Barcelona: An Urban History of Science and Modernity, 1888–1929

The four decades between the two Universal Exhibitions of 1888 and 1929 were formative in the creation of modern Barcelona. Architecture and art blossomed in the work of Antoni Gaudí and many others. At the same time, social unrest tore the city apart. Topics such as art nouveau and anarchism have attracted the attention of numerous historians. Yet the crucial role of science, technology and medicine in the cultural makeup of the city has been largely ignored. The ten chapters of this book recover the richness and complexity of the scientific culture of *fin de siècle* Barcelona. The authors explore a broad range of topics: zoological gardens, natural history museums, amusement parks, new medical specialities, the scientific practices of anarchists and spiritists, the medical geography of the urban underworld, early mass media, domestic electricity and astronomical observatories. They pay attention to the agenda of the bourgeois elites but also to hitherto neglected actors: users of electric technologies and radio amateurs, patients in clinics and dispensaries, collectors and visitors of museums, working class audiences of public talks and female mediums. Science, technology and medicine served to exert social control but also to voice social critique. *Barcelona: An Urban History of Science and Modernity, 1888–1929* shows that the city around 1900 was both a creator and facilitator of knowledge but also a space substantially transformed by the appropriation of this knowledge by its unruly citizens.

Oliver Hochadel is a Tenured Historian of Science at the Institució Milà i Fontanals, Consejo Superior de Investigaciones Científicas (IMF-CSIC), Barcelona. His research focuses on the relationship between science and its publics. Book publications include *El mito de Atapuerca. Orígenes, ciencia, divulgación* (2013), *Playing with Fire. Histories of the Lightning Rod* (edited with Peter Heering and David Rhees, 2009) and *Öffentliche Wissenschaft. Elektrizität in der deutschen Aufklärung* (2003).

Agustí Nieto-Galan is Associate Professor of History of Science, and Director of the Centre d'Història de la Ciència (CEHIC) at the Universitat Autònoma de Barcelona. He has written widely on the history of chemistry and natural dyestuffs, and on the history of science popularization (eighteenth to twentieth centuries). He is currently working on several aspects of popular science and urban history of science in Europe. His book *Science in the Public Sphere* is scheduled to appear in 2016.

Science, Technology and Culture, 1700–1945

Science, Technology and Culture, 1700–1945 focuses on the social, cultural, industrial and economic contexts of science and technology from the 'scientific revolution' up to the Second World War. Publishing lively, original, innovative research across a broad spectrum of subjects and genres by an international list of authors, the series has a global compass that concerns the development of modern science in all regions of the world. Subjects may range from close studies of particular sciences and problems to cultural and social histories of science, technology and biomedicine; accounts of scientific travel and exploration; transnational histories of scientific and technological change; monographs examining instruments, their makers and users; the material and visual cultures of science; contextual studies of institutions and of individual scientists, engineers and popularizers of science; and well-edited volumes of essays on themes in the field.

Also in the series

Entrepreneurial Ventures in Chemistry: The Muspratts of Liverpool, 1793–1934
Peter Reed

Science Policies and Twentieth-Century Dictatorships
Spain, Italy and Argentina
Edited by Amparo Gómez, Antonio Fco. Canales and Brian Balmer

From Local Patriotism to a Planetary Perspective
Impact Crater Research in Germany, 1930s–1970s
Martina Kölbl-Ebert

Francis Watkins and the Dollond Telescope Patent Controversy
Brian Gee, edited by Anita McConnell and A.D. Morrison-Low

Acid Rain and the Rise of the Environmental Chemist in Nineteenth-Century Britain
The Life and Work of Robert Angus Smith
Peter Reed

Matter and Method in the Long Chemical Revolution
Laws of Another Order
Victor D. Boantza

Meeting Places: Scientific Congresses and Urban Identity in Victorian Britain
Louise Miskell

Barcelona

An urban history of science and modernity, 1888–1929

Edited by
Oliver Hochadel and
Agustí Nieto-Galan

Routledge
Taylor & Francis Group
LONDON AND NEW YORK

First published 2016
by Routledge

2 Park Square, Milton Park, Abingdon, Oxfordshire OX14 4RN
711 Third Avenue, New York, NY 10017

Routledge is an imprint of the Taylor & Francis Group, an informa business

First issued in paperback 2017

British Library Cataloguing in Publication Data
A catalogue record for this book is available from the British Library

Library of Congress Cataloging in Publication Data
Names: Hochadel, Oliver, 1968– editor. | Nieto-Galan, Agustí, editor.
Title: Barcelona : an urban history of science and modernity, 1888–1929 / edited by Oliver Hochadel and Agustí Nieto-Galan.
Description: 2016. | Series: Science, technology and culture, 1700–1945 | Includes bibliographical references and index.
Identifiers: LCCN 2015024494 | ISBN 9781472434197 (hbk. : alk. paper) | ISBN 9781315568713 (ebk)
Subjects: LCSH: Public spaces—Spain—Barcelona—History. | Science—Spain—Barcelona—History. | Civilization, Modern.
Classification: LCC HT185 .B37 2016 | DDC 307.7609467/2–dc23
LC record available at http://lccn.loc.gov/2015024494

ISBN: 978-1-4724-3419-7 (hbk)
ISBN: 978-0-8153-6674-4 (pbk)

Typeset in Bembo
by Apex CoVantage, LLC

Contents

List of Figures and Maps

Figures

Maps

Notes on contributors

Ferran Aragon holds a PhD in Developmental Biology from the Universitat Pompeu Fabra, Barcelona (2008) and a Master in History of Science from the Universitat Autònoma de Barcelona, Universitat de Barcelona (2012). He is interested in the intersections between science and ideology, especially in the coproduction of nation, nationalism and scientific authority. His current research is about the Natural History Museum of Barcelona and its relation with the conservative Catalan nationalists at the beginning of the twentieth century. He designed a prosopographic relational database that is currently used by a small team of researchers and that will later on be opened to the entire community.

Mònica Balltondre is a Postdoctoral Research Fellow at the Faculty of Psychology and member of the Centre d'Història de la Ciència (CEHIC) at the Universitat Autònoma de Barcelona. She was a visiting scholar in Fiocruz, Brasil, and at the University of California, Berkeley, USA. Her research focuses on the history of psychology, sixteenth-century women's spirituality (with the book: *Éxtasis y visiones. La experiencia contemplativa de Teresa de Ávila*, 2012) and on the history of the relationship between science and religion, with several publications in journals such as *History of the Human Sciences*, *History of Psychology* and *Dynamis*.

Jordi Ferran holds a degree in Physics from the Universitat de Barcelona (1995) and a PhD in History of Science from the Universitat Autònoma de Barcelona (2013). His research focuses on the popularization of everyday technologies in the twentieth century. He published several articles about electricity and the 1929 Barcelona International Exposition in journals such as *Annales historiques de l'électricité* (2006). His book (with Mireia Fernández-Ardèvol) *Un viatge per la fractura digital. Mòbils, SMS, Blogs i Internet a l'Àfrica* (2007) received the XXIII Humbert Torres Prize for the Popularization of Science.

Álvaro Girón Sierra is a tenured historian of science at the Institució Milà i Fontanals, Consejo Superior de Investigaciones Científicas (IMF-CSIC), Barcelona. His research focuses on the relation between anarchist culture and evolutionary thought. On this topic he published two books: *Evolucionismo y anarquismo en España, 1882–1914* (1996) and *En la mesa con Darwin. Evolución y revolución en el movimiento libertario en España (1869–1914)* (2005) as well as numerous articles in journals and edited volumes.

Andrea Graus holds a PhD in History of Science from the Universitat Autònoma de Barcelona (2015). Her thesis explores psychical research and spiritism in Spain at the turn of the twentieth century from a European transnational perspective. Her research interests include the history of psychology, the interplay between science and religion and the history of hypnotism. On the latter topic she recently published an article in *Studies in History and Philosophy of Science Part C: Studies in History and Philosophy of Biological and Biomedical Sciences.*

Meritxell Guzmán holds a degree in Audiovisual Communication from the Universitat Autònoma de Barcelona (2003) and a Master in History of Science from the Universitat Autònoma de Barcelona, Universitat de Barcelona (2013). Her research deals with knowledge popularization on the radio in twentieth-century Catalonia, with special emphasis on the case of Radio Barcelona from 1923 to 1930.

Oliver Hochadel is a tenured historian of science at the IMF-CSIC, Barcelona. His research focuses on the relationship between science and its publics, for example, in the case of electricity in the German enlightenment, the history of the zoo in the nineteenth century and the history of human origins research. Book publications: *Öffentliche Wissenschaft. Elektrizität in der deutschen Aufklärung*, 2003; *El mito de Atapuerca. Orígenes, ciencia, divulgación,* 2013. Edited books: (with Ursula Kocher) *Lügen und Betrügen: Das Falsche in der Geschichte von der Antike bis zur Moderne,* 2000; (with Peter Heering and David Rhees) *Playing with Fire. Histories of the Lightning Rod,* 2009.

Àlvar Martínez-Vidal is a Senior Lecturer in History of Science at the Universitat de València and a researcher at the Institut d'Història de la Medicina i de la Ciència 'López Piñero' since 2010. Previously he had spent two decades at the Universitat Autònoma de Barcelona, where he was one of the founders of the CEHIC. In a long-time joint venture with José Pardo-Tomás, he studied the so-called Spanish 'novator' movement in the Early Modern period and the overlaps between medicine and nationalism in the case of Catalonia during the first decades of the twentieth century. He edited the book *L'Hôpital Varsovie (1944–1950). Médecine, exil et résistance* (Toulouse, 2011). He is currently working with Alfons Zarzoso on the emergence of medical specialties in Barcelona.

Jorge Molero-Mesa is a Senior Lecturer in History of Science and a member of the board of the CEHIC at the Universitat Autònoma de Barcelona. He is also the director of the History of Science PhD Programme at the same university, and a member of the editorial board of the journal *Dynamis*. His research focuses on the social history of medicine (nineteenth and twentieth centuries). He has published extensively on the history of social diseases (tuberculosis, malaria), Spanish health policy and administration, Spanish colonial medicine in Morocco and the relationship between medicine and society from a class perspective.

Agustí Nieto-Galan is Associate Professor of History of Science, and Director of the Centre d'Història de la Ciència (CEHIC) at the Universitat Autònoma de Barcelona. He has written widely on the history of chemistry and natural dyestuffs, and on the history of science popularization (eighteenth to twentieth centuries). He is currently working on several aspects of popular science and urban history of science in Europe. His most recent book is *Science in the Public Sphere* (2016).

José Pardo-Tomás holds a PhD in History (Universitat de València). Since 1994 he is a tenured historian of science at the IMF-CSIC, Barcelona. He was a visiting scholar at the universities of Padua (Italy), Humboldt (Berlin, Germany), Bordeaux (France) and the Universidad Nacional Autónoma de México. He works on social and cultural history of medicine, natural history and scientific books. His most recent publications include: *Salvadoriana. The Cabinet of Curiosities of Barcelona* (2014).

Antoni Roca-Rosell is a Senior Lecturer in History of Science at the Universitat Politècnica de Catalunya. In 2010 he was elected corresponding member of the International Academy of History of Science, and in 2014, full member of the Institut d'Estudis Catalans. His field of research is the history of science and technology in Catalonia and Spain, including the history of engineering and technical education (eighteenth to twentieth centuries). He published 180 papers and 15 books, among them, as editor, *Josep Comas i Solà. Astrònom i divulgador* (2004).

Pedro Ruiz-Castell lectures History of Science at the Universitat de València. After being awarded a DPhil at the University of Oxford, he worked for several years as Head of the Department of Research and Documentation at the Spanish Museo Nacional de Ciencia y Tecnología in Madrid and as a Postdoctoral Research Fellow at the Universitat Autònoma de Barcelona. His research focuses on the history of physics and allied sciences in the nineteenth and twentieth centuries – paying special attention to the history of astronomy and astrophysics in Spain – science in the public sphere, and scientific instruments. Among his publications stand out his book entitled *Astronomy and Astrophysics in Spain (1850–1914)* (2008) and several articles published in journals such as *Annals of Science, The British Journal for the History of Science, Dynamis* and *History of Science.*

Jaume Sastre-Juan is a post-doc researcher at the Centro Interuniversitário de História das Ciências e da Tecnologia (CIUHCT) – Universidade de Lisboa. His main research interests within the history of technology are the politics of display in science and technology museums and the politics of technological fun in "interactive" museums and amusement parks. He holds a degree in Philosophy from the Universitat de Barcelona (2005) and a PhD in History of Science from the Universitat Autònoma de Barcelona (2013), with a dissertation on the politics of technological display at the New York Museum of Science and Industry (1912–1951). He has taught as an adjunct lecturer under conditions of exploitation at the Universitat de Barcelona (2012–2015) and also at the Universitat Oberta de Catalunya. He is a member of the co-operative of historians Allaqqat.

Carlos Tabernero teaches History of Science at the Universitat Autònoma de Barcelona and is a member of the CEHIC. After his PhD in Biology he worked as a molecular biologist at the National Cancer Institute in the United States and as a researcher at the Internet Interdisciplinary Institute (IN3) of the Universitat Oberta de Catalunya. His current research interests include media and the processes of popularization of science, medicine and technology in the twentieth century; screen (film and television) studies; and the representation and articulation of life sciences (natural sciences, biology, and medicine–health) in media in contemporary Spain. His publications include articles in journals such as *Dynamis, Arbor, Host, Observatorio (OBS) Journal* and the *Journal of Medicine and Movies*, as well as contributions to books.

Jaume Valentines-Álvarez holds a degree in Engineering from the Universitat Politècnica de Catalunya (2002), a degree in History from the Universitat de Barcelona (2005) and a PhD in History of Science from the Universitat Autònoma de Barcelona (2012). He has worked as an archivist and a curator in universities and museums. As a researcher he published on topics dealing with the social construction of technology and the technological construction of societies during the twentieth century. His PhD dissertation focused on engineering expertise, technocracy and nationalism between 1929 and 1939. After being 'usefully unemployed' (in Illich's terms) for two years, he is now a post-doc researcher at the Inter-University Centre for the History of Science and Technology in Lisbon. He is a member of the co-operative of historians Allaqqat.

Laura Valls holds a degree in Biology from the Universitat de Barcelona (2000) and a master in History of Science from the Universitat Autònoma de Barcelona, Universitat de Barcelona (2009). Her doctoral research focuses on the popularization program of the Junta de Ciències Naturals de Barcelona at the Parc de la Ciutadella and on the material culture of objects. On these topics she published papers in journals such as *Geo Crítica. Cuadernos Críticos de Geografía Humana* and *Afers: fulls de recerca i pensament*. Currently she works at the Department of Scientific Culture of the CSIC in Barcelona, where she coordinates different projects and outreach activities.

Alfons Zarzoso is curator of the Museu d'Història de la Medicina de Catalunya since 2000. He curated medical exhibitions and developed website-related projects on nineteenth and twentieth centuries medical press and publicity as well as exiled physicians. He also lectures in History of Medicine at the Universitat Autònoma de Barcelona. His research focuses on the material culture of medicine and the visual representations of science and medicine in modern Spain, and their role in the definition and delimitation of medical specialties such as anatomy, obstetrics and plastic surgery. He is the author of three books and a number of papers in international journals. He is currently working with Àlvar Martínez on the emergence of medical specialties in Barcelona.

Preface

After having completed this book, we, the 18 authors, look at Barcelona differently. We thought we knew the city most of us live in pretty well. Yet the sheer richness of Barcelona's scientific culture around 1900 we helped each other to discover provides us with a lasting thrill. For quite a number of years many members of our local history of science community had come across a host of different 'urban' aspects in their research projects. What seems obvious now needed time to mature. In our local seminars, but also in many of the papers presented at international conferences the role of the urban space as an active agent in shaping scientific knowledge became more and more explicit. We realized that a common project addressing the interactions between science and the city in Barcelona was called for to bring those numerous individual research agendas together. In times of scarce research funds and our local research institutions being savaged by 'the crisis', we needed to find new models of collaboration to embark on a large-scale project and to overcome traditional disciplinary boundaries. In March 2012 we founded together with José Pardo-Tomás and Alfons Zarzoso the 'Barcelona Laboratory Science and the City. Scientific Culture in the City, 1868–1939: Sites, Actors and Practices'. This Laboratory was to serve as a base to launch a number of projects directed both at local and international audiences, at scholars and the general public: the gathering of relevant sources, the building of databases, the organization of academic meetings and the fostering of international collaborations, research projects and subsequent publications but also outreach activities, such as guided city tours, exhibitions and public lectures.

One of the projects of our Laboratory is this book. Its time frame is traditional, as it addresses the four decades between the two international exhibitions in 1888 and 1929, the formative period of modern Barcelona. Yet it was our intention to present new case studies and topics and to explore little known sources, urban spaces and actors. This ambitious endeavour required sheer manpower and we were fortunate to be able to count on so many colleagues from our local community. For each topic chosen we asked two historians – often a senior and a junior scholar – to research and write the article together. We, the two editors, also committed ourselves to pen the introduction together. In other words, this book was conceived from the beginning as the project of a team.

Pre-circulated papers of increasing length were discussed at three workshops held at the Institució Milà i Fontanals (IMF-CSIC) in 2013 and 2014. We all cherished the fruitful exchange seeing the project – and the book – becoming more intriguing and intellectually stimulating each time. Apart from the authors many colleagues were present at these workshops contributing with constructive criticisms and comments that substantially improved the volume. We are grateful to Annette Mülberger, Fernando Vidal, Miquel Carandell, Clara Florensa, Jon Arrizabalaga, Xavier Roqué, José Luis Oyón, Enric March and many others. An advanced draft of each article was sent out to two referees (see the individual chapters for their names). We would like to thank all 22 of them for most valuable criticism and numerous helpful suggestions. We are particularly grateful to the anonymous referee of the entire manuscript for his or her scrupulous reading and precise suggestions for improvement, that reflected an impressive understanding of *all* the topics addressed. And of course we would also like to thank the editors of the series for their support and input.

In recent years we presented parts of this project at several international workshops and conferences. We organized the sessions: 'Science in the Public Sphere: Barcelona 1888–1936', at the 4th European Society for the History of Science Conference, in Barcelona in November 2010; 'Technological Cityscapes: Barcelona around 1900', at the 39th ICOTEHC Conference, in Barcelona in July 2012; 'Urban Peripheries? Science in "Second Cities" around 1900', at the 9th STEP Conference in Lisbon; 'Ciència i ciutat: una història urbana de la ciència: Barcelona 1888–1929', at the 13th Trobada d'Història de la Ciència i de la Tècnica, Sant Feliu de Guíxols, both in September 2014. In June 2014 and thanks to the support of Neale Watson, we held the Third Watson Seminar on the History of Material and Visual Science in Barcelona: 'How to Write an Urban History of Science: New Approaches and Case Studies'. We benefited enormously from the discussion with our colleagues from abroad. They helped us to sharpen our historiographical framework and to design a comparative approach for the urban history of science. We thank all the speakers and participants of these conference sessions and workshops for their stimulating contributions.

The project has been also funded by the ICREA-Acadèmia Research Prize, which one of us (ANG) was awarded in the period 2010–2015. The Spanish Ministry of Economy and Competitiveness contributed by financing the following research projects: (HAR2009–12918-C03–02) (subprograma HIST): 'Science and Expertise in the Public Sphere: Barcelona (1888–1992)'; (HAR2012–36204-C02–02) 'Scientific Authority in the Public Sphere in Twentieth-Century Spain'; and (HAR2013–48065-C2–1-P) 'Science and the City. Natural History, Biology and Biopolitics in the Divided City: Barcelona and Buenos Aires (1868–1936)'. In addition, our History of Science research group in Barcelona: 'Science, Technology and Medicine in Modern Catalonia (18th–20th centuries)', has been funded since 2009 by the Catalan Government as a 'Grup de recerca consolidat i finançat' (SGR2009–887) (2014 SGR 1410).

A note on spelling: around 1900 the use of the Catalan language progressively became more widespread in the public sphere, in politics, administration, and the media, a sphere that had been dominated nearly exclusively by the Spanish language since the eighteenth century. What is more, Catalan grammar was not standardized until the 1920s. Therefore the period covered by our volume is characterized by a certain linguistic 'confusion', for example, with respect to the spelling of the first and second names of historical actors, of place names and institutions. Our 'solution' is to reflect as far as possible the spelling of the Spanish and Catalan primary sources – and thus the mix of languages and spellings.

Much urban history of science remains to be explored. We very much hope that the case studies of this volume and the historiographical framework in which they are developed will stimulate further research – on Barcelona and many other cities. The comparative potential of this historiography seems enormous to us. It would require though not only transdisciplinary but even more so transnational collaborations.

Oliver Hochadel and Agustí Nieto-Galan
Barcelona, February 2015

Barcelona historical timeline

1860 The end of the medieval city: the construction of the Eixample (Ildefons Cerdà)
1868 Spanish Liberal Revolution (La Gloriosa)
1872 Foundation of the Ateneu Barcelonès
1873 First Spanish Republic
1875 Monarchic Restoration (Restauración)
1877 Public debates on Darwinism at the Ateneu Barcelonès
1880 First Catalanist Conference
1881 Foundation of the Ateneu Obrer de Barcelona (Workers' Athenaeum)
1886 Valentí Almirall, *Lo catalanisme* (federal, liberal Catalanism)
1888 **Universal Exhibition** at the Parc de la Ciutadella
International scientific conferences (medicine, spiritism)
Founding Conference of *Unión General de Trabajadores* (UGT), socialist
1892 Josep Torres i Bages, *La tradició catalana* (Catholic, conservative Catalanism)
1893 Anarchist bombing at the Liceu Opera Theatre
New sewage system and hygienic programmes (Pedro García Faria)
1895 First telephone wire Barcelona-Madrid
1896 First cinema at the Rambla
1897 Old adjacent towns (Sants, Gràcia, Les Corts, Sant Gervasi, Sant Martí, Sant Andreu) incorporated into Barcelona
1898 Spanish colonial crisis. Loss of Cuba and the Philippines
1899 First electric tramways
1900 Opening of the Parc Güell
Modernisme movement
1901 Francesc Ferrer Guàrdia opens the Escuela Moderna
Funicular at the Tibidabo mountain
Foundation of La Lliga Regionalista (conservative Catalanism)
1903 First Congrés Universitari Català
1904 Opening of the Fabra Obsevatory at the Tibidabo mountain, Josep Comas Solà (director)
1905 First Electric railway to Sarrià
First Congrés d'Higiene de Catalunya

1906	Noucentisme movement
	Enric Prat de la Riba, *La nacionalitat catalana*
	Solidaritat Catalana as a Catalanist political movement
1907	First automobile
	Foundation of the Institut d'Estudis Catalans
1908	Opening of the art nouveau Palau de la Música
1909	The Tragic Week – 'La Setmana Tràgica' (a violent social revolt)
	International Conference on Esperanto
1910	Founding Conference of the Confederación Nacional del Trabajo (CNT), anarchist
1911	Opening of Saturno Park
1913	First Conference of Doctors of Catalan language (Congrés de Metges en Llengua Catalana)
1914	Yellow fever plague
	Mancomunitat de Catalunya as an autonomous provincial government
1915	Public works at the Montjuïc mountain
1916	Francesc Carreras Candi, *La ciutat de Barcelona*
1918	The Spanish Flu
1919	Workers' long strike at La Canadenca
1920	King Alfonso XIII's visit to the city's textile factories
1921	More than hundred deaths caused by social and political violence
1923	Visit of Albert Einstein
	General Primo de Rivera's coup d'état
1924	Radio Barcelona begins to broadcast
	Opening of the first underground line: Plaça Catalunya-Lesseps
1925	Repression of Catalanist societies
1926	Death of Antoni Gaudí
	Opening of Jorba stores
1927	Inauguration of Plaça Catalunya
1928	Zeppelin flies around the city
1929	**International Exhibition** at the Montjuïc mountain
1930	Crisis of Primo de Rivera's dictatorship
1931	End of the Monarchy: the Second Spanish Republic begins
	Recovery of the Catalan Autonomous Government: Generalitat de Catalunya
1936	General Franco's coup d'état
	Spanish Civil War begins
	Anarchist-communist revolution in Barcelona
1938	George Orwell, *Homage to Catalonia*
	Fascist bombing of Barcelona's civil population
1939	Franco's victory, and massive Catalan exile, the end of the republican regime, and the beginning of a new dictatorship

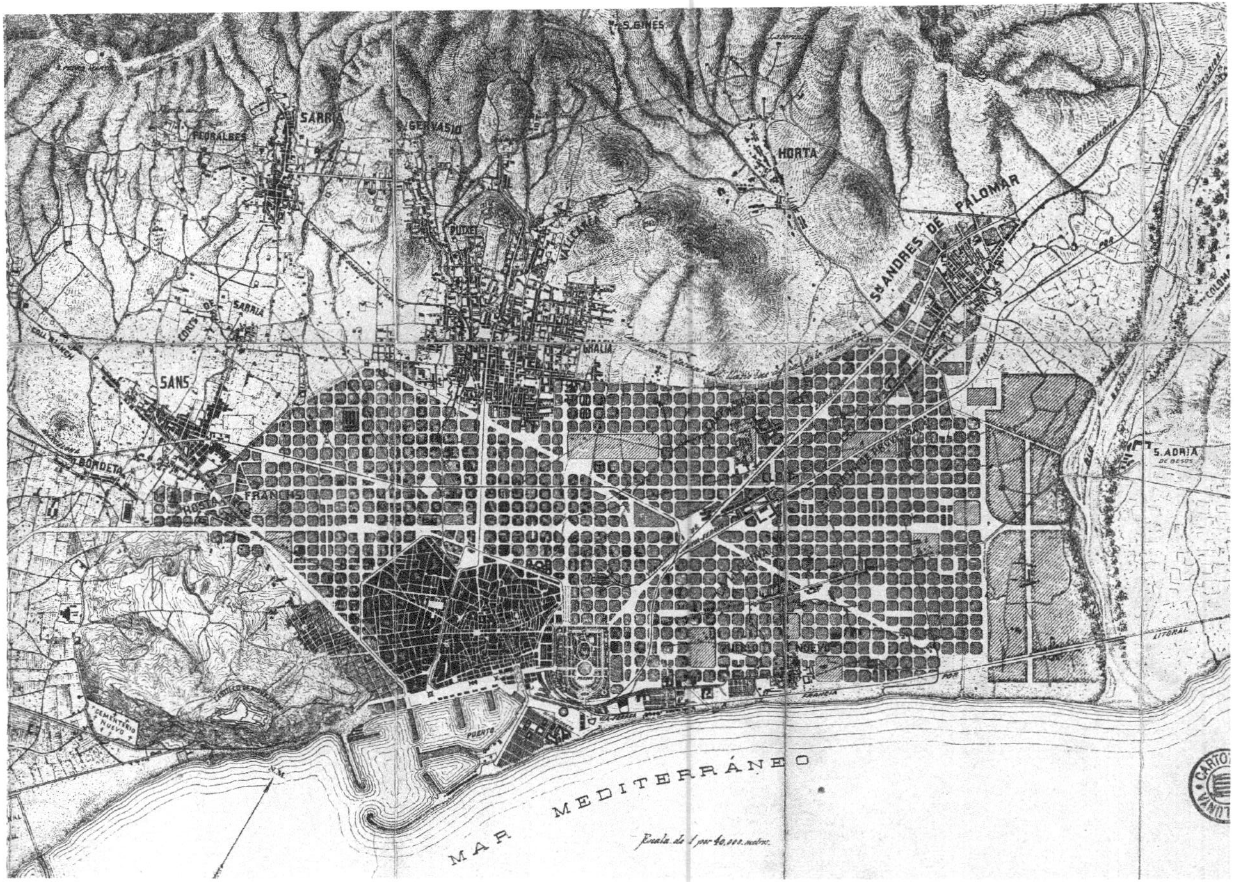

Map 1 Barcelona in 1888.

1 Introduction

Oliver Hochadel and Agustí Nieto-Galan[1]

At the beginning of his novel *City of Marvels*, Spanish writer Eduardo Mendoza, sketches with a few brush strokes a scientific panorama of Barcelona in the nineteenth century:

> [...] Barcelona was always at the forefront of progress. In 1818, the *first* regular stagecoach service in Spain went into operation between Barcelona and Reus. The *first* experimental gaslight system was installed in the courtyard of the Palace of La Lonja, housing the Chambers of Commerce, in 1826. In 1836, the *first* steam-powered motor went into operation [...] Spain's *first* railroad was built to link Barcelona and Mataró, dating from 1848. The *first* electric power station was likewise built in Barcelona, in the year 1873. The gap between Barcelona and the rest of the peninsula was enormous, and the city made an overwhelming impression on the newcomer.[2]

With an image of a city marked by technological innovation and brimming with energy and promise, Mendoza sets the stage for the meteoric career of his protagonist. *City of Marvels* begins in 1887 when the young Onofre Bouvila, barely a teenager, comes from the Catalan countryside to the bustling city. His first job is to hand out anarchist pamphlets to workers in the Parc de la Ciutadella, the site of the first Universal Exhibition in Barcelona to be celebrated the following year. As the story unfolds, Onofre rises from abject poverty to power and wealth propelled by his relentless criminal energy. Mendoza's novel ends in May 1929 on the very day the second major exhibition, the International Exhibition, opens this time located on the other side of town, at the foot of Montjuïc. In a spectacular ascent, Onofre rises vertically into the

1 We would like to thank in particular Dorothee Brantz and Mitchell G. Ash for numerous helpful suggestions and most valuable advice. We also benefited substantially from constant discussion with all the contributors of this volume.

2 Eduardo Mendoza, *City of Marvels* (London: Collins Harvill, 1990), p. 12 [our emphasis]. Original: *La ciudad de los prodigios* (Barcelona: Seix Barral, 1986). Mendoza takes some poetic license with respect to historical data.

air with a seemingly miraculous machine, leaving the masses speechless only to disappear into the Mediterranean Sea. He had urged a Catalan inventor and a German engineer to collaborate in order to develop this flying contraption. The machine has roughly the shape of a helicopter – the *dernier cri* of aeronautic technology at the time. Yet it lacks a propeller – a 'wonder' as Mendoza writes, tongue in cheek.

City of Marvels, first published in Spanish in 1986, was written amidst the great expectations created by the Olympic Games which Barcelona was to host in 1992. Once again, we might say, the city was eager to catch up with 'modernity' as it had done a century earlier, casting off the grey vestiges of the oppressive Franco regime (1939–1975). For Mendoza this era between the two international exhibitions in 1888 and 1929 serves as a mirror for his own present. Despite 'being first' on so many accounts, the Catalans – their ruling class of bourgeois industrialists – were unable to assume a leading political role in Spain. Instead they were being bamboozled by 'Madrid', the Spanish central power, who was reaping the cream of the two exhibitions for the greater glory of Spain. And – couched in the sometimes funny, sometimes bitter sarcasm of his 'mock historical novel' – Mendoza sees the same danger lurking on the horizon with the Olympics of 1992 in the making.[3]

In spite of Mendoza's gloomy prognosis that a major international event such as the Olympic Games would be high-jacked once more by the Spanish state, Barcelona was able to gain the world's attention and to maintain it. The Games did not only convert the city into a major tourist destination. Still in the Olympic year of 1992, art critic and journalist Robert Hughes lamented in his acclaimed book *Barcelona* the lack of international attention for the city and its history.[4] Yet more than two decades later, the political and social history of Barcelona in our Mendozian timeframe (1888–1929) and until the end of the Spanish Civil War (1939) is much trodden turf. In those closing years of the nineteenth century and in the first decades of the twentieth century, rapid industrialization, the dramatic changes in the physiognomy of the city (its expansion and the new architecture) and the increasing social unrest with all its violent manifestations have attracted the interest of social historians, urban scholars, historians of architecture and of art, just as Hughes had demanded.[5] Barcelona is known for its modernist art (art nouveau) and its architecture. Architects such as Antoni Gaudí, Lluís Domènech i Montaner and Josep Puig i Cadafalch and painters such as Joan Miró, Pablo Picasso and Antoni Tàpies are nowadays crucial for the marketing

3 Joan Ramon Resina, *Barcelona's Vocation of Modernity. Rise and Decline of an Urban Image* (Stanford: Stanford University Press, 2008), pp. 182–5.

4 Robert Hughes, Barcelona (New York: Vintage Books, 1992).

5 See for instance: Pere Hereu et al., *Arquitectura i ciutat a l'Exposició Universal de Barcelona de 1888* (Barcelona: UPC, 1988); Edmond Valles, *Història gràfica de la Catalunya comtemporània.* 3 vols (Barcelona: Edicions 62, 1974–1976); Jordi Casassas, *Intel·lectuals, professionals i polítics a la Catalunya contemporània (1850–1920)* (Sant Cugat del Vallès: Amelia Romero, 1989); Alexandre Cirici Pellicer, *1900 en Barcelona: modernismo, modern style, art nouveau, jugendstil* (Barcelona: La Polígrafa, 1967); Pere Gabriel (ed.), *El Noucentisme* (Barcelona: Edicions 62, 1996).

of the city.[6] More than that, historians from near and far have discovered Barcelona, from among other labels, as the 'world capital of anarchism' in the first third of the twentieth century as an intriguing object of investigation.[7] Yet in our perception this thriving research on *fin-de-siècle* Barcelona is incomplete. This ever-increasing amount of scholarship rarely addresses at all the role of science (as science, technology and medicine) in this epoch.[8]

What we would like to borrow from *City of Marvels* is not only the timeframe but also Mendoza's perceptiveness about the role of science. Nowadays it is a commonplace that the years between 1888 and 1929 were formative in the creation of 'modern' Barcelona.[9] Yet what Mendoza clearly grasps is that scientific progress was both a driving agent and a tangible result of this process. *City of Marvels* abounds with references to technological innovation indicating a seemingly successful modernization. These four decades between the two international exhibitions seem to us an ideal historical window to pursue our project on the scientific culture of Barcelona.

At that time, the city was in a permanent state of construction. Between 1888 and 1929, Barcelona, the 'Catalan Manchester', as it was often referred to, grew substantially, to roughly one million inhabitants. Industrial growth demanded a new labour force and shaped its urban geography and architecture with new transport networks and infrastructures. The second industrial revolution brought electricity and chemistry to the already powerful textile industry run by steam. A substantial cluster of textile factories grew in highly industrialized areas such as Poble Nou, but also in the surrounding towns such as Sant Martí de Provençals and Sant Andreu del Palomar. In the search for cheaper sources of energy, other factories moved to the countryside next to the rivers to profit from hydraulic power. This network of circulation of raw materials, workers and finished products, shaped the scientific culture of Barcelona for decades.[10]

6 See for instance: Cirici Pellicer, *1900 en Barcelona*; Teresa-M. Sala (ed.), *Història de la cultura catalana: Barcelona 1900* (Barcelona: Edicions 62, 2007).

7 Relevant works in Spanish and Catalan are: Pere Solà, *Els ateneus obrers i la cultura popular a Catalunya (1900–1939). L'Ateneu Enciclopèdic Popular* (Barcelona: La Magrana, 1978); José Luis Oyón, *La quiebra de la ciudad popular: espacio urbano, inmigración y anarquismo en la Barcelona de entreguerras, 1914–1936* (Barcelona: El Serbal, 2008). In English: Angel Smith (ed.), *Red Barcelona. Social Protest and Labour Mobilization in the Twentieth Century* (London/New York: Routledge, 2002); Chris Ealham, *Anarchism and the City: Revolution and Counter-Revolution in Barcelona, 1898–1937* (Oakland: AK Press, 2010).

8 In the remainder of this introduction, we will only use the term science (scientific) understood in its broadest sense, including technology and medicine: John V. Pickstone, *Ways of Knowing. A New History of Science, Technology and Medicine* (Manchester: Manchester University Press, 2000).

9 For a textbook-like synthesis, see Alejandro Sánchez (ed.), *Barcelona, 1888–1929. Modernidad, ambición y conflictos de una ciudad soñada* (Madrid: Alianza, 1994).

10 For a history of industrialization on a regional level, see: Norman Pounds, *A Historical Geography of Europe* (Cambridge: Cambridge University Press, 1990). See also Jordi Nadal (ed.), *Atlas de la industrialización de España: 1750–2000* (Barcelona: Fundación BBVA, 2003). See also: Horacio Capel (ed.), *Las Tres Chimeneas: implantación industrial, cambio tecnológico y transformación de un espacio urbano*

Figure 1.1 A general view of Barcelona and its harbour during the opening of the 1888 International Exhibition.

At the turn of the century, Barcelona's landscape of scientific institutions also changed dramatically. To name but a few of the most relevant ones: In 1882, the Museu Martorell, the first public museum (dedicated to natural history and related fields) opened in the Parc de la Ciutadella. In 1887, the Laboratori Microbiològic Municipal was founded, not least to fight or even better prevent the recurrent epidemics. The rise of Catalan nationalism at the very end of the nineteenth century led to a host of new institutions. In 1899, the Institució Catalana d'Història Natural was founded, in 1907 the Institut d'Estudis Catalans. The latter was an ambitious project for a new 'national' (Catalan) research centre embracing all fields of knowledge.[11]

Yet the dynamic of the local scientific culture resulted – as in many other cities at the time – from a combination of this 'official' or 'elite' science with a vibrant 'public' or 'civic' science.[12] Obviously these two kinds of science are often intrinsically connected and hard to separate. In terms of civic science, Barcelona was home to a wide variety of activities and institutions: public lectures, exhibitions, university extension courses, collections of living wild animals, amateur astronomical observations, international conferences on popular but controversial fields such as spiritism, athenaeums and pedagogically highly innovative schools. Yet this thriving scientific culture, which often crossed social barriers, has been largely ignored by recent historiography. This is the raison d'être of this book: to write a new, genuine urban history of science of Barcelona.

Urban science

The thesis of this book is that the history of Barcelona between 1888 and 1929 cannot be properly understood without accounting for the role of science. So far there have been only some valuable but isolated case studies on the scientific culture of the city.[13] Our goal is more than simply to add 'a bit of history of science', but to provide a new perspective on the cultural makeup of the city

barcelonés (Barcelona: FECSA, 1994); Mercè Tatjer, 'La industria en Barcelona (1832–1992)'. *Scripta Nova*, 10 (2007), http://www.ub.edu/geocrit/sn/sn-218-46.htm (last accessed 26 January 2015); *150 anys d'història de la Maquinista Terrestre i Marítima, S.A. i de MACOSA: de la revolució industrial a la revolució tecnològica* (Barcelona: Dos Punts Documentació i Cultura, 2009); Pere Colomer, *Barcelona, una capital del fil: Fabra i Coats i el seu model de gestió, 1903–1936* (Barcelona: Ajuntament de Barcelona, 2014).

11 For a more complete list of scientific institutions, see: Alexandre Galí, *Història de les institucions i del moviment cultural a Catalunya, 1900–1936*, vol. 16: 'Acadèmies i Societats científiques' (Barcelona: Fundació Alexandre Galí, 1986); Antoni Roca-Rosell, 'Les possibilitats d'una producció científica catalana: entorn de l'acció de la Mancomunitat de Catalunya', *Recerques: història, economia, cultura*, 14 (1983): 81–95.

12 For the concept of civic science, see Lynn K. Nyhart, *Modern Nature: The Rise of the Biological Perspective in Germany* (Chicago/London: The University of Chicago Press, 2009).

13 There are some 'city guides' such as Xavier Duran and Mercè Piqueras, *Passejades per la Barcelona científica*. 2nd ed. (Barcelona: Ajuntament de Barcelona, 2003), which deal in a very general way with the history of science in Barcelona. For explicit historical scholarship, see: Matiana González-Silva and Néstor Herran, 'Ideology, elitism and social commitment: alternative images of science in two *fin de siècle* Barcelona newspapers', *Centaurus*, 51 (2) (2009): 97–115. For the physical sciences, see the

and its attempts at modernization. In fact, we do not intend to write a 'separate' history of science in Barcelona around 1900. It would make no sense to neatly disjoin these spheres (art, architecture, science, medicine, politics and so on). They need to be understood as a seamless web with numerous intersections. This is obviously a tremendous historiographical challenge and runs the danger of stating truisms such as 'everything is connected'. Therefore we need to be more precise: what is lacking is not so much a history of Barcelona that also mentions the history of the municipal natural history museum; or the conflictive reception of Darwinism; or the visit of a foreign luminary such as Albert Einstein.[14] What is lacking is an *urban* history of science of Barcelona. Such a history would have to focus on the specific conditions of knowledge production and circulation in these critical decades around 1900.

Starting in the mid/late nineteenth century, many new institutions, formats, media and technologies enter the urban stage. To name some obvious examples: public museums, urban parks, international exhibitions, electric lighting, private clinics, amusement parks, but also newspapers (only becoming a mass medium in the late nineteenth century) and illustrated journals, evening schools and new places of sociability such as the athenaeum, the cinema and eventually radio. Barcelona's *fin-de-siècle* medical doctors, patients and their clinics, public health programmes, anarchist and spiritist circles, amateurs, users, local experts and technological networks formed part of a complex cultural mesh. By means of a selection of actors, sites and scientific practices shaped in this specific urban context, we believe that the overall historical account of the city between 1888 and 1929 can be significantly enriched. Our basic assumption is that the urban space is both: a creator, incubator and facilitator of these practices of knowledge production and circulation but also an object substantially transformed by these practices. As urban scholars have argued, 'rather than a passive container of institutions and practices, urban space was a complex material and symbolic environment that was shaped by – and that in turn shaped – institutions in historically specific ways'.[15]

---

paper by Antoni Roca-Rosell and Xavier Roqué, 'Physical Science in Barcelona', *Physics in Perspective*, 15 (4) (2013): 470–98.

14 For the natural history museum (the Museu Martorell), see: Alicia Masriera (ed.), *El Museu Martorell, 125 anys de Ciències Naturals (1878–2003)*, vol. 3 (Barcelona: Monografies del Museu de Ciències Naturals, 2006) – and Chapter 3 in this volume; for the reception of Darwinism, see Agustí Camós, 'Darwin in Catalunya. From Catholic Intransigence to the Marketing of Darwin's Image', in *The Reception of Charles Darwin in Europe*, edited by Eve-Marie Engels and Thomas F. Glick (London/New York: Continuum, 2008), 400–12; for Einstein's visit to Barcelona, see: Thomas Glick, *Einstein in Spain. Relativity and the Recovery of Science* (Princeton: Princeton University Press, 1988), pp. 112–19; see also: Roca-Rosell, Roqué, 'Physical Science in Barcelona'.

15 Sven Dierig, Jens Lachmund and J. Andrew Mendelson, 'Introduction: Toward an Urban History of Science', *Osiris*, 18 (2003): 1–19, p. 5.

Our attempt to write a new, genuine urban history of science inscribes itself into the recent 'urban turn' in the history of science.[16] This urban turn is a consequence of a more general trend labelled the 'spatial turn'.[17] The insight that the actual site where knowledge is produced matters has taken centre stage in numerous research projects in the last two or three decades. Historians of science have not only looked at 'traditional' sites of knowledge production such as laboratories, observatories and collections (cabinets and museums). They have also shifted their attention to sites such as the field, zoological and botanical gardens, hospitals, churches, princely courts, pubs and coffee houses, libraries, lecture theatres, salons and many more. This spatial turn has also led historians of science to look at 'sites' on a larger scale, such as the city, encompassing many of the places above.[18]

A spatial approach should tell us to what extent the specific urban setting of Barcelona shaped the practices of our actors and led to the emergence of the media and institutions related to science, technology and medicine. And the other way round: in how far these practices changed the urban environment. It is such a dialectical approach that is required in order to analyse the scientific practices in a particular quarter (Barrio Chino or Eixample), in a specific urban public place (the Parc de la Ciutadella, the central Plaça Catalunya or the Tibidabo mountain) and even in the whole of the city. It is of significance whether anarchists met in the quarter of Sants or the spiritists in the quarter of Gràcia. It matters where the amusement parks were located, where the Barcelonese visited museums and the zoo, and where the sick sought medical advice and treatment. In our volume, a detailed look at the myriad of sites will help us to understand the ways in which scientific knowledge was transferred and appropriated by different actors.

16 For a helpful review of the existing scholarship, see Martina Hessler, 'Technopoles and Metropolises: Science, Technology and the City. A Literature Overview', in Mikael Hård and Thomas J. Misa (eds), *The Urban Machine. Recent Literature on European Cities in the 20th Century*, A 'Tensions of Europe' electronic publication (July 2003) (www.tc.umn.edu/~tmisa/toe20/urban-machine), 57–82. Dorothee Brantz discusses the (all too tentative) connections between urban history, environmental history and the history of science in 'The Natural Space of Modernity: A Transatlantic Perspective on (Urban) Environmental History', in Ursula Lehmkuhl and Hermann Wellenreuther (eds), *The Historian's Nature: Comparative Approaches to Environmental History* (Oxford: Berg Publishers, 2007), 195–225; and in 'The Urban Discovery of Nature: Science, Zoos, and Natural History Museums', paper presented at the Third Watson Seminar on the Material and Visual History of Science 'How to write an urban history of science. New approaches and case studies', Barcelona, 6 June 2014.

17 To name but two of the most important contributions to the spatial turn: Jon Agar and Crosbie Smith, *Making Space for Science: Territorial Themes in the Shaping of Knowledge* (London: Macmillan, 1998); David N. Livingstone, *Putting Science in its Place: Geographies of Scientific Knowledge* (Chicago: The University of Chicago Press, 2003).

18 Charles W.J. Withers and David N. Livingstone, 'Thinking Geographically about Nineteenth-Century Science', in Charles W.J. Withers and David N. Livingstone (eds), *Geographies of Nineteenth-Century Science* (Chicago/London: The University of Chicago Press, 2011), 1–19; Diarmid A. Finnegan, 'The spatial turn: Geographical approaches in the history of science', *Journal of the History of Biology*, 41 (2) (2008): 369–88, in particular pp. 374–8.

In the following 10 chapters of this book, we reconstruct a host of specific sites. When we speak of a site, we usually think of a very concrete physical place than can be located on a map. It has an address and a material existence, such as an academy, an observatory, a laboratory, an athenaeum or a clinic. Yet this 'localized' understanding is insufficient to get hold of the scientific culture of Barcelona in its entirety. Many sites might only be occupied for a certain period of time. An anarchist meeting place might only be used for a month or two and be clandestine. The spiritists too might gather in changing venues. In these cases, site becomes a fluid category, part of dynamic urban networks in which people, objects and ideas constantly circulated and contributed to the making of a specific scientific culture.

In fact, the insight that the city, an assembly of so many small-scale sites, is a dynamic arena of scientific practices, and therefore requires and merits a specific perspective dates back a long time. Robert Kargon's seminal book on nineteenth-century Manchester was first published in 1977.[19] Yet it seems to us – as part of the spatial turn in history of science – that the city has acquired increased prominence as an object of research since roughly the turn of the millennium. In 2003, a special volume of *Osiris* was dedicated to *Science and the City*. Sven Dierig, Jens Lachmund and Andrew Mendelson suggested investigating the interactions between the city itself and the sites of production and circulation of knowledge. It is precisely at the intersection between urban expertise, cultural representations, places of knowledge and everyday practices that a truly urban history of science can emerge.[20]

Another important contribution is the edited volume *Urban Modernity* that came out in 2010.[21] Miriam Levin and others focused on the role of social elites (including scientists) in urban growth, international exhibitions and museums. Their case studies were Paris, London, Berlin, Chicago and Tokyo. This is representative for current scholarship on science and the city. So far historians of science have mostly focused on the metropolis. There are, of course, quite a number of individual articles dealing with specific aspects of the scientific

19 Robert Kargon, *Science in Victorian Manchester: Enterprise and Expertise* (Baltimore: Johns Hopkins University Press, 1977) (new edition, Piscataway: Transaction Publishers, 2009). Roy Porter, *London, a Social History* (London: Hamish Hamilton, 1994), is another example.

20 They distinguish four main interconnections, co-productions between science and the city: (1) The intersection between scientists and politicians gave rise to what was known as *urban expertise*; (2) Science played a crucial role in the *cultural representation of the city*: new literary genres (science-fiction and so on), daily papers, photography, films, marketing; scientific activities were deeply embedded in the social and material infrastructures of the city; (3) that is to say specific *places of knowledge* can only be properly understood through their role within the urban context; (4) There was a significant level of interaction between science and urban *everyday life*. Dierig, Lachmund and Mendelson, 'Introduction'.

21 Miriam Levin, Sophie Forgan, Martina Hessler, Robert Kargon and Morris Low, *Urban Modernity. Cultural Innovation in the Second Industrial Revolution* (Cambridge, Mass.: The MIT Press, 2010).

culture of smaller cities in this formative period around 1900.[22] Yet what are missing – as far as we can see with very few exceptions – are comprehensive and systematic studies. The existing scholarship on smaller cities has left little impact on the overall picture. We need to ask to what extent can results of studies on science in the metropolis be transferred to other urban contexts,[23] or whether a different interpretative framework is required to approach places such as Lyon, Marseille, Milan, Hamburg, Glasgow, Budapest, Amsterdam, Copenhagen and Antwerp and, of course, Barcelona. Are there different things to be learned from such supposed 'second cities'?

To repeat: the general historiography has neglected the scientific culture of Barcelona around 1900. Or rather, scientific practices are taken as givens, amalgamated as 'scientific and technological progress', as an *explanans* and never as an *explanandum*. And within history of science and the growing scholarship on urban science, cities such as Barcelona have not received distinct and in-depth treatment. As David Livingstone has argued: 'Bristol science, Manchester science, and Newcastle science are not the same as science in Bristol, science in Manchester, or science in Newcastle. The place-name adjectives in these designations attest to scientific practices that were constituted in different ways and different urban cultures'.[24] We shall therefore try to pinpoint the genuine and distinctive features of the urban history of 'Barcelona science'.

A genuine scientific culture

Despite the increased interest of historians of science in the urban history of science, it is our perception that much of this scholarship focused in one way or another on the elites of a given city: on leading scientists and engineers but also on town planners, architects, organizers of exhibitions, promoters of public health services, in short: experts of all sorts. It goes without saying that this focus on scientific elites is indispensable when dealing with science and the city around 1900. Yet this book proposes a different approach, a complementary one and possibly more than that. The ten following chapters assembled here ask how the scientific culture affected and changed the everyday life of the city dweller, how these Barcelonese citizens perceived and appropriated the changes and developments in

22 For nineteenth-century 'peripheral' cities such as Madrid and Lisbon, see: Tiago Saraiva, *Ciencia y ciudad. Madrid y Lisboa, 1851–1900* (Madrid: Ayuntamiento de Madrid, 2005); Antonio Lafuente and Tiago Saraiva, 'The urban scale of science and the enlargement of Madrid (1851–1936)', *Social Studies of Science*, 34 (4) (2004): 531–69.

23 Maiken Umbach used the concept of 'second cities' in order to advance a comparative perspective within history (not within history of science): Maiken Umbach, 'A Tale of second cities: Autonomy, culture, and the law in Hamburg and Barcelona in the late nineteenth century', *The American Historical Review*, 110 (3) (2005): 659–92.

24 Livingstone, *Putting Science*, p. 108.

their urban habitat induced by science, technology and medicine and how they themselves participated in a variety of ways in this new scientific culture.

In this sense this book attempts to combine two burgeoning historiographies, the scholarship on science and the city with the scholarship on popular science. 'Popular science', or rather the intricate relationship between science and its publics has proven to be a very fertile field in the understanding of the production of knowledge. To sum up some of the most important insights of this research in a bullet point way: The public of science is not a passive recipient. It actively appropriates knowledge in its own, often idiosyncratic way. It is crucial to stress the diversity and plurality of publics. Science popularization should not be described as a top-down model but rather as 'knowledge in transit', as an interactive process between scientists and their audiences. Historians of science now ask how knowledge is communicated, validated or rejected, appropriated or contested and so on in order to explain how a claim may gain acceptance.[25]

Combining these two historiographies of the urban history of science and of popular science, thus implies identifying the numerous audiences of science in a specific city. In the urban jungle of actors, objects, sites and events, there are lots of micro-histories which deserve further attention. In the same way that historians defend the epistemological value of a particular case study as opposed to big historical narratives, particular episodes *in the city* might shed new light on the complexities of the urban history of science. We will ask: How do certain scientific practices and technologies change everyday life? But also: how does the latter influence the former?

This quest leads us to the methodological question as to which sources we could use to reconstruct the perception and appropriation of science in the city around 1900. This question has kept a large number of historians of science busy: How to give this variety of publics, the students, the audiences of public talks, the users of technologies, the patients in the clinics and dispensaries, the radio amateurs and the instrument makers, the animal keepers and the technicians their proper voice? Testimonies from these groups are more often than not hard to come by. If they show up at all, they are mostly indirect, presented by a different voice situated on a higher level, an expert, a physician or a university professor. In other words: the historian is confronted with a stark asymmetry in the sources. The ambition of this book is to expand the existing attempts to try and reconstruct these voices, which generally have only a faint or strongly distorted echo in the material available. The contributors of this volume have identified a host of new primary sources such as city guides, medical guides, specialized journals, the archives of publishing houses, local societies and institutions, personal diaries, travel reports, but also reportage, novels, satirical magazines, advertisements as well as photos, caricatures, maps and architectural plans.

25 James A. Secord, 'Knowledge in Transit', *Isis*, 95 (4) (2004): 654–72. For a revision of 'popular science' as a historiographical category, see: *Isis*, 100 (2) (2009), 'Focus: Historicizing 'Popular Science'. For a general overview, see Agustí Nieto-Galan, *Science in the Public Sphere* (London/New York: Routledge, 2016).

This vast range of different practitioners played a crucial role in the struggle for scientific authority and the idiosyncratic appropriation of expert knowledge. Barcelona was the city of mediums, cheap telescopes, radio stations, underground sexual hygiene, electricity users and anarchist notions of health; an urban space in which citizens appropriated scientific culture in complex and varied ways, more or less explicitly, as a strategy, or perhaps simply as a tactic – in Michel de Certeau's terms – to resist the cultural hegemony of the elite and its science.[26]

Following this framework, we have organized the case studies of our volume under three different headings: 1. 'Control – Elite Cultures' gathers the Parc de la Ciutadella as a paradigmatic place for popular science and its agenda of disciplining (Hochadel & Valls), the political lining of natural history in the public sphere (Aragon & Pardo-Tomás), the new hegemony of laboratory medicine emerging in the Eixample district (Zarzoso & Martínez-Vidal), and the elite's control of the technology of amusement parks all over the city (Sastre-Juan & Valentines-Álvarez). 2. 'Resistance – Counter-Hegemonies' includes anarchist science and its places across the urban fabric (Girón Sierra & Molero-Mesa), spiritist practices and its social impact on the city (Balltondre & Graus), and the underground medical culture of the Barrio Chino (Zarzoso & Pardo-Tomás). 3. 'Networks – Experts and Amateurs' shows how technology spread through the city and how professionals, businessmen, but also aficionados and clients interacted. These last three case studies are devoted to the emergence of an 'urban astronomy' with its rich amateur culture (Roca-Rosell & Ruiz-Castell), the impact of the radio and its users (Guzmán & Tabernero) and popularization and domestication of electric lighting (Ferran & Nieto-Galan).

Readers, visitors, patients and users of new technologies appropriated what they were told or advised to do in their own way. They question or defy authorities, come up with their own interpretations, build their own devices or apply heterodox medical treatments. They also contribute to the legitimization of new scientific disciplines and specialties (in medicine, urban planning, sanitary reforms, natural history, astronomy and so on) and to the making of scientific authority in the urban public sphere. In what follows, we shall sum up very briefly what these ten chapters might contribute to a better understanding of science and the public in an urban, *fin-de-siècle* context.

At the very same time the city became a place for social control but also for social critique. On the one hand the major projects and institutions of this new urban culture of science aimed at educating and 'moralizing' the masses, and thus controlling them.[27] Manifestations of this attempt to achieve cultural hegemony and to exert social control are the two international exhibitions, the public museums, the Parc de la Ciutadella, and even amusement parks.

26 Michel de Certeau, *The Practice of Everyday Life*. 2nd ed. (Berkeley/Los Angeles: University of California Press, 2003) (1st French edition, 1980).

27 Stephen Jacobson, 'Interpreting Municipal Celebrations of Nation and Empire: The Barcelona Universal Exhibition of 1888', in William Whyte and Oliver Zimmer (eds), *Nationalism and the Reshaping of Urban Communities in Europe, 1848–1914* (London: MacMillan, 2011), 74–109.

Figure 1.2 City Council members visiting the Natural History Museum in the Parc de la Ciutadella, May 1917.

The Junta, the Board of Natural Sciences, transformed the new urban Parc de la Ciutadella into a place for the popularization of natural history. The Junta was caught between the notion of an ideal visitor, well behaved, educated and eager for self-improvement and the bugbear of the 'real' visitor, allegedly only interested in amusing himself or worse torturing animals at the zoo or poking at the enormous mammoth made of stone (Hochadel & Valls).

These leanings were also very tangible in the well-documented donations to the first public museum in Barcelona, the Museu Martorell, founded in 1882, which allows us to sketch the astonishing variety of publics. The donors range from the foremost aristocrats and businessmen to travellers, military officers and avid naturalists to the 'simple' citizen with no training in natural history finding a fossil on his Sunday stroll on Montjuïc. This sketch also shows the change in the social milieus of the donors in the decades between roughly 1880 and 1920: broadly speaking the 'cosmopolitan' aristocrats gave way to middle-class naturalists exploring the natural history of the region. The people of Barcelona knew that the Martorell was the place to bring strangely shaped eggs or a rare mineral. Yet that does not mean that the museum was a 'democratic' space. Particularly from 1900 onwards, it was geared towards a middle-class audience with specific ideological leanings: Catholic, conservative and Catalanist (nationalist) (Aragon & Pardo-Tomás).

In our Mendozian timeframe from 1888 to 1929, medical practice also underwent radical changes in Barcelona. It was precisely in the late 1880s that in the new district of the Eixample the first specialized practices ('clinics') emerged. The 'new age' of laboratory medicine with its host of new instruments and technologies manifested itself in these private clinics and in their specific functional architecture. This movement left the old city and its hospitals behind in a double sense: geographically but also epistemologically distancing itself from the 'old-fashioned' university-based medicine of the time. The physicians running these new 'clinics' needed to forge a new 'public', principally from among the bourgeoisie. They managed to create new demands advertising their 'modern' treatments. It was this group of patients that also provided empirical evidence, 'material' for scientific publications in these new fields of medical specialties. Thus these new practices shaped the identity of patients but also of the physicians (Zarzoso & Martínez-Vidal).

Technology had a profound impact on the everyday life of the Barcelonese in numerous respects. It also changed the way they spent their leisure time. In the early twentieth century, a number of amusement parks were set up in different parts of the city and outside of it, following a specific social geography. Each one of these parks provided 'technological fun' for a specific class of people ranging from the working class to the affluent. While the visitors were screaming in the roller coaster, they were in fact subjected to new regimes of pleasure immersed in an ideology of technological progress (Sastre-Juan & Valentines-Álvarez). Conservative Catalanism and Catholic science were the crucial ingredients of a cultural hegemony established around 1900. As in many other European urban contexts, the bourgeois, industrial city and its cultural hegemony sustained by the ideology of progress were confronted with popular values rooted in republican, working class, socialist and anarchist ideas.

In Barcelona, the 'divided city', torn apart by the increasing social conflict, 'counter-hegemonies' such as anarchism emerged. They not only promoted their own moral codes but also their understanding of what constitutes scientific knowledge.[28] For anarchists science and its popularization served as a mean of emancipation, most notably in their specific reading of Darwin. Anarchists in Barcelona had their own journals and their own meeting places (even if only transient), where they educated themselves. Barcelona was the city of popular athenaeums (ateneus populars), in which workers attended evening schools (including courses on science) after their long working hours. Health-related issues were particularly high on their agenda. (In a sense, bourgeois and anarchist views overlapped in this respect: science served the goal of personal improvement, intellectually *and* morally). For anarchists, the notion of an 'expert' and of authority was by definition problematic. Working-class audiences forcefully questioned the competence of invited speakers, thereby proving how far they had risen in their self-civilizing mission (Girón Sierra & Molero-Mesa).

Barcelona was also home to other associations (republican, freethinkers, masons), which competed with conservative Catholic organizations for cultural hegemony in education and in the public sphere. Starting in 1895, a popular library, the Biblioteca Arús, provided books and journals (including popular science) for the

28 For an analysis of resistance and counter-hegemonies in history of science, see: Agustí Nieto-Galan, 'Antonio Gramsci revisited: Historians of science, intellectuals, and the struggle for hegemony', *History of Science*, 49 (4) (2011): 453–78.

working classes, informed by anarchist, republican and masonic values.[29] Yet in the very same year, because of his lectures on Darwinism, the Catholic authorities shut down Odón de Buen's chair in natural history at the University of Barcelona.[30]

Spiritists should be understood as a 'counter-hegemonic' movement as well in particular in the Barcelonese form. They too challenged the existing epistemological hegemony and very consciously pursued heterodox practices and beliefs. Just like the anarchists, the spiritists were resolutely anti-Catholic, setting up their own journals, associations and schools in order to implement their social and moral agenda. They even started their own clinic treating a large number of patients with heterodox practices such as hypnosis and magnetism. Women played a central role in the spiritist movement in Barcelona, both as actors (media, authors, public speakers) and as target audience. Despite their complete lack of formal education, female media such as Amalia Domingo were able to become influential figures in Barcelona's public arena (Balltondre & Graus).

In stark contrast to Eixample, built upon hygienic principles, stands the Barrio Chino. The quarter today known as Raval was marked by poverty, prostitution, the entertainment industry – and medically speaking – by tuberculosis and venereal diseases. For the bourgeoisie, it represented both a symbol of moral decay as well as an irresistible lure. The Barrio Chino was characterized by constant attempts to medicalize its inhabitants. Yet as a close analysis shows, this medicalization went far beyond official attempts to contain the danger the Chino represented to public health. It reveals the plurality of – often self-declared – medical experts constantly manoeuvring between treatment, health (and sexual) education, commerce and entertainment bordering on pornography (Zarzoso & Pardo-Tomás).

In many fields science had not been fully professionalized in Barcelona around 1900. As elsewhere in Spain, amateurs of all sorts played a crucial role. We already mentioned the naturalists of the Museu Martorell. Many of them had no formal training in the natural sciences yet became the leading scholars for example in geology, palaeontology and speleology. A 'classic' field for amateurs is astronomy and the scientific culture of Barcelona provides a particularly instructive example. Even celebrated astronomers of international renown such as Josep Comas i Solà were strictly speaking amateurs. The widespread interest of the public and the support of wealthy patrons led to the establishment of well-equipped observatories, the founding of associations, the proliferation of telescopes and other instruments made possible by an increasing number of specialized workshops as well as valuable research in observational astronomy. The

29 Montserrat Comas and Víctor Oliva (eds), *Llum entre ombres: 6 biblioteques singulars a la Catalunya contemporània* (Vilanova i la Geltrú: Organisme Autònom de Patrimoni Víctor Balaguer, 2011).

30 Agustí Nieto-Galan, 'A Republican natural history in Spain around 1900: Odón de Buen (1863–1945) and his audiences', *Historical Studies in the Natural Sciences*, 42 (3) (2012): 159–89.

Barcelonese astronomers were furthermore highly active in popularizing their own work reaching thus an even larger public (Roca-Rosell & Ruiz-Castell).

Amateur astronomers were able to build communities of their own in the city. In these networks of sociability, they shared skills and exchanged information, objects and instruments in constant tension with institutional science and official expertise. This is similar to the case of early radio users in the 1920s. There was a sizeable community of radio enthusiasts building their own devices and voicing their ideas and concerns in the journal of Radio Barcelona, *RMB*. In other words, they were asserting their own expertise in the epistemological space they shared – not without friction – with accredited experts. Radio Barcelona itself also became a medium for science popularization. Astronomy and health-related issues were by far the most popular topics of the radio lectures, thus reflecting the scientific culture of the city (Guzmán & Tabernero).

Just like the radio antennas on the rooftops, the spread of electric lighting altered Barcelona's cityscape and reshaped the everyday life of its citizens. The nocturnal city became a new 'site', full of life, overcoming the routines dictated by natural light: from the 'electric' night parties at the 1888 Exhibition, to the new facilities at the operating rooms of the clinics (Zarzoso & Martínez-Vidal), and the electrical lights at the fun parks (Sastre-Juan & Valentines-Álvarez). The sophisticated display of lights and colours at the 1929 Exhibition was widely considered a triumph of electricity in the city. Part of this large-scale event was the Exposición de la Luz. This 'Exhibition of Light' was a mix of techno-scientific

Figure 1.3 Dr Josep Agell (1882–1973) and his co-workers at the laboratory of the Escola de Directors d'Indústries Químiques, August 1916.

popularization and commercial promotion of the new, electricity-based technologies aimed at the private consumer. New spaces such as showrooms appeared in the city skilfully displaying the virtues of electrical appliances. These initiatives rested on the emerging alliances between companies, engineers, advertisers and designers and created new experts as well as new publics, that is, (mostly female) customers (Ferran & Nieto-Galan).

Despite the evident male hegemony of the urban scientific culture of Barcelona at the time, this is only one example where women contributed in a variety of ways to the circulation of scientific knowledge in public as well as private sites in the city. Women emerge as donors of natural history specimens to public collections (Aragon & Pardo-Tomás), as bourgeois patients in the clinics of Eixample (Zarzoso & Martínez-Vidal), or as prostitutes in the Chino resisting programmes of medicalization (Zarzoso & Pardo-Tomás), as audiences of public lectures of astronomy (Roca-Rosell & Ruiz-Castell), as users of electrical appliances at home (Ferran & Nieto-Galan), as spiritist media (Balltondre & Graus) or as workers in textile factories.[31]

To be sure, much of the evidence cited in theses cases is indirect and remains fragmentary if not distorted. One may even mockingly say that our knowledge, no, our ignorance of 'the public', is made even more obvious by these approaches which leave so many questions unanswered. Yet that would be a mistake. Clearly acknowledging the large holes, which remain with respect to the publics of science, these methodologically varied attempts help us to paint a far richer and more interesting picture of the discursive processes in the scientific culture of Barcelona around 1900.

Urban modernity

Urban planner Peter Hall's optimistic understanding of cities as innovative milieus, as ideal contexts for overcoming cultural inertia and stimulating creativity will probably always be counterbalanced by historian of technology Lewis Mumford's vintage but still powerful vision of cities as inevitable sources of environmental degradation, social division and even cultural barbarism.[32] In a sense, these antagonistic visions of the modern city were already formulated in our Mendozian timeframe in the decades before and after 1900. Just like in any other city, our historic actors in Barcelona put forward both highly optimistic and sceptical views. Political and economic elites appropriated international exhibitions in order to legitimize their own interests. Technophilic discourses radiating with optimism celebrated the city as a paradigmatic space

31 Cristina Borderías, 'Women's work and household economic strategies in industrializing Catalonia', *Social History*, 29 (3): 2004, 373–83.

32 See Peter Hall, *Cities in Civilization* (New York: Pantheon, 1998) and Lewis Mumford, *The City in History* (New York: Harcourt, Brace & Co., 1961)

of modernity.[33] Railways, tramways, electric lighting, radio, sewage, power stations and public health campaigns established new ways of expertise and public trust in science, and changed dramatically the everyday life of normal citizens. Miriam Levin has recently stressed for example how institutions and social elites shaped city life in Paris with urban plans to legitimize discourses of linear progress.[34] Similarly Sophie Forgan has examined the way in which urban elites shaped London's life and culture, being full of enthusiasm for technological modernity.[35] Nevertheless, resistance against and criticism of the 'modern' city was also frequent, taking different forms in different local contexts. The technological progress and the rise of new expertise was apparently inevitably accompanied by the danger of pauperism, moral decadence, prostitution, epidemics, poor housing, noisy factories and never ending social conflicts.[36] In fact, as Mikael Hård and Markus Stippak have pointed out, different social groups construct diverging images of the city, in which modernity as a universal concept becomes complex, varied and never homogeneous. From a politically conservative point of view, the city seems to be a dangerous place, full of social unrest and political struggles, in contrast to the 'stable' countryside. It appears to be a site of degeneration and decadence, simply a repulsive place. But from a more progressive, liberal position, the late nineteenth-century industrial city soon became a paradigmatic place for social reforms (schools, public services, social housing, traffic systems, water works), innovation, creativity and amusement.[37]

In Barcelona, in the period under study, debates on the meaning of modernity, linked the scientific practices intimately with the city in a variety of ways. If we take for instance special events such as the 1888 and 1929 international exhibitions, it is easy to see how conservative elites were in general in favour of and even enthusiastic about the projects. By contrast, members of the liberal middle class expressed more reluctance, while the working class organizations were often the most critical.[38] At the 1888 Exhibition, elite discourses on the supposed benefits of the public were counterbalanced on different fronts.[39] Dissenting voices such as the writer and journalist Josep Yxart (1852–1895) very

33 Mikael Hård and Markus Stippak, 'Discourses on the modern city and urban technology, 1850–2000. A review of recent literature', in Hård and Misa, *The Urban Machine*, 35–56.

34 Miriam Levin, 'Dynamic Triad: City, Exposition and Museum in Industrial Society', in *Urban Modernity*, 1–12, pp. 11–12.

35 Sophie Forgan, 'From Modern Babylon to White City: Science, Technology and Urban Change in London, 1870–1914', in *Urban Modernity*, 75–132.

36 Hård and Stippak, 'Discourses', p. 36. See also: Clifford Geertz, *Local Knowledge* (New York: Basic Books, 1983); Bruno Latour, *Nous n'avons jamais été modernes. Essai d'anthropologie symétrique* (Paris: La Découverte/Poche, 1997).

37 Hård and Stippak, 'Discourses', p. 44.

38 Agustí Nieto-Galan, 'Scientific "marvels" in the public sphere: Barcelona and its 1888 International Exhibition', *HoST – Journal of History of Science and Technology*, 6 (2012): 33–63.

39 Jacobson, 'Interpreting Municipal Celebrations'. For the plurality of voices surrounding the first world exhibition in London, see: Louise Purbrick (ed.), *The Great Exhibition of 1851. New Interdisciplinary Essays* (Manchester: Manchester University Press, 2001).

much doubted the alleged modernity of the project. He gave a long list of reasons: the local and national economic crisis that the city was suffering at the time; the industrial weakness of the nation; the proximity in time to the 1889 Paris Exhibition; but also the lack of experience in organizing such an ambitious event; the lack of a reliable urban transport and communication network, together with the decay of the old town and its difficult renewal.[40] Equally, the workers', socialist and anarchist associations denounced the supposed modernity of the exhibition as a 'bourgeois' display of luxury goods and waste, and criticized the false rhetoric of peace, the commodification of knowledge and the corruption of the prize juries.[41] At the International Exhibition of 1929, for instance, Jorba, one of the city's most popular department stores, organized a public display on the 'progress of the city' in its own pavilion. A comparison between the way of living in 1888, at the time of the first international exhibition, and that of 1929 showed the proper path towards progress and modernity: education, public transport, housing, labour conditions, business and trade, all seemed to have achieved a new stage of 'modernity'.[42]

The chapters of this book are replete with examples of how the discourses of the elites shaped their own idea of modernity: there is the policy of medicalization, that is, the projects of hygiene and public health in socially desolate areas such as the Barrio Chino; the emergence of new medical specialities and chirurgical practices; the pursuit of a utilitarian agenda of applied natural history as in the case of Francesc Darder, the founder of the Barcelona Zoo; the vision of the engineers of new electrical grids as agents of social change and irresistible progress, and a kind of 'mechanical optimism' as a metaphor for the new industrial order in the amusement parks. In many of these examples, urban growth went hand in hand with what was understood as progress induced by science, technology and medicine.

But modernity also took other genuine forms. Take for instance a group of naturalist-priests at the Seminari de Barcelona who eagerly pursued natural history. Within a framework of biblical temporality, scholars such as Jaume Almera and Norbert Font explored and described 'Catalan' nature.[43] But this conservative, apparently anti-modern programme was not opposed to the idea of a utilitarian science as a crucial ingredient of industrial progress. Darder and his programme of applied natural history fitted very well into this kind of conservative modernity.

40 Joseph Yxart, 'La exposición por fuera', in *Ateneu Barcelonés, Conferencias públicas relativas a la Exposición Universal de Barcelona*, (Barcelona: Busquets y Vidal, 1889) 117–42, p. 123.

41 *El Productor, Periódico Socialista*, 20 April 1888, 4 May 1888, 14 December 1888; see Nieto-Galan, 'Scientific "marvels"', p. 45.

42 *Las Noticias*, 19 May 1929, *Suplemento: Barcelona y sus exposiciones, 1888–1929*.

43 Jordi Bohigas, '*Per Deú i per la Ciència*'. *L'Església i la Ciència a la Catalunya de la Restauració (1874–1923)* (PhD thesis, Universitat de Girona, 2011). Santos Casado, *Naturaleza Patria. Ciencia y sentimiento de la naturaleza en la España del regeneracionismo* (Madrid: Marcial Pons, 2012), p. 195.

Between 1888 and 1929, Barcelona became a complex technological network in which people, animals, commodities, electrical power and light circulated widely. Animal traction coexisted with electrical tramways and undergrounds and later automobiles. Gas, water, sewage and electricity were progressively extended through neighbourhoods, streets, buildings and apartments.[44] Public works became part of the everyday technological culture, not only in the massive transformation of the Parc de la Ciutadella and the Montjuïc area for the two international exhibitions, but also in the building of new railways and streets, such as Via Laietana in the centre of the Gothic quarter. A sort of industrial, machine-like modernity – which included amusement parks located on two symbolic mountains of the city, Tibidabo and Montjuïc – took hold of the city. The spread of electric lighting altered the habits of the everyday life of the Barcelonese. The sophisticated display of lights and colours at the 1929 Exhibition was specifically designed to convince its visitors of the transformative power of electricity and industrial progress. For the urban middle classes, the 'triumph' of electric lighting clearly indicated 'modernity'. When Radio Barcelona started broadcasting in 1924, it also began to shape the everyday life of the Barcelonese – another hallmark of this urban modernity so typical of the 1920s.

But there were other concepts of modernity in Barcelona that often clashed with the above one. We have already mentioned the struggle between hegemonic bourgeois science in its particular Barcelonese form of Catholic and Catalanist conservatism, and the 'counter-hegemonies' of republicans, freethinkers and in particular, anarchists but also spiritists. There were republicans such as the Darwinist Odón de Buen and alternative spaces such as Francesc Ferrer's Escuela Moderna. This educational project was based on different 'modern' values such as freethinking, secularism, coeducation and a monistic explanation of the universe. Its aim was to substitute religious worldviews by the new 'objectivity' of modern science. Ferrer was executed in October 1909, accused of (intellectually) instigating the riots in Barcelona in late July that year. In the so-called 'Setmana Tràgica' (Tragic Week), protesters attacked and burnt down Catholic churches and monasteries across the city. The uprising was brutally struck down.[45] In a similar way, spiritist circles campaigned in favour of a new positivist religion, a system of beliefs that challenged the moral authority of the Catholic Church. Spiritists considered their own approach to moral values and religion as 'modern'.

44 Mercedes Arroyo, *La industria del gas en Barcelona (1841–1933): innovación tecnológica, territorio urbano y conflicto de intereses* (Barcelona: El Serbal, 1996); Joan Carles Alayo, *L'Electricitat a Catalunya: de 1875 a 1935* (Lleida: Pagès, 2007). On the industrial vitality of the city see, Mercè Tatjer: *Barcelona. Ciutat de fàbriques* (Barcelona: Albertí, 2014).

45 For an overview see: Joan Connelly Ullman, *The Tragic Week: A Study of Anticlericalism in Spain, 1875–1912* (Cambridge, Mass.: Harvard University Press, 1968); Jordi de Cambra Bassols, *Anarquismo y positivismo: el caso Ferrer* (Madrid: Centro de Investigaciones Sociológicas, 1981); Pere Voltes Bou, *La Semana trágica* (Madrid: Espasa Calpe, 1995).

Figure 1.4 Electric lighting at the 1929 International Exhibition, designed by engineer Carles Buigas.

In the *City of Marvels*, as in many other western cities, the rhetoric of modernity was also intimately tied to the nineteenth-century building of the modern nation.[46] But the case of Barcelona was in many ways peculiar. In the public debates about the pros and cons of the two international exhibitions in 1888 and in 1929, the tension between the conflicting nationalisms – the Spanish and the Catalan one – was very tangible. In the Restoration period (1875–1931), the Spanish liberal state, with its central administrative centre in Madrid was weak. A considerable concentration of industrial urban elites on the periphery, most notably in Barcelona, demanded their share of political influence.[47] In addition, from the 1860s onwards, a cultural movement named 'Renaixença' sought a 'rebirth' of Catalan culture through the recovery of the Catalan language and literature. This movement strived for economic, political and cultural autonomy from Spain. Around 1900 Catalanism also extended to science. Many actors linked pure and applied knowledge to the progress and modernity of the Catalan nation. Scientific Catalanism attempted to shape a scientific culture – including here again science, technology and medicine – in a nationalist sense,

46 On the topic of science and nationalism, see: Carol E. Harrison and Ann Johnson (eds), *National Identity. The Role of Science and Technology*. *Osiris*, 24 (2009).

47 Joan-Lluís Marfany, *La cultura del catalanisme: el nacionalisme català en els seus inicis* (Barcelona: Empúries, 1995); Josep Termes, *(Nou) resum d'història del catalanisme* (Barcelona: Base, 2009).

using a modern scientific Catalan language and working in an academic context in which research and its applications to industry did not depend on Spanish centralism. Medical doctors, but also engineers and emerging professional scientists, contributed significantly as professional groups to the construction of a Catalan national identity.[48] The International Exhibition on Montjuïc had been conceived in a sense as the culmination of this. Yet this project was 'high-jacked' – in the view of Eduardo Mendoza and many others – by Primo de Rivera and his coup d'état in 1923 and the subsequent dictatorship.

It is the Spanish king and Primo de Rivera who inaugurate the exhibition in May 1929. And during the opening celebration – at least in the novel *City of Marvels* – they are spellbound as they watch Onofre Bouvila shooting up into the air with his propeller-less helicopter. Glancing down on Barcelona from the sky Onofre 'said with emotion, "how beautiful you are! And when I think what it was like when I saw it for the first time! [...] The country began right there, the houses were puny little things, and these teeming suburbs were just villages [...] The future is already my past"'.[49] Having said that Onofre crashes into the Mediterranean Sea. The navy rushes out immediately but neither remains of him nor of his miraculous machine are found by its divers or its 'radar'. Onofre simply disappears. Mendoza even 'quotes' the obituaries of several newspapers: '"The 1888 World's Fair marked the beginning of his career, and this fair of '29 its end"'.[50]

Already in the perception of many contemporaries, the second exhibition of 1929 marked the end of an epoch. Modernity had firmly settled into Barcelona. Yet what it meant and what it implied was more controversial than ever. The 1930s were marked by a series of political upheavals: the birth of the Second Spanish Republic and the establishment of Catalan autonomy in 1931; the outbreak of the Civil War and the anarchist revolution in Barcelona in July 1936, and finally, in 1939 the fall of Barcelona and the beginning of Franco's dictatorship. One may well describe these conflicts as a violent clash of different, in fact, incompatible notions of modernity. Yet it seems that the integral part that science and technology played in these antagonistic political ideologies has disappeared from sight just as Onofre Bouvila's flying contraption vanished into the sea without leaving a trace. This book will attempt to recover the urban culture of science of the *City of Marvels*.

48 For a more specific definition of 'Medical Catalanism', see: Enrique Perdiguero, José Pardo-Tomás and Àlvar Martínez-Vidal, 'Physicians as a public for the popularisation of medicine in interwar Catalonia: The *Monografies Mèdiques* series', in Faidra Papanelopoulou, Agustí Nieto-Galan and Enrique Perdiguero (eds), *Popularizing Science and Technology in the European Periphery* (Aldershot: Ashgate 2009), 195–217. For a definition of 'Technological Catalanism', see for instance: Jaume Valentines, *Tecnocràcia i catalanisme tècnic a Catalunya als anys 1930: els enginyers industrials, de l'organització del taller a la racionalització de l'estat* (PhD thesis, Universitat Autònoma de Barcelona, 2012); Antoni Roca-Rosell and Vicent Salavert, 'Nacionalisme i ciència als Països Catalans durant la Restauració', *Afers*, 18 (46) (2003): 549–63.

49 Mendoza, *City of Marvels*, pp. 414–15.

50 Ibid., pp. 415–16, quote p. 416. Once more Mendoza takes some poetic license here: The radar was, of course, invented later.

Part I

Control – elite cultures

2 Civic nature

The transformation of the Parc de la Ciutadella into a space for popular science

Oliver Hochadel and Laura Valls[1]

On 12 December 1907 the reproduction of a life-sized mammoth was inaugurated with great fanfare in the Parc de la Ciutadella just north of the city centre.[2] The prehistoric beast was made of armoured concrete, at the time a brand-new construction material. Yet for its makers, conservative Catholic naturalists, the creature with the enormous tusks represented a species that was lost through the biblical flood and could thus be read as an anti-evolutionary statement. At the same time the naturalists emphasized that it was a 'Catalan' mammoth, representing all the fragmentary mammoth fossils found in Catalonia since 1883. The massive sculpture standing over six metres tall was supposed to educate but maybe even more to impress. Yet its makers were afraid that the visitors to the Parc might climb on the mammoth (or worse) and therefore made sure to keep them at bay by erecting a fence around it.

Between 1872 and 1917 the Parc de la Ciutadella was transformed into a space for popularizing natural history. Its agenda was informed by a set of ideologies that were neither homogenous nor static. The mammoth embodies this peculiar mix of ideas: the construction of a 'Catalan' nature in tune with the rising political Catalanism of the time; a Catholic and concordist approach to science, insisting on the compatibility of revealed religion and natural history; the modernization of the country by popularizing the natural sciences in combining instruction and entertainment.

Other elements of the Parc such as the Museo Zootécnico, the Zoological Garden and its Fish Laboratory were characterized by an instrumental understanding of nature. Nature was to be exploited and more than that, to be actively transformed through specific animal technologies for the benefit of the economy and thus society as a whole.

1 We would like to thank Helen Cowie and Christina Wessely for their most helpful reviews. Ferran Aragon and José Pardo are our indispensable 'Parc partners'. Many authors of this volume contributed in different ways to this chapter. Research was supported by the Spanish Ministry of Economy and Competitiveness, grant HAR2013–48065-C2–1-P.

2 For reasons of brevity and easy identification, we will henceforth refer to the Parc de la Ciutadella as 'Parc'. It was originally mostly referred to as 'parque de Barcelona'. The specification 'de la Ciutadella' came into use much later. Due to historical circumstances and concrete context, the official name also changed between 'parque' (Spanish) and 'parc' (Catalan).

Recent historiography on nineteenth and early twentieth century science frequently uses the word 'civic', as in 'civic science' or 'civic pride'. Looking at institutions such as the natural history museum, the zoological gardens and associations of naturalists, these studies emphasize the involvement of interested, committed and often highly skilled citizens in the construction of an urban culture of natural history – largely outside universities and academies of science.[3]

With our Barcelonese case study we would like to build on this research. We entitled this chapter 'Civic Nature' in order to focus on the different uses and understandings of nature. Nature was civic in a triple sense: 1. Educational: The Parc with all its different features was supposed to civilize its visitors, that is, not only teach them natural history but also educate them to be better citizens. 2. Patriotic: At the same time many of the naturalists active in the Parc pursued a nationalist programme eager to present (read: construct) 'Catalan' nature. 3. Economic: Much of the Parc agenda is marked by a distinctly utilitarian way of doing science that may best be summed up in the formula of applied natural history. While the Barcelona Parc was highly typical of its time with respect to its civilizing mission, the admixture of nationalism and an explicit economic objective point to a peculiar blend of ideologies. In order to fully understand this 'civic nature', we will have to situate the Parc in its broader historical context.

This chapter will therefore first describe the emergence of the public park in the nineteenth century, a prime site for science popularization – and for disciplining its visitors. We will then zoom in on the Parc de la Ciutadella. Among the numerous institutions founded and activities launched between 1872 and 1917 (see Figure 2.1), we will look at three in greater detail: the zoological garden and its acclimatization programme; the nationalist appropriation of the mammoth and of a geological open-air exhibit of stone samples; and finally the activities aimed at the breeding of fish such as the Festa del Peix and the exhibition on pisciculture.[4]

A place for science – public parks in the nineteenth century

In many respects the Parc and its ideological underpinnings are typical for the second half of the nineteenth century. In numerous cities, municipal authorities pursued a policy of urban renewal – the 'greening' of the city – to counter

3 The most important study in this respect is Lynn K. Nyhart, *Modern Nature: The Rise of the Biological Perspective in Germany* (Chicago/London: The University of Chicago Press, 2009). Also see her earlier article 'Civic and economic zoology in nineteenth-century Germany: The "living communities" of Karl Möbius', *Isis* 89 (4) (1998): 605–30 and Denise Phillips, *Acolytes of Nature: Defining Natural Science in Germany, 1770–1850* (Chicago/London: The University of Chicago Press, 2012).

4 For a first exploration of this topic, see Laura Valls, 'El museo de ciencias naturales de Barcelona (1882–1917): popularización de las ciencias naturales dentro y fuera del museo', *Geo Crítica. Cuadernos Críticos de Geografía Humana* 15 (918) (2011).

the effects of industrialization.[5] The rapid growth in population of urban centres, the disappearance of commons and the private character of existing parks made it increasingly difficult for the average city dweller to have any access to nature. This situation alarmed reformists and governments because – as they saw it – workers in particular would spend their leisure time engaged in morally reproachable activities such as gambling and drinking. Public parks, free of charge, were supposed to provide a social remedy. And more than that: for the hygiene movement of the time and its demands for a healthier environment in cities struggling with tuberculosis and the occasional outbreak of cholera, the public park was also a medical remedy.[6]

The new public parks came with a specific socio-political programme: 'recreation should be associated with improvement'.[7] In a time of growing social tensions in increasingly industrialized urban centres, a policy was formulated which argued for the social control of the working classes by means of 'civilized' entertainment and recreation, which the public park could provide. The idea was to provide a 'class-free' space, yet conflicts about proper conduct and 'appropriate' activities were rather the norm.[8] As Hazel Conway observes: 'Some saw parks as places where the classes could mix, but on middle- and ruling-class terms – "gentility mongering places"'. And later she writes: 'Parks were one manifestation of the rise of modern institutions to control the physical and social processes of urbanisation [...] recreation was increasingly controlled by commercial or local authorities'.[9]

Recreation in the urban green may come in different forms, for example, walking, sports, rowing on artificial lakes or concerts. Within this agenda to regulate the behaviour of the public and to educate it, science, or rather the popularization of science, played a pivotal role. More often than not a major public park would also be home to one or several of the three main 'popular' scientific institutions at the time: the natural history museum, the zoological garden and the botanical garden. And quite often a major park provided the venue for an international exhibition.

In Vienna, for example, the Schönbrunn Park hosts to the present day, the Schönbrunn Menagerie (founded in 1752, open to the public since 1778) and the Botanische Garten, until 1918 both property of the Austrian Emperor. In

5 There is a vast literature on the public park in the nineteenth century. To quote but a few relevant books: Hazel Conway, *People's Parks: The Design and Development of Victorian Parks in Britain* (Cambridge: Cambridge University Press, 1991); Roy Rosenzweig and Elizabeth Blackmar, *The Park and the People: A History of Central Park* (Ithaca/London: Cornell University Press, 1992); Heath Massey Schenker, *Melodramatic Landscapes: Urban Parks in the Nineteenth Century* (Charlottesville: University of Virginia Press, 2009).

6 Diego Armus, *La ciudad impura. Salud, tuberculosis y cultura en Buenos Aires, 1870–1950* (Buenos Aires: Edhasa, 2007), Chapter 1.

7 Conway, *People's Parks*, p. 29.

8 Dorceta E. Taylor, 'Central Park as a model for social control: Urban parks, social class and leisure behavior in nineteenth-century America', *Journal of Leisure Research,* 31 (4) (1999): 420–77.

9 Conway, *People's Parks*, pp. 206, 222.

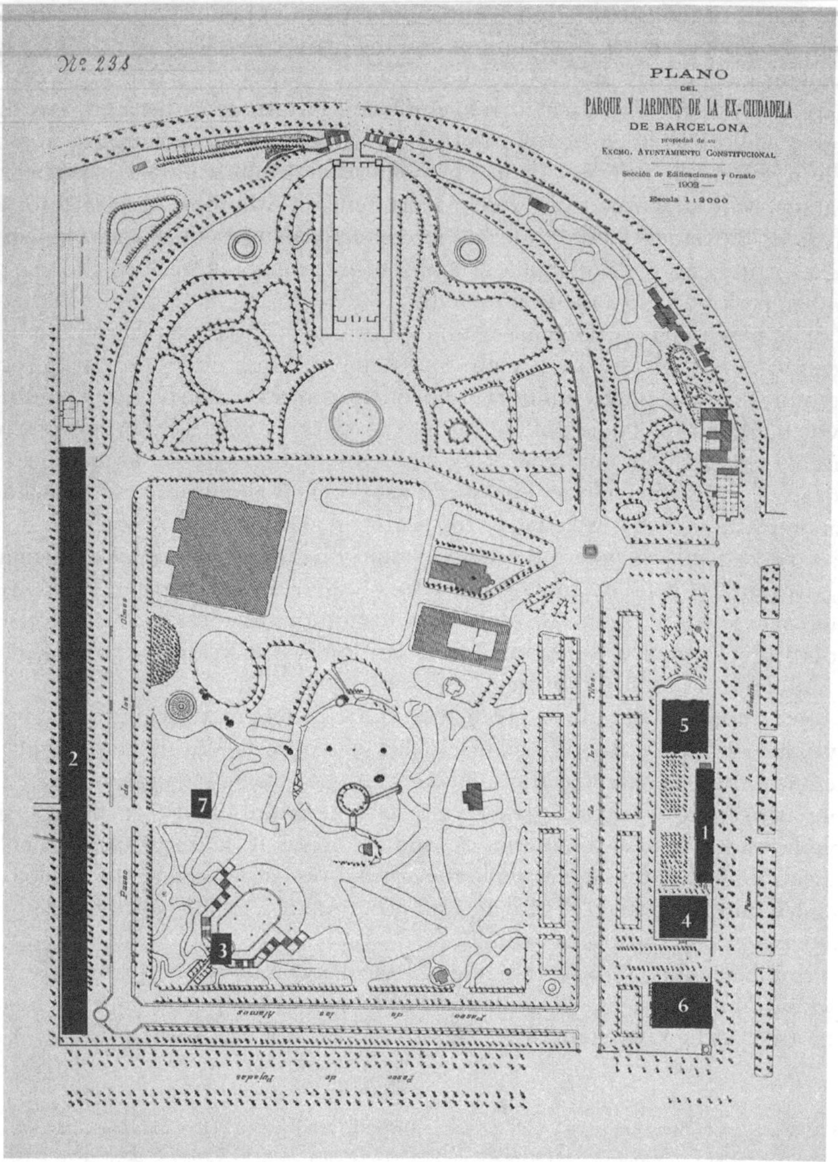

Figure 2.1 Parc de la Ciutadella with installations devoted to natural history (1882–1917), located in a map of 1902: 1. Museu Martorell, 2. Parque Zoológico, 3. Cascada Monumental: Aquarium, 4. Hivernacle (greenhouse), 5. Umbracle ('shade-house'), 6. Castell dels Tres Dragons: Pisciculture Exhibition (1912) and Museu de Catalunya (1917); and 7. Mammoth Sculpture.

Paris, all three institutions (Muséum d'Histoire naturelle, Ménagerie and Jardin des Plantes) are still assembled within one park in the centre of the city (dating back to the late eighteenth century). In Dublin the zoo (opened in 1831) is still situated in the ample Phoenix Park, also the venue of the 1907 International Exhibition. The Central Park in New York (inaugurated in 1873) had a small zoo (already opened in 1864), while the American Museum of Natural History is right at the side of Central Park (opened in 1877). The Barcelonese case fits this pattern very well. Hence science, or in more concrete terms natural history, was nearly always part of the educational agenda of any major public park from the mid-nineteenth century onwards. Yet as we shall see, the Parc in Barcelona is remarkable due to the sheer mass and density of institutions and activities pursuing natural history and its popularization.

The creation of the Parc de la Ciutadella

In 1872, Barcelona City Council approved a plan to turn the site of the former military citadel just outside the city walls into a public park. The fortress – demolished in 1868 – had been for nearly a century and a half a symbol of 'Bourbon oppression' (read: the Spanish monarchy and the central government of Madrid) of the city. After the legal handover of the terrain in 1869, the master-builder Josep Fontserè (1829–1897) won the call for tenders advertised by the city.[10] The intention was to offer a space for learning, leisure and rest for – at least in theory – all citizens. Yet for Fontserè, just like for most proponents of parks prior to the 1880s, the crucial 'target group' was adult males. Men were considered as more likely to have their health impaired in the city due to the strain of professional life in business, factories and public life.[11]

Fontserè's scheme, that is, the elements he had in mind for the Parc, was eclectic. His concept of nature as it materialized in the Parc oscillated between the romantic and the utilitarian and instructive. The first architectural landmark to be built was the enormous Cascada in the northern corner of the Parc, started in 1875 and inaugurated in 1881. The cascade hosted a grotto with picturesque arches as entries, the walls and ceilings were marked by sprawling rock

10 For the early history of the Parc, see: Adolfo Florensa, 'José Fontserè y el Parque de la Ciudadela', in *Miscel·lània Fontserè Barcelona* (Barcelona: Gustavo Gili, 1961), 175–82; Miquel Vidal, 'El Parque de la Ciudadela, entre el Romanticismo y la "Renaixença"', *Arquitecturas Bis. Información gráfica de actualidad*, 40 (1981), 14–17; Ramon Grau and Marina López, 'La gènesi del Parc de la Ciutadella: projectes, concurs municipal i obra de Josep Fontserè i Mestre (1868–1885)', in *El pla de Barcelona i la seva història. Actes del I congrés d'història del Pla de Barcelona* (Barcelona: La Magrana, 1984), 441–67; Manuel Arranz, Ramon Grau and Marina López, *El Parc de la Ciutadella. Una visió històrica* (Barcelona: Ajuntament de Barcelona-L'Avenç, 1989); Stephanie Kickum, 'Auf dem Weg zu internationaler Bedeutung: Parkanlagen in Spanien', in Angela Schwarz (ed.), *Der Park in der Metropole. Urbanes Wachstum und städtische Parks im 19. Jahrhundert* (Bielefeld: Transcript, 2005), 161–218, pp. 193–209.

11 José Fontserè, *Proyecto de un parque y jardines en los terrenos de la ex-ciudadela de Barcelona* (Barcelona: Narciso Ramírez y Compañía, 1872), p. 5; Terence Young, 'Modern Urban Parks', *Geographical Review – Thematic Issue: American Urban Geography*, 85 (4) (1995): 535–51, p. 538.

formations. A nearby artificial lake with an island completed this romantic part of the Parc enabling the visitor to experience 'real nature'. Yet from the very beginning, the Parc was also supposed to provide a space for study, instruction and the development of industry. Fontserè's plan included sites for storing seeds, greenhouses in order to grow specific botanical specimens, a school, a museum of botany as well as a cabinet of mineralogy and zoology and a zoological garden.[12] In 1882, the Museu Martorell was inaugurated in the Parc, the first public museum in Barcelona, devoted to natural history and archaeology.[13]

The coexistence of two concepts of nature, romantic and utilitarian, in Fontserè's plan may be typical of the time, at least in Catalonia. Roughly speaking, the romantic notion of nature is predominant in the middle of the nineteenth century and wanes towards the end of the century while the utilitarian notion is clearly on the rise at that stage.

In 1888 the city of Barcelona hosted the Exposición Universal in the grounds of the Parc.[14] This major event changed the face of the Parc by adding new architectural elements such as the Umbracle, the Hivernacle (both for plants, a 'shade-house' and a greenhouse) and the Café-Restaurant, known as the Castell dels Tres Dragons.

In order to manage and extend the facilities of the Parc devoted to natural history, the City Council created in 1893 the 'Junta técnica de los Museos de Ciencias Naturales, Parque Zoológico y Jardín Botánico'. It is telling that this Junta (henceforth Junta técnica) actually bore the trinity of 'popular science park' (natural history museum, zoo, botanical garden) in its name. The Junta técnica was made up of representatives from the City Council and various scientific institutions in the city, such as the University, the Reial Acadèmia de Ciències i Arts de Barcelona and the Institut Agrícola Català de Sant Isidre. As we shall see, an important member of this Junta técnica was the poly-faceted Barcelonese naturalist, Francesc Darder (1851–1918), the founding director of the zoo. His lifelong credo was to develop and exploit the economic opportunities provided by animals of all sorts. Yet the Junta técnica had little leeway not least because of financial restrictions. This changed in 1906 with the creation of the Junta Autònoma de Ciències Naturals (henceforth simply referred to as Junta), which replaced the Junta técnica. The new Junta reflected the rise of political Catalanism and marked a turning point in the agenda setting for the Parc with its special emphasis on the natural history of Catalonia and thus national identity.[15]

12 Fontserè, *Proyecto*, pp. 12–16.

13 See Chapter 3 of this volume.

14 There is an enormous amount of literature on this exhibition: For a history of science point of view, see Agustí Nieto-Galan, 'Scientific "marvels" in the public sphere: Barcelona and its 1888 International Exhibition', *HoST – Journal of History of Science and Technology*, 6 (2012): 33–63.

15 For the general context, see Antoni Roca-Rosell and Vicent L. Salavert, 'Nacionalisme i ciència als Països Catalans durant la Restauració', *Afers: fulls de recerca i pensament*, 18 (46) (2003): 549–63.

In his prosopographical studies of the Barcelonese naturalists of the first decade of the twentieth century, Ferran Aragon identified a closely-knit network of Catholic and Catalanist scientists.[16] These like-minded scholars, among them the priest Norbert Font (1874–1910), the canon Jaume Almera (1845–1919) and Artur Bofill (1846–1929), the director of the Museu Martorell, also dominated the new Junta. The ideological leanings of this Junta (in its majority, not in its entirety) were Catalanist, conservative and Catholic. Their conservative convictions rested on a conciliatory understanding of science and Catholic faith as being perfectly compatible with each other. It was the geologist Norbert Font, disciple of Almera, who formulated the programme of the new Junta. In this agenda he proposed to convert the Parc into a 'veritable park of culture'.[17] In what follows we shall outline the different institutions promoted by the Junta técnica and later on by the Junta. The question is what concept of civic nature drove their initiatives?

Acclimatizing animals – the Barcelona Zoo

Fontserè's project already toyed with the idea of building a 'casa de fieras' (a house for the beasts).[18] Yet for most of the 1870s and 1880s, there were mainly cows roaming the greens. In the late 1880s the city already used the Parc as a 'parking space' for exotic animals. Live gifts from foreign politicians and businessmen such as caimans, crocodiles and pheasants were deposited there. In 1886, Francesc Darder demanded that the Parc be reformed in order to provide more attractions than merely the 'monotonous vegetation'.[19] Darder had started to work as a taxidermist in the Museu Martorell in 1883, that is, he was well familiar with the Parc. What he had in mind was a zoological garden. And indeed, on 24 September 1892 the Parque zoológico de aclimatación y naturalización was inaugurated on the far side of the Parc.

Darder became the first director and the name of the zoo reflected his objectives. The idea to acclimatize economically useful species from different climates to the European one was thriving in the mid-nineteenth century, in particular in France.[20] Exotic animals were supposed to supply meat, eggs, wool and feathers as well as traction force to national economies. Yet already by 1870,

16 Ferran Aragon, *Anàlisi prosopogràfica del Catòlic-Catalanisme Científic (1904–1910)* (Master Thesis, Universitat Autònoma de Barcelona, 2012).

17 Norbert Font, 'Junta autònoma de Ciències Naturals', *La Veu de Catalunya*, 18 September 1907 (evening edition), 1–2, and 'Junta autònoma de Ciències Naturals II', *La Veu de Catalunya*, 2 October 1907 (morning edition), 1. Quote from the second article.

18 Fontserè, *Proyecto*, p. 15.

19 Francesc Darder, 'Variedades', *El Naturalista*, 3 (1886), 19. Reprinted in *La Vanguardia*, 4 March 1886, 4–5. Darder was a veterinarian, taxidermist, and author of natural history journals and books. A thorough investigation of his manifold activities is a desideratum.

20 On the acclimatization movement of the nineteenth century, see: Michael A. Osborne, *Nature, the Exotic, and the Science of French Colonialism* (Bloomington: Indiana University Press, 1994).

this movement had lost its momentum because it had largely failed to be economically viable and had gone out of fashion as a research programme, at least in Europe.[21] There had been a Spanish variant of the 'acclimatization movement' spearheaded by Mariano de la Paz Graells that was closely associated with the French one. Yet its institutional base was feeble, the acclimatization zoo that Graells founded in 1858 in the grounds of the Jardín Botánico of Madrid had to close down after the revolution of 1868 with virtually nothing to show for in terms of successful and profitable acclimatization.[22]

Therefore it may appear as an historical oddity to found an acclimatization zoo as late as 1892. Yet upon closer inspection, there is a clear connection. Darder saw himself explicitly in the tradition of the French acclimatizers and of Graells. In 1865 the Barcelonese banker Lluís Martí-Codolar started his collection of exotic animals on a private estate in Horta, at the time still outside the borders of the city. Already from the early 1870s, Darder was acting as an advisor to Martí-Codolar, taking care of the animals and purchasing new ones. Eventually this very collection of roughly 160 animals became the nucleus of the municipal zoo when financial difficulties led Martí-Codolar to sell it to the city of Barcelona in the spring of 1892.[23] Darder's particular focus was on poultry but he also advocated the acclimatization and breeding of camels and llamas. During the first two decades of its existence, the Barcelona zoo held regular sales and auctions. One could purchase animals (dogs, wild boar, goats) and in particular birds (poultry, pheasants, ducks, pigeons) as well as fertilized eggs.

Unlike most other late nineteenth-century zoological gardens, the Barcelona zoo did not charge an entry fee until the end of the 1920s.[24] Because the subsidies of the City Council were insufficient to cover the running costs, Darder resorted to the sale of animals and their products.[25] This was not atypical for the time. The zoo of Antwerp was particularly business-minded and became a centre for the international trade in animals in the later part of the nineteenth

21 Warwick Anderson, 'Climates of opinion: Acclimatization in nineteenth-century France and England', *Victorian Studies,* 35 (2) (1992): 135–57; Osborne, *Nature*, p. 127.

22 Santiago Aragón, *El Zoológico del Museo de Ciencias Naturales de Madrid. Mariano de la Paz Graells (1809–1898), la sociedad de aclimatación y los animales útiles* (Madrid: Museo Nacional de Ciencias Naturales, CSIC, 2005), pp. 142–88 and particularly pp. 199–200.

23 The literature on the early history of the Barcelona zoo is sparse but see: Emili Pons, *El parc zoològic de Barcelona. Cent anys d'historia* (Barcelona: Edicions 62, 1992), pp. 48–53; Ramon Alberdi and Rafael Casasnovas. *Els Jardins de Martí-Codolar. La Granja Vella* (Barcelona: Ajuntament de Barcelona, Districte d'Horta-Guinardó, 1999), pp. 46–53; Rossend Casanova, 'Francesc d'Assís Darder i l'origen del Parc Zoològic de Barcelona', *Revista de Catalunya Barcelona,* 142 (1999): 36–41; Daniel Venteo, 'La Ciutadella i el Zoo, una història ciutadana paral·lela', in *Parc del zoo, el cor de la Ciutadella* (Barcelona: Ajuntament de Barcelona, 2009), 15–61.

24 Pons, *El parc zoològic*, p. 50.

25 Darder complained regularly to his superiors that he had barely enough money to feed the animals; for example, AMCNB (Arxiu històric, Museu de Ciències Naturals de Barcelona), 3/6, 18 March 1902; also Pons, *El parc zoològic*, p. 64. The first guide to the zoo also mentioned the prices for the animals on display, see *Catálogo del Parque zoológico municipal* (Barcelona: Henrich, 1897).

Figure 2.2 Avi and his admirers: the elephant was the most popular animal in the zoo.

century. In a few years it was able to finance half of its budget through the sale of animals. The Jardin d'acclimatation in Paris, the Schönbrunn Menagerie (Vienna) and other zoos sold surplus animals as well as feathers, eggs and even excrements to anybody willing to pay the money.[26]

In contrast to these major European zoos, the Barcelona zoo was very small, only covering about two hectares of the grounds of the Parc (today it covers 13 hectares, that is, almost half the surface of the Parc).[27] The Barcelonese loved the elephant Avi (see Figure 2.2) but all in all there were few spectacular animals (read: large quadrupeds). In 1912, in his 600-page volume on the history of the modern zoo, Gustave Loisel dedicates exactly one sentence to the Barcelona zoo, stating that is was 'of no interest'.[28] As the biologist Ramón Turró put it in his obituary of Darder in 1918, people wanted exotic-looking birds, 'lions, tigers and panthers, who would roar night and day'. According to Turró, Darder disappointed the expectations of everybody, the general public, the media and the City Council with his practically-minded approach and his shunning of the show.[29] Yet this assessment might be overstated. Darder's call for an applied natural history and the acclimatization of exotic animals certainly did not fall on deaf ears. It is echoed, for example, in a report of the Junta técnica presented

26 Gustave Loisel, *Histoire des ménageries de l'Antiquité à nos jours* (Paris: Doin, 1912), vol. 3, pp. 269, 290–1; Roland Baetens, *The Chant of Paradise. The Antwerp Zoo: 150 Years of History* (Tielt: lannoo, 1993), pp. 94–5; Osborne, *Natura*, p. 122; for Vienna, see Gerhard Heindl, *Start in die Moderne. Die Entwicklung der kaiserlichen Menagerie unter Alois Kraus* (Wien: Braumüller, 2006), p. 39.

27 Casanova, 'Francesc d'Assís Darder', pp. 37, 39.

28 Loisel, *Histoire des ménageries*, p. 110.

29 Ramón Turró, 'En Francesc Darder', *La Veu de Catalunya*, 16 April 1918 (morning edition), 9.

to the City Council in 1899. Its members once more advocated the development of the scientific facilities of the Parc, mentioning the acclimatization of new species at the zoo and the creation of an aquarium. The target group of these efforts were the workers struggling to make a living. The idea was that they would grow plants and raise useful animals themselves at home in order to obtain cheap and healthy food. The report explicitly states: 'The acclimatization of domestic animals, whose keeping should be cheap and easy, would help to resolve one of the questions which contemporary society tends to dwell on most'.[30] In other words: the aim was to dispel the social tensions created by industrialization and urban poverty. In this sense the Parc and its institutions were conceived of as producers and distributors of both knowledge and actual specimens.

The inauguration of the Museo Zootécnico (Museum of Animal Technologies) in the very same year 1899 illustrates once more how Darder and the Junta técnica conceived of nature as a resource to be exploited economically. The objects, among them stuffed animals, skeletons, fur, feathers and so on originated from the zoo and from Darder's private collection. There were sections dedicated to comparative anatomy, embryology, hunting and fishing, as well as a library and a reading cabinet. The Museo Zootécnico was supposed to show potential industrial applications of natural history. There is very little information on how these materials were actually exhibited – and even less to indicate how far this collection led to any useful inventions.[31]

And more generally: There is little evidence that these initiatives advocated by Darder and promoted by the Junta did indeed mitigate the lot of the lower classes. In a more cynical reading one may qualify these attempts as predominantly rhetorical, as a tranquilizer for the anxieties of the bourgeois classes due to the social conflict rampant at the time.

A fatherland made of stone: The mammoth and the petrographic collection

Political Catalanism became the major political force in Barcelona in the general elections of 1901. In November 1905 its party, the Lliga Regionalista – conservative, Catalanist and Catholic – achieved a landslide victory in the municipal elections. As mentioned above, this shift in politics led to the instalment of a new Junta in 1906, now bearing a Catalan name: Junta Autònoma de Ciències

30 Ayuntamiento Constitucional de Barcelona, *Informe relativo a la organización y sucesivo desenvolvimiento de los Museos y Parque histórico-naturales de Barcelona* (Barcelona: Junta Técnica de los Museos de Ciencias Naturales, Parque Zoológico y Jardín Botánico, 1899), p. 25.

31 *Anuario Estadístico de la Ciudad de Barcelona* (Barcelona: Henrich, 1902), pp. 309–10; Rossend Casanova, 'El museo zootécnico de Barcelona. Uso y abuso del patrimonio efímero tras la Exposición Universal de 1888', in *Arte e identidades culturales* (Oviedo: Universidad de Oviedo, 1998), 425–32, p. 428; Nieto-Galan, 'Scientific "marvels"', p. 61.

Naturals.[32] This new Junta was, unlike its predecessor, endowed with a budget of its own – and it had a clear agenda. The sculpture of the mammoth (see Figure 2.3) inaugurated in December 1907 was the first of a series of reproductions of prehistoric large mammals from Catalonia that were to be built all over the Parc. The project was inspired by the dinosaur sculptures exhibited since 1854 in the gardens of the rebuilt Crystal Palace in Sydenham (London).[33] The aforementioned naturalists, Jaume Almera and Artur Bofill, had seen these sculptures in 1888, while participating at the Fourth International Geological Congress. They were fascinated by the visual power of these archaic creatures. Their idea to adapt a similar project for the Parc was stalled for 20 years until economic and political conditions allowed for it.

Around 1900 there was a revival of sculpting of three-dimensional life-sized dinosaurs. One of the reasons for this was a new building material. Reinforced concrete made the reconstructions easier and cheaper than 50 years before.[34] The 18 life-sized reproductions of mostly dinosaurs made by the sculptor Josef Pallenberg in 1911 for the Zoological Garden of Carl Hagenbeck in Stellingen near Hamburg are a splendid example of this revival.[35]

Reinforced concrete was introduced into Catalonia in the 1890s and at first its principal use was private (water deposits, laundries, sewers and so on).[36] In the 1900s it was deployed in public works such as viaducts, bridges and canalization but still not in the city. The Parc offered a highly visible site to display the astonishing properties of this brand new material. Thus, the Junta gave 'life' to an extinct animal using a distinctly modern technique: following a small-scale model of a mammoth made of wood by the sculptor Miquel Dalmau, the company of Claudi Duran, which specialized in works of iron and concrete, produced the life-sized sculpture.

The Junta had specifically chosen to begin with the mammoth because it was also the first species of prehistoric mammal to be unearthed in Catalonia. (As it turned out, it was the only sculpture of the planned series of extinct animals that was actually built.) In 1883, a piece of a tusk of a mammoth was dug up in the quaternary stratum of the river Llobregat. More fossil discoveries were to follow, though no complete mammoth was found.

Nevertheless the Junta presented the sculpture in the Parc as a 'Catalan' mammoth, an autochthonous being preserved in the soil of the fatherland and now resurrected. The project of sculptures also included large relief maps representing geological landscapes. With the help of these media, the Junta wanted to

32 Josep Maria Camarasa, *Cent anys de passió per la natura. Una història de la Institució Catalana d'Història Natural: 1899–1999* (Barcelona: Institució Catalana d'Història Natural, 2000), p. 40.

33 AMCNB 15/1, 15 September 1906; James A. Secord, 'Monsters at Crystal Palace', in Soraya de Chadarevian and Nick Hopwood (eds), *Models. The Third Dimension of Science* (Stanford: Stanford University Press, 2004), 138–69.

34 Secord, 'Monsters', p. 161.

35 Herman Reichenbach, 'Carl Hagenbeck's Tierpark and Modern Zoological Gardens', *Journal of the Society for the Bibliography of Natural History,* 9 (4) (1980): 573–85, p. 579.

36 'Nueva industria', *Resumen de arquitectura,* 1 January 1895, 7.

Figure 2.3 An antediluvian creature protected by a fence – the mammoth in the Parc.

show a 'natural' origin of Catalonia, clearly distinct from Spain and conceived of by God from the beginning of time.[37] Yet apart from this nationalist appropriation of the mammoth, the creature was also presented as an 'antediluvian animal'. The reconstruction represented a species that belonged to a world that

37 Font, 'Junta autònoma II'; Jordi Bohigas, *'Per Déu i per la ciència'. L'Església i la ciència a la Catalunya de la Restauració (1874–1923)* (PhD thesis, Universitat de Girona, 2011), p. 278.

had been lost through the Flood. Although only implicitly, the Junta – in its majority conservative Catholics – did take a stand against the theory of evolution. They situated the mammoth in the framework of biblical temporality. The plan of the Junta to exhibit several sculptures of 'Catalan' and 'antediluvian' creatures fused a nationalist and a biblical concept of 'nature'. Despite their extinction in the Flood, animals such as the mammoth retained their national identity.[38]

At the same time as the installation of the mammoth, the Junta launched another patriotic project in the Parc: the display of 40 large stones from all over the country in front of the Museu Martorell.[39] This open-air exhibition included big slabs of limestone, gypsum, basalt, granite, porphyry and slate. Although still in the making, the collection aspired to be a comprehensive catalogue of Catalan geology. The shape of the stone samples was standardized in order to give unity to the collection. They were usually extracted from a quarry and cut into a given size (between 1.5–2 metres by 0.5–0.6 metres). In the Parc they were placed upon pedestals (0.7 metres wide) and exhibited as 'raw materials' with a polished side. The blocks were marked with signs informing the visitor of the Parc about the type of stone, the location and its origin. The aim was to show the richness and utility of the geological resources –, that is, of 'mother nature' – for the modernization of the country. The project had an explicitly 'regional' character and as the Junta boasted, it had no equal elsewhere.[40]

The petrographic collection grew significantly during the following months. This was a most welcome opportunity for the Junta, which was eager to show the results of its activity. By the end of June 1908, the collection already numbered nearly 100 stone samples. The acquisition of such an extensive collection in such a short time was accomplished with the help of commercial and industrial elites of the Catalan bourgeoisie. The majority of rocks proceeded from different quarries or private lands from all over Catalonia. The industrialist and politician Eusebi Güell (an important patron of Antoni Gaudí), the architect Pere Falqués and the politician Salvador de Samà – known by his noble title Marqués de Marianao – figured among the donors of the petrographic collection. Not surprisingly, building and railroad companies such as the Sociedad de Obras y Construcciones and the Compañía de Ferrocarriles Madrid-Zaragoza-Alicante, also donated several blocks.

The donation of a selected piece of rock to the Museu Martorell not only demonstrated that the donor was a 'connoisseur' of the natural world,[41] it also bore testimony to his patriotic zeal and commitment to Barcelona and

38 On the history of the mammoth, see Julio Gómez-Alba, 'El mamut y la colección petrológica de grandes bloques del Parque de la Ciudadela (Barcelona, España)', *Treballs del Museu de Geologia de Barcelona*, 10 (2001): 5–76, pp. 7–11; and Laura Valls, 'El mamut de la Ciutadella (1907). Biografia d'un objecte urbà', *Afers: fulls de recerca i pensament*, 30 (80/81) (2015): 243-68.

39 Gómez-Alba, 'El mamut', provides an exhaustive account of the provenance of the stones. Most but not all of the samples were from sites in Catalonia.

40 AMCNB, 25/14, 1910.

41 Rossend Casanova, *El Castell dels Tres Dragons* (Barcelona: Museu de Ciències Naturals/Institut del Paisatge Urbà i la Qualitat de Vida, 2009), p. 67.

Catalonia.[42] The message the Junta wanted to convey was clear: nature was a resource that could be and had to be exploited for the benefit of the country. The donors whose names appeared on the plaques of the blocks knew that only too well.

Exhibiting fish in the city

Unlike the acclimatization of land animals, the introduction of certain species of fish was high on the agenda of naturalists at the end of the nineteenth century. The International Fisheries Exhibition in London in 1883 was a highly visible event in the debate – both scholarly and practical – on what were the proper methods of fish farming.[43] The new Junta also took to these ideas. Keeping up an impressive pace, the Aquarium at the top of the Cascada Monumental was inaugurated already on 23 June 1908. The exhibition of seals, acquired from the Netherlands, certainly would have increased the attraction of the new installation.[44] The Aquarium was part of the existing zoo and completed the collection of living animals. During the following months, many species of fishes were put in the basins of the Aquarium, the cascade itself and the small lake nearby. This was the starting point of a larger project spearheaded by Darder.

In February 1909, the Laboratori Ictiogènic (Fish Laboratory) was founded on the grounds of the zoo. The breeding programme included both indigenous as well as exotic species, such as the green tench of Russia, the red tench of Mongolia and the Japanese carp 'Hi-Goi'. The plan was to repopulate the rivers and lakes of Catalonia, which had been severely affected by pollution and overfishing. The species to be introduced were supposed to improve the aquatic resources of the nation by increasing and varying production. The Junta distributed fish and fish eggs produced in the Laboratori free of charge to any Catalan municipality with a river or a lake in its vicinity. To celebrate the introduction of fish into local fresh waters, the town supported by the Junta, organized a popular festivity called Festa del Peix (festival of fish).[45]

The first town to celebrate such a festival was Banyoles in October 1910, followed by Terrassa and Manresa in 1911, and other Catalan towns until 1915. The Festa included different activities such as popular lectures by Darder and other naturalists, the singing of the Himne del Peix (Hymn of the Fish)[46], a fishing contest and a regional exhibition on 'Pisciculture and Fishery'.

On 20 December 1912, the great 'Pisciculture and Fishery Exhibition' was inaugurated in the building known as the Castell dels Tres Dragons in the western

42 Valls, 'El museo de ciencias naturales', p. 15.

43 See, for example, James Ramsey Gibson Maitland, *On the culture of Salmonidae and the Acclimatization of Fish* (London: W. Clowes and Sons, 1883).

44 AMCNB, 22/18, September 1908.

45 The topic has not yet been studied in any depth. For a useful first step in collecting material, see Georgina Gratacós, 'La festa del peix a Banyoles el 1910, una recerca documental i fotogràfica', *Jornadas imatge i recerca,* 12 (2012).

46 Joan Vidal, *L'Estany de Banyoles* (Girona: La Económica, 1925), p. 63.

corner of the Parc. With this highly visible event, the Junta wanted to publicize the results of repopulating the Catalan fresh waters. Outside the building the visitors could contemplate living animals and objects related to the reproduction and farming of fish. A great variety of different species were put on display in huge aquariums. A laboratory of artificial fertilization proved to be a major attraction.[47] Once inaugurated, the exhibition was enriched with an exhibit of aquatic plants and another one showing small mammals typical of the fluvial ecosystems.

Inside the building, objects of a different nature were on display: 'not how to breed fish, but on the contrary, how to kill them'.[48] There was a repertory of all classes of artefacts representative of fishery (nets, rods, traps) as well as instruments related to the 'fish industry': impermeable clothes for fishermen, pictures, instruments for research, a small collection of books, hydraulic machines, collections of fluvial fauna and so on. This collection of objects proceeded from more than 30 companies and individuals.

At the same time, a whale was exhibited on the Avenida del Paralelo, the heart of Barcelona's amusement industry.[49] The huge whale had been washed ashore some months earlier on the coast of Sant Feliu de Guíxols. The satiric magazine *L'Esquella de la Torratxa* poked fun at this coincidence, imagining the escape of the whale from Paralelo to the exhibition in the Parc.[50] The joke soon became reality (see Figure 2.4). In March 1913 the organizers of the exhibition acquired the whale in order to display it outside the Castell dels Tres Dragons.[51] In the Bar Piscatorio inside the building, seafood such as oysters, caviar, salmon, trout and eel was served.[52] Organized events not only included lectures on pisciculture but also popular dances, activities for children and concerts of the municipal band. Thus the exhibition became a success drawing large crowds. In the first three months alone, 50,000 tickets were sold.[53] Initially scheduled until 30 June 1913, the exhibition was eventually prolonged until 31 December 1913.

Pisciculture was the auxiliary industry of agriculture most promoted by the Junta. But other 'cultures' were also featured in the Parc by means of temporary exhibitions. The first exhibition on sericulture was inaugurated on 6 June 1909 in order to foster the silk industry in Catalonia.[54] A professional society called Fomento de la Sericicultura Española had proposed the exhibition, which was staged annually, supported by the Junta, until 1911 and again in 1914.

The Junta expanded its scope by adding new areas and agendas. It went beyond the walls of the Museu Martorell and used the Parc as an extension to the city, attempting to occupy different buildings in the Parc. The celebration of

47 'Exposició de piscicultura, acte inaugural", *La Veu de Catalunya*, 21 December 1912 (morning edition), 2.

48 Ibid.

49 'Balena de 13 metres', *La Veu de Catalunya*, 17 December 1912 (morning edition), 4.

50 'Última hora', *L'Esquella de la Torratxa*, 27 December 1912, 844.

51 'Exposició regional de piscicultura i pesca', *La Veu de Catalunya*, 19 March 1913 (evening edition), 7.

52 Casanova, *El Castell dels Tres Dragons*, p. 69.

53 'Exposició regional de piscicultura i pesca'.

54 AMCNB 94/002, 27 April 1909, p. 207.

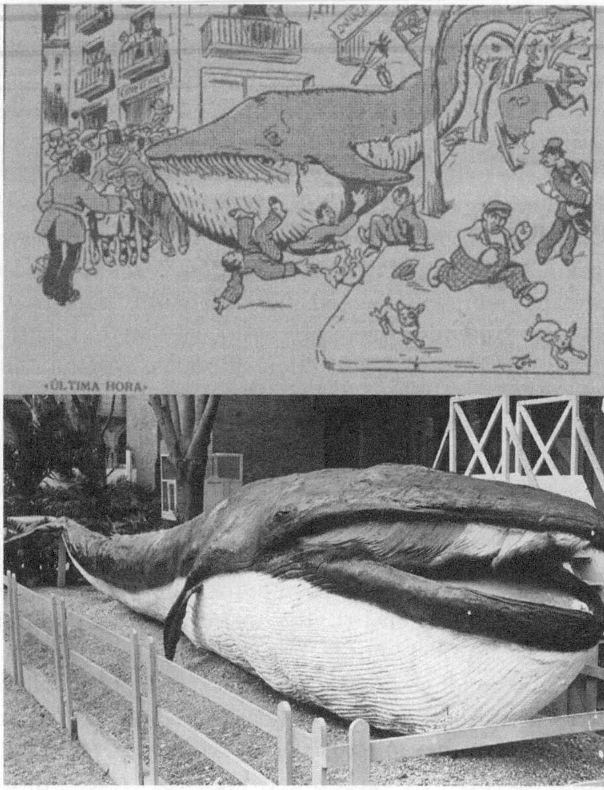

Figure 2.4 The stuffed whale 'travelled' from the Paralelo to the Pisciculture Exhibition (1913).

the exhibition on 'Pisciculture and Fishery' in the Castell dels Tres Dragons was one such attempt. And there were more to come, once more triggered by political changes. The creation of the 'Mancomunitat de Catalunya' in 1914, uniting the four Catalan provinces administratively into a 'Catalan Commonwealth' eventually also had an impact on the Parc. In 1916 a new Junta mixta de Ciències Naturals was created as the Joint Board of the Barcelona City Council and the Diputació de Barcelona (the government of the province Barcelona). The Junta mixta incorporated several facilities that were built for the Exposición Universal of 1888 such as the Umbracle, the Hivernacle and the Aquarium into its programme for the promotion of the natural sciences in the Parc. Most importantly, the Castell dels Tres Dragons became the seat of the new Museu de Catalunya de Ciències Naturals in 1917. This marked the completion of the transformation of the Parc de la Ciutadella into a space for popular science as well as for research in natural history very much anchored in the rise of political Catalanism in the early twentieth century.

Imagined citizens, real visitors

What can we say about the success of these 'out-reach' endeavours? Did the Juntas impact the audience they were aiming at? To what extent did visitors to the Parc assimilate the way nature and science were presented? These are difficult questions whose answers demand a more roundabout route. Due to the limited character of the sources, we are able to say much more about how the different Juntas conceived of the public and how they wanted to educate it.

As mentioned above, the idea behind the creation of the municipal park was to create a space that served as a 'meeting point for all social classes', as Darder put it in 1886.[55] Hence this idea – or rather this ideal – of social inclusion was current in Barcelona as well. But here, as well as in London or Copenhagen, New York or Buenos Aires, the public park was an institution permeated by bourgeois norms. True, its intended audience encompassed the lower classes as well as the upper and middle classes. But the aim was clearly to immerse them in the values of the latter: civilized behaviour, rational recreation, education, self-improvement and self-discipline.[56] What impact did this 'civic nature' of the Parc and its bourgeois values have on their target audience?

Even basic information such as visitor numbers is in many cases impossible to obtain. What the sources offer are rather isolated comments such as the following by Darder in early 1886: 'It is with regret that we observe that for a year, the gardens and parks have been every day less visited with the exception of Thursdays and Sundays'.[57] This seemed indeed to be a major flaw in the public park scheme in general. How were the workers to benefit from the educational facilities and the clean air of the urban greens if they had to work six days a week? And how were they supposed to travel across town given that the working-class quarters were mostly not in the proximity of the park?[58]

Let us review the interaction between the promoters of the Parc agenda and their publics in the three case studies we just expounded, starting with the zoo. As has been shown with respect to the nineteenth-century zoo in general, the principal way to learn anything about the actual behaviour of its visitors is to look at the transgressions of the existing rules and norms – because these violations of regulations generated sources, internal reports and public comments. Scholars have repeatedly argued that one may think of the zoo, or public parks more generally, as institutions for disciplining behaviour. The visitor is supposed

55 Darder, '*Variedades*'.

56 The case of popular astronomy – 'a useful tool to discipline and preach morality to citizens' – is analogous; see Chapter 9 in this volume.

57 Darder, '*Variedades*'.

58 Conway, *People's Park*, pp. 60–1.

to learn how to treat the animals (no poking with umbrellas, no throwing of stones), to keep on the paths and not to litter.[59]

In the case of the Barcelona zoo, it only took two weeks after its inauguration on 24 September 1892 for the first complaints to be voiced in public. According to newspaper reports, visitors threw lit cigarettes at Avi, the elephant of the zoo, and even a glass bottle to see what the pachyderm would do with it.[60] The deaths of some flamingos and a white swan were attributed to deliberate poisoning.[61] Complaints about the incessant attempts of visitors to feed animals seem to constitute the 'eternal' problem of the zoo.[62] No different in Barcelona: Already on 4 October 1892, signs were put up admonishing the public not to feed the animals.[63]

Yet there was also an alternative to the image of the zoo visitors as in need of instruction and at least potentially deviant troublemakers, an image that is predominantly painted by conservative newspapers such as *La Vanguardia* and *El Diario de Barcelona.*[64] On the pages of the republican and anticlerical *El Diluvio,* we encounter a public that is active and attentive, demands its 'rights' and does behave properly when it visits the zoo and the Parc.[65] It does not seem farfetched to hear echoes of the debate about male suffrage, introduced in Spain the previous year in 1891. Should the 'masses' be kept in check or should 'public opinion' prevail? And instead of casting a critical eye on the general public, *El Diluvio* asks for explanations regarding the cost of the animals and the management of Darder as director of the newly inaugurated zoo.[66]

Turning to our second case study: The installation of the mammoth in 1907 reflects once more the ambivalence of the Junta and its networks of Barcelonese

59 On the zoo disciplining its visitors (what they should and should not do), see Christina Wessely, *'Künstliche Tiere etc.'. Zoologische Gärten und urbane Moderne in Wien und Berlin* (Berlin: Kadmos, 2008), pp. 66–73; Oliver Hochadel, 'Im Angesicht des Affen. Die Besucher des Tiergartens im 19. Jahrhundert', in Sabine Nessel and Heide Schlüpmann (eds), *Zoo und Kino* (Frankfurt/M: Stroemfeld, 2012), 29–48, pp. 30–2.

60 The incident with the cigarette is related in *Diario de Barcelona,* 29 September 1892 (morning edition), 11316; 'Justicia elefantisiaca', *El Diluvio,* 29 September 1892 (morning edition), 8221; *La Vanguardia,* 3 October 1892, 1. Also see Pons, *El parc zoològic,* pp. 51–2.

61 *El Diario de Barcelona,* 14 October 1892, 11974; Pons, *El parc zoològic,* p. 52.

62 Oliver Hochadel, 'Vor den Gitterstäben: Die Besucher der Menagerie Schönbrunn im 19. und frühen 20. Jahrhundert', in Mitchell G. Ash and Lothar Dittrich (eds), *Menagerie des Kaisers – Zoo der Wiener. 250 Jahre Tiergarten Schönbrunn* (Wien: Pichler, 2002), 158–79, pp. 173, 176.

63 *La Vanguardia,* 4 October 1892, 2; Pons, *El parc zoològic,* p. 52.

64 For an instructive comparison on how *La Vanguardia* and *El Diluvio* represent science, see Matiana González-Silva and Néstor Herran, 'Ideology, elitism and social commitment: Alternative images of science in two *fin de siècle* Barcelona newspapers', *Centaurus* 51 (2) (2009): 97–115.

65 'La coleccion zoológica del Parque', *El Diluvio,* 22 September 1892 (morning edition), 3 (8012); also see *El Diluvio,* 25 September 1892 (morning edition), 8111. On the role and place of *El Diluvio* in the public sphere of Barcelona, Stephen Jacobson, 'Interpreting Municipal Celebrations of Nation and Empire: The Barcelona Universal Exhibition of 1888', in William Whyte and Oliver Zimmer (eds), *Nationalism and the Reshaping of Urban Communities in Europe, 1848–1914* (Basingstoke: Palgrave Macmillan, 2011), 74–109, pp. 98–9.

66 'Que se aclare', *El Diluvio,* 13 October 1892 (morning edition), 8692–3.

naturalists and civil servants. The 'antediluvian' animal was intended to teach visitors to the Parc about the prehistory of Catalonia. Yet at the same time, the Junta felt the mammoth needed to be protected from likely damage and therefore erected a protective barrier around it (during the works, first wooden panels, then barbed wire, and after the inauguration a metal fence).[67] The public was to be educated but also to be kept on a short leash. As the members of the Junta put it: the fence was necessary due to the ignorance and lack of culture of people who could not appreciate the importance of the work.[68] The magazine *La Esquella de la Torratxa* satirized this patronizing attitude that ascribes to the public the nature of an animal: the fence around the reproduction of the mammoth was to ensure that 'no one would hurt it'.[69] And only a few months later, with the petrographic collection taking shape outside the museum in early 1908, the Junta saw itself faced with the very same problem. In order to protect the samples of different stones and the labels attached to them against 'a certain type of public', it proposed to put a fence around the collection as well.[70]

These examples from the zoo, the mammoth and the petrographic collection show the fear of, or at least the uneasiness toward, the 'masses'. The 'masses' in this discourse were constructed as the constitutive opposite of the bourgeois individual, civilized and educated, receptive to what the Parc had to offer as a school of 'civic nature'. The contradictions in the agenda of the Parc as a public space for science are obvious: Its content was designed to educate and maybe even more to pacify the 'masses' (read: the workers) yet at the same time the Junta was afraid and put up panels with prohibitions and fences in order to protect itself.

Our third case, the exhibition on pisciculture points in a different direction. In their rhetoric Barcelonese naturalists would frequently defend the frontier between the spectacular and the instructive. Yet the case of the 'travelling whale' shows that the border between popular culture – Paralelo – and bourgeois popular education – the Parc – was porous. The organizers of the exhibition did not miss out on the opportunity to attract new audiences by buying and displaying such a spectacular specimen. In fact, the various incentives on offer – seafood, concerts, activities for children – indicate a growing awareness that certain concessions were called for in order to satisfy the desire of the 'real visitor' for some good entertainment. The exhibition on pisciculture may even be interpreted as a learning process for the Junta.

Conclusions

In the summer of 2014, the Natural History Museum of Barcelona and the Barcelona Zoo were digging for their roots. They organized a public lecture series

67 Gómez-Alba, 'El mamut', p. 8.
68 AMCNB, 15/001, 31 August 1907, pp. 20–1.
69 *La Esquella de la Torratxa*, 13 September 1907, 603.
70 AMCNB, 94/002, 24 February 1908, p. 117.

entitled 'La Ciutadella, Barcelona's first science park' in an attempt to connect the past with the present.[71] In 1907 Norbert Font chose a different key word, he envisioned a 'veritable park of culture'. For Barcelona's *fin-de-siècle* naturalists, the Parc was certainly more than just about science. The time between the approval of Fontserè's plan in 1872 and the inauguration of the Museu de Ciències Naturals de Catalunya in 1917 was marked by several agendas with distinct ideological underpinnings. These programmes were characterized by their different notions of nature – and their common notion of the public.

Fontserè's eclectic plan reflected both his (older) romantic understanding of nature as manifested in the grottos of the Cascade as well as a utilitarian one with its focus on applied natural history. It was this utilitarian understanding of investigating and using nature as a resource that is common to all the different 'Parc agendas'. In its purest form we find it in the Junta técnica (1893–1906), and in the person of Francesc Darder, his zoo and his acclimatization efforts. As we have seen, in a radically different political context, the creation of the new Junta in 1906, focusing on the natural history of Catalonia (and thus national identity) marks a turning point.

Yet these two agendas were largely compatible. Many of the activities promoted by the Junta were based on a utilitarian concept of nature, in line with the programme of the preceding Junta técnica.[72] The large-scale programme to enrich the Catalan fresh waters with 'foreign' species of fish between 1908 and 1915 clearly shows this. It has been argued that doing applied science may be considered a specific feature of Catalan science at the time.[73] This is certainly the case with the Parc. Both the Junta técnica and the Junta were eager to demonstrate that nature as a resource could be controlled, improved and exploited in order to modernize Catalonia. What was needed was research in applied natural history.

Yet at the same time, the citizens needed to be won over to support this 'civic nature'. The agenda that the two Juntas promoted aimed at more than simply increasing the knowledge of the visitor to the Parc about the natural world. The citizens were essential in order to fully realize the utilitarian programme. They were needed as contributors to the collections of the Parc, most notably to the Museu Martorell and the Petrographic Collection. They were supposed to buy animals and animal products at the zoo, in order to breed and raise animals themselves, be it on a domestic or even on a commercial scale. They were encouraged to come up with their own business ideas inspired by the Museo Zootécnico or by one of the exhibitions. Some of the bigger fish caught by 'ordinary people' in the spring of 1913 in the lake of Banyoles went straight to the exhibition on pisciculture in Barcelona. They were presented as direct proof

71 See http://barcelonacultura.bcn.cat/descobreix/la-ciutadella,-el-primer-parc-cientific-de-barcelona#.VK0WNSfz9PR (last accessed 7 January 2015).

72 See the programmatic article by Font, 'Junta autònoma II', where the Catalanist and the utilitarian agenda are neatly fused.

73 Roca-Rosell/Salavert, 'Nacionalisme', pp. 552, 555.

of the success of the Festa de Peix and the introduction of the exotic species into Catalan fresh waters.[74]

These attempts to recruit the ordinary citizen, thus ascribing at least in principle an active role to him, stand in stark contrast to the patronizing attitude of both the Junta técnica and the Junta. The way the facilities of the Parc offered knowledge was 'top down'. The experts in applied natural history told the presumably ignorant citizen what to do (and, for example, in the case of the mammoth also what to believe). The Juntas found themselves in a seemingly contradictory position. On the one hand, they depended on the general public for their agenda to have a noticeable impact. Yet on the other hand they were suspicious about the lack of education and proper behaviour of the 'masses'. They felt they needed to protect the animals of the zoo and the mammoth in the Parc from evildoers. This boils down to the opposition of an imagined citizen versus a real visitor.[75] The former is eager to participate in the projects of applied natural history proposed by the Juntas in the Parc. The latter tends to seek amusement rather than instruction. It seems that particularly with the exhibitions in the Parc from 1909 onward, the Junta tried to combine instruction with entertainment. The acquisition of the large stuffed whale by the organizers of the exhibition on pisciculture in March 1913 is the most striking example of this attempt. And the Festa del Peix hosted in smaller Catalan towns became a highly popular event, at least judging by the extensive press coverage. Yet there was always competition for public attention even within the Parc. This competition became particularly fierce in 1911 when the amusement park Saturno opened in the heart of the Parc, drawing the masses within earshot of the Museo Martorell and the zoo.[76]

The Saturno only provided technological fun for 15 years in the Parc, yet the stone mammoth is still there. More than 100 years after its inauguration, the huge sculpture nowadays provides a much-used photo opportunity. On Sundays one may even have to queue in order to position oneself behind the enormous trunk – a trunk that is vertically bent so that it easily serves as a seat for a child. The fence around it is long gone. It seems like an ironic twist of history that this creature, this strange mix of old and new, of defending biblical chronology and advocating modern political Catalanism, of conservative ideology and state-of-the-art technology, is nowadays an eye-catcher for tourists from all over the world, and is thus probably more entertaining than instructive.

74 See the substantial documentation in Francesc Darder and Jerónimo Darder, *Piscicultura: Crónica piscatoria* (Barcelona: Hijos de Domingo Casanovas, 1913)

75 This split between an ideal and real visitor was typical at the time. For example, the Viennese naturalist and short-term zoo director Friedrich Knauer (1850–1926) was struggling along the same lines; see Hochadel, 'Im Angesicht', pp. 30–1.

76 See Chapter 5 of this volume.

3 Reconstructing the Martorell

Donors and spaces in the quest for hegemony within the natural history museum

Ferran Aragon and José Pardo-Tomás[1]

In 1883 the Duchess of Medinaceli (1858–1936), head of one of the oldest aristocratic lineages in Spain, gave to the recently inaugurated Museu Martorell a series of crystals extracted from the salt mines in Cardona, province of Barcelona. Donating to the Martorell was considered a patriotic act and received publicity in the press.[2] But in the case of Medinaceli's donation, the symbolism of the act goes deeper, as it poses the question of property and tension between elites. Indeed, seven years after the donation by the Duchess of Medinaceli, the Martorell received another donation from the salt mines in Cardona. This time the donor was the economist and liberal politician Miquel Elias. Years ago, Elias had published a report containing a compilation of historical arguments and citations to the recently approved Mining Law to defend the rights of people other than the Duke to extract and commercialize Cardona salt.[3] If in 1883 the Duchess gave 74 samples of Cardona salt to the Martorell, in 1890 Elias donated 159. This example illustrates the role that the symbolic act of donating to a museum like the Martorell can have. And how politics can make its way into the museum through the objects that it receives.

In the present chapter we will analyse who the donors who shaped the content of the museum were, what kind of objects they donated, how these were exhibited (or not), and how the museum's different sections evolved and grew in consonance with the political and social events of the period. We will follow the development of these donors and donations during the four initial decades of the museum. While reconstructing this evolution we will see how the practice and popularization of natural sciences acted as tools for the cultural construction

1 The authors would like to thank Irina Podgorny, Samuel Alberti and Morgan Meyer for their comments and suggestions, which helped to improve the chapter substantially.

2 Arxiu del Museu de Ciències Naturals de Barcelona (hereafter AMCNB), C08 E36, C06 Exp.01: *La Vanguardia*, 17 February, 1883, 1; *Diario de Barcelona*, 31 July 1889; *Diario de Barcelona*, 1 February 1890; *Diario de Barcelona*, 6 December 1890; *Diario de Barcelona*, 9 May 1891.

3 Miguel de Elias, *Memoria sobre el criadero de sal gema de Cardona* (Barcelona: Imprenta de Ramírez, 1854). The report was dedicated to the Asociación General de Minería de España, a private association aimed to defend the interests of the emerging group of investors devoted to the exploitation of mines in Spain that had been growing since the expansion of railways.

of an elitist, Catalanist, Catholic and conservative project of society. The role of donations and donors in the making of collections of natural history museums appears frequently in the literature. Yet to treat the donors as central agents for the political and social definition of the museum has hardly ever been attempted.[4] This approach is only possible due to the extraordinary richness of the sources preserved in the archive of the Museu de Ciències Naturals.

Our analysis runs from the legacy of the founding collection in 1878 to 1924, when the museum underwent its third reorganization and the dictatorship of Primo de Rivera abolished the Mancomunitat, the autonomist project of the Catalan government. We divide this period in four roughly 10-year phases. In the first phase, the time of Miquel Elias and the Duchess of Medinaceli, the Martorell was a heterologous repository of natural history, also containing numismatics, archaeology and ethnology; from 1891 the museum was exclusively devoted to natural history and emphasized the role of regional collections; from 1906 the museum grew quickly, thanks mainly to Catalanist forces in the local government who created an autonomous board for the development of the Martorell; from 1916 the museum reached a new stage of development, acquiring a second building in the park, creating laboratories and professionalizing several of its collaborators.

From an eclectic museum to a natural history space, 1878–1891

In 1878 Francesc Martorell (1822–1878), a merchant and stockbroker, died and left his private library and collections of numismatics, archaeology and natural history (especially malacology) to the Barcelona Council, all of which came with a sum of money that the Council was to use for the creation of a library and a public museum to exhibit the collections.[5]

Aspiring to the cosmopolitan and civic style of the European metropolises, the Council of Barcelona built the Museu Martorell in the recently created public urban park (Parc de la Ciutadella). The museum was officially inaugurated in September 1882 by the city's mayor, Francesc Rius i Taulet (1833–1889). The architecture of the building was inspired by the monumentality of the 'temples of science' in neoclassical style that were built in the European and American metropolises during the last decades of the nineteenth century.[6] It is a

4 See, however, the interesting perspectives in Argentinian cases in: Irina Podgorny and Maria M. Lopes (eds), *El desierto en una vitrina. Museos e historia natural en la Argentina, 1810–1890* (México: Limusa, 2008).

5 We lack a serious biographical study of Francesc Martorell and his economic activities. A source to start with is Artur Bofill, *Francisco Martorell y Peña: apuntes biográficos* (Barcelona, 1905).

6 This stylistic choice was not exclusive to big national museums, but also to regional ones, including La Plata (Argentina), opened in 1885, which also adopted a monumental appearance reminiscent of Greek temples, as a thinly veiled allusion to 'temples of knowledge': Maria M. Lopes and Irina Podgorny, 'The shaping of Latin American museums of natural history', *Osiris*, 15 (2000): 108–18. On

Figure 3.1 The facade of the Museu Martorell.

two floor plus basement construction occupying a total surface of 1,288 square metres. The main floor is divided in three parts: two lateral bodies and a central one, the atrium, in the front part of which a set of four Doric-style columns support an attic with the city's coat-of-arms on its frontispiece (see Figure 3.1).

In 1883, the patron's collections, furniture and library were installed in the museum. Visits were initially being restricted to Thursdays and Sundays from 10 am to 6 pm. In this manner, the first display had the air of the rich and cultured stockbroker: in the twin wards and the hall, his luxurious shelves and cupboards showed the magnificence of his collections. This eclectic collection was shown as follows: in the left ward were the zoology, botany and geology sections; in the hall were the fossils; while the right ward housed the archaeology and numismatics collections, with some space reserved for temporary exhibits of other objects and topics.[7] Above each ward, and surrounding the entire perimeter,

architecture and natural history museums, see Susan Sheets-Pyenson, *Cathedrals of Science: The Development of Colonial Natural History Museums during the Late Nineteenth Century* (Kingston: McGill-Queen's University Press, 1988); Sophie Forgan, 'Bricks and Bones: Architecture and Science in Victorian Britain', in Peter Galison and Emma Thompson (eds), *The Architecture of Science* (Cambridge, Mass.: Massachusetts Institute of Technology, 1999), 181–208.

7 For example, exhibitions organized by elitist artistic associations such as the Centro de Acuarelistas. In their application they appealed to the director's 'interest always shown in the development of the Art': AMCNB, C05 E15.

there was a balcony with metallic showcases where it seems that the library was initially located. Alternating with the walls of these showcases was a series of large windows to illuminate the wards, which, like the internal spaces, had no artificial lighting. In the central body of the building, after crossing the hall, a door led to a modest staircase up to the offices and working areas.

The initial eclecticism of the museum, however, stemmed not only from the passion for collection of its patron but also from the assumption by the municipal authorities that the museum had to keep the kind of Renaissance spirit shown by high-society men, which was reflected by their interest in diverse fields of knowledge. This idea would seem to be confirmed by the appointment of Manuel Martorell, Francesc's brother, as director of the museum, which reveals the desire of the director and municipal authorities to be associated with the aristocracy and the city's upper-class bourgeoisie. A lawyer with a passion for art, Manuel Martorell was also an amateur entomologist and the owner of a geological collection that he gave to the Martorell just after being appointed. He was to be the 'honorary' – that is, unpaid – director of the Museum from 1880 until his death 10 years later. The other members of the small staff were: Enric Grau, curator, who donated his zoological collection; Eusebi Palomera, janitor and head of security; and Francesc Darder (1851–1918), taken on in 1883 as 'dissector'.[8] Grau resigned in 1887 and was replaced by Artur Bofill (1844–1929), who became the director of the museum in 1893.

The key actors in the development of the Martorell during this period, however, are not to be found in the museum, but outside of it. What most influenced the changes in the use and distribution of spaces in the early years of the museum were contributions made by donors, rather than its internal policies. Indeed, the tendency of donors to bring natural objects to the Martorell must have been essential to its commitment towards natural history ahead of other possible avenues that the museum could had gone down.[9]

Donating objects to a museum was an appropriate way of enhancing its reputation for a local elite that was willing to demonstrate civic commitment, cosmopolitanism and wealth. The museum regularly issued local press releases about the donations that it received and thus made donation a public, 'disinterested' and patriotic act. The presents that members of the highest social classes gave to the museum were representative of their way of living and had the stamp of colonial power on them. In this first period, the presence among the donors of the noble and upper-class bourgeoisie classes was remarkable, but was to become much less so. A few examples will suffice to illustrate the main features of a typical donor's profile in these years.

Salvador Samà (1861–1933), Marqués de Marianao, was a landowner and a dynastic politician, being mayor of Barcelona in the periods 1905–1906 and

8 See Chapter 2 of this volume.

9 From 1878 to 1891, about 80 per cent of the donation events found in the registers are related to natural specimens and the remaining 20 per cent is a compilation of archaeological, historical, ethnological and even artistic objects. AMCNB, C10 E01.

1910–1911. He had his own botanical and zoological garden located 100 kilometres from Barcelona, created in a romantic style by Josep Fontserè (1829–1897), who had also designed the Parc de la Ciutadella. Samà used to donate to the Martorell animals from his living collection that had died, which were stuffed and exhibited in the museum.[10]

Emilia Serrano (1845–1922), Baronesa de Wilson, devoted part of her life to travelling all around Latin America in search of ruins of pre-Columbian civilizations. In 1894, she donated an archaeological–ethnographical and a natural history collection to museums in Barcelona, the latter of which went to the Martorell, with 'products of the regions that one day were precious jewels belonging to our Spain and now constitute a beautiful group of nationalities'. In the letter that accompanied her donation, she explained that the reason for making these gifts to the city was the special predilection that she felt for Barcelona, where she had lived at different times after coming back from America.[11]

Barcelona citizens with diplomatic, military or commercial duties in other countries, normally Spanish colonies or former colonies, constituted a source of exotic objects. Barón de la Barre, who in 1899 was ambassador in Colombia, sent from this country a collection of 42 birds and another of 24 butterflies.[12] Manuel Porcar, delegate of Ecuador at the 1888 Universal Exhibition and mayor of Barcelona from 1891 to 1892, gave the museum 23 flasks with Ecuadorian vipers.[13] Leandro Mella, captain of a frigate, left the museum on his death in 1899 a collection of 266 molluscs and other objects – among them a turtle shell and a tooth from a cachalot – from the Philippines, Guinea and India.[14]

Finally, turning from the colonies to the Barcelona elite, we must mention hunters. Legal hunting was both an expensive pursuit and a social activity that denoted high social status. Many members of the bourgeois elite tried to emulate this aristocratic practice. Leading industrialists, for example, were among these hunters who donated some of their trophies to the Martorell collections: Mariano Rifà, owner of forges, who forged the lamps and fences of Parc de la Ciutadella; Camil Fabra (1833–1902), Marqués d'Alella, owner of textile factories, who also had political duties in some of the dynastic parties; and Joseph Richardson, who headed the English Barcelona Besòs Waterworks Ltd., which managed the water supply for the city.

10 These donations included different species of monkeys, for example a *Cebus* from Brazil; birds, like hawks and pheasants, and mammals, like a deer from the Philippines. AMCNB, C06 E01; C08 E56; C09 E43; C06 E01; C02. E04; C09 E05.

11 AMCNB, C09 E09; C29 E01; C17 E01. The quotation is in C01 E35. Serrano collected objects which aroused her curiosity, but she was not a connoisseur of natural history; among the objects collected were several insects, vegetables, minerals and birds pending classification, a task that was to be performed in the museum, mainly by Artur Bofill.

12 AMCNB, C01 E11.

13 AMCNB, C08 E74.

14 AMCNB, C09 E54.

As mentioned above, natural history donations were much more frequent than other kinds of gifts. It is not surprising, therefore, to see how when the museum planned its first reorganization for the Universal Exhibition of 1888, the mineralogy and botany collections, initially in the left ward, were moved to the right ward, being forced to share it with the archaeological and numismatic collections, whereas the zoological collections were expanded to fill the whole left ward.[15] After the Universal Exhibition, several collections of minerals, rocks, Iberian wildlife, as well as forest and crafted products were given to the Martorell.

But the event that finally sealed the exclusive dedication of the museum to natural history was the acquisition in 1891 of a reference fossil collection comprising 12,000 samples, the so-called Baron collection, bought in Paris from the merchant Félix Pisani.[16] The acquisition was controversial; some councillors considered it a waste of money, largely because there was not a clear plan for the Martorell, the specific functions and objectives of the museum never having been defined.[17] When the acquisition was eventually accepted, the collections of archaeology and numismatics were moved outside the park to the former Pavilion of Fine Arts of the Universal Exhibition, and the Martorell became the natural history museum of the city. The museum acquired 20 new showcases to house their collections and redistributed the space as follows: left ward for zoology, right ward for geology and palaeontology; a herbarium from Montserrat mountain on the first floor, together with the library and the director's office; on the second floor were the housekeeper's lodge and in the basement were a storage room and a taxidermy laboratory.[18]

Collecting the fatherland, 1892–1906

Despite the extra space gained in 1891 due to the removal of the archaeological and numismatics collections, complaints about lack of space followed almost immediately.[19] The conversion of the Martorell to a Natural History Museum was, however, irreversible. This particular path having been cleared, the next steps in the museum's development would progressively orientate it towards more regionally focused Catalan collections and displays.[20] It is worth noting that in spite of the significance of colonial and exotic donations discussed in

15 *Guía itineraria y descriptiva de Barcelona, de los alrededores y de la Exposición Universal* (Barcelona, 1888).

16 Acquired after a favourable report by Artur Bofill for 6,000 French francs. AMCNB, C08 E23.

17 The museum's archive contains a summary of the debate that was published in the press. *El Diluvio*, 8 April 1891. AMCNB, C06 E01.

18 Julio Gómez-Alba, 'El Museo de Geología de Barcelona: desde su fundación a la Junta de Ciencias Naturales (1878–1905)', *Treballs del Museu de Geologia de Barcelona*, 1 (1990): 7–34.

19 Laura Valls, 'El Museo de Ciencias Naturales de Barcelona (1882–1917): popularización de las ciencias naturales dentro y fuera del museo', *Biblio 3W. Revista bibliográfica de Geografía y Ciencias Sociales*, 15 (918) (2011) n° 918, http://www.ub.edu/geocrit/b3w-918.htm; AMCNB, C02 E21 (last accessed 12 February 2015).

20 This emphasis on the homeland is for example also typical for some German natural history museums of the time, especially the Altona one: Lynn K. Nyhart, *Modern Nature. The Rise of the Biological Perspective in Germany* (Chicago: The University of Chicago Press, 2009), pp. 268–78.

the previous section, objects from Catalonia and its bordering regions had an important presence in the museum from the beginning. Indeed, the proportions of exotic and regional contributions would not substantially change in this second period, the former accounting for around 25 per cent and the latter growing from 40 to 50 per cent.[21] However, it was during this second period when the museum rulers defined new policies for the regional orientation of the Martorell. These guidelines would be reinforced and clearly start making their effect felt later on, especially from 1907. There is an obvious relation with the formulation of the regional programme and political changes in Barcelona. It was during these years that the political parties close to the monarchy were defeated and substituted by republican ones, and later by them in alliance with Catalanists in the Barcelona Council.

In 1893, shortly after the republicans won the elections, a technical board for the management of the Martorell was created, the so-called Junta técnica del Museo de Ciencias Naturales y Jardines Zoológico y Botánico.[22] Modest Casademunt – a trader of glassware, pottery and porcelain closely linked to Barcelona apothecaries – chaired the Junta técnica as deputy mayor. Its members were: the aforementioned Francesc Darder, the museum's former taxidermist who only a few months earlier had been named director of the Zoo; Odón de Buen (1863–1945), professor of natural sciences at the university; Canon Jaume Almera (1845–1919), director of the Institute of Geology and Paleontology in the Seminary of Barcelona, the school for Catholic priests; the engineer Luis M. Vidal (1842–1922) as president of the Academy of Sciences and Arts; the presidents of the Academy of Medicine and the College of Apothecaries; the professor of natural sciences at the Barcelona Public High School; and the Mining and Mountains District chief of engineers.[23] In 1899, the Junta técnica published a report stating their main objectives:

> To popularize knowledge about the organic and mineral forms that exist in different countries of the Globe.
>
> To show the multiple economic applications that man can make of natural beings.
>
> To give knowledge about the flora, fauna and gea[24] of the Catalan-Balearic region.[25]

21 The other 25 per cent corresponded to Spain and its colonies (16 per cent approx.), Europe (7.6 per cent) and multiple origin collections (1.3 per cent).

22 Depending on a Commission of the Barcelona City Council.

23 Artur Bofill, 'Memòria sobre l'origen i desenrotllo del Museu Martorell', in Junta de Ciències Naturals, *Anuari 1916* (Barcelona, 1916), p. 39.

24 Gea is a Spanish word (from Gaea, the Greek Goddess of the Earth) that refers to the set of beings belonging to the inorganic kingdom of a country or region.

25 The report was written by Odón de Buen, Silvi Thós (1843–1911) and the botanist and apothecary Frederic Trèmols (1831–1900), professor of inorganic chemistry in the Pharmacy School at the University of Barcelona: Gómez-Alba, 'El Museo de Geología', p. 9.

The aims set out in the report were quite ambitious, but the Junta técnica had to face serious obstacles to achieving them, namely economic scarcity and a series of internal political tensions in the Council, which combined to limit public funding and slow the development of the museum. Nevertheless, in spite of this lack of resources, many donations continued to flow into the museum, in addition to a remarkable number of visitors.[26]

There were no professional studies of natural sciences in Catalonia when the Martorell was founded,[27] though there were a few professors based in the schools of Medicine and Pharmacy at the University of Barcelona. Most of those who repeatedly contributed with donations to the museum were amateurs. This group of naturalists showed a high level of social homogeneity among them. They were mostly middle class and bourgeois professionals, young men holding degrees mainly in pharmacy, medicine, engineering or teaching, but also in law and humanities.

Two important ideological tenets characterized this group of men. First of all, they were fervently Catholic; indeed, some of the most prominent local natural history practitioners of the moment were priests, and, as we will see, they had very close ties with the Martorell. The second tenet was an explicit commitment to the building of the Catalan nation through science. This group of Catholic, conservative and Catalanist scientists were part of a broader phenomenon in Catalan society, which was the search for platforms for social intervention by a new generation of middle-class professionals. These people, energized by a feeling of class discontent, or a sense of 'status incongruity', to use Alberto M. Banti's term,[28] believed that the old ways of doing politics in the old Spanish national framework were discredited, and that a new national project could not get off the ground without them assuming a preeminent role in society. The new politics would have to be pursued in the public sphere and therefore political activism needed to act from all flanks of public life including the media, scientific associations and institutions.[29]

From the last quarter of the nineteenth century, and especially from the beginning of the twentieth, the group of Catholic and Catalanist scientists constituted a community and generated a network of associations and institutions that supported not only their scientific activity, but also their ideological programme. The most important of these was the Institució Catalana d'Història Natural (ICHN), funded as a private association of Catholic naturalists in 1899, and re-funded in 1904, under the motto 'Nulla usquam fide et ratio vera

26 For example, in September 1899, the museum had 3,500 visitors: *Informe de la Junta técnica* (Barcelona, 1899).

27 The chair of natural sciences was not created until 1889 and given to Odón de Buen, but Barcelona University did not have a school of natural sciences until 1911.

28 Alberto M. Banti, 'Italian Professionals: Markets, Incomes, States and Identities', in Maria Malatesta (ed.), *Society and the Professions in Italy, 1860–1914* (Cambridge: Cambridge University Press, 1995), pp. 223–54.

29 This has been ignored or poorly studied in much of the literature on Catalan nationalism.

dissensio esse potest'.[30] Most of the recurrent donors of the Martorell were members of the ICHN. As the group of Catholic and conservative scientists were traditionally related to political Catalanism, the best moments in terms of financing and institutionalization of the group coincided with Catalanist forces being in power.

The Martorell was a crucial element in the institutional network of the Catholic and Conservative scientists group. Indeed, Artur Bofill, who directed the museum for 30 years and constantly contributed to the museum with personal donations, fits this sociological and ideological profile perfectly. Bofill was a wealthy lawyer in Barcelona, whose position allowed him to travel all around Europe and North Africa collecting species and acquiring experience in malacology and palaeontology. Bofill was closely related to representatives of the scientific apologetics in Spain, a branch of the Catholic Church devoted to the defence of Catholic beliefs in the scientific sphere.[31]

Although the Martorell, unlike private associations, had no explicit commitment to Catalan nation-building in these years, the plan of 1899 already recognized, as mentioned above, that one of the aims of the museum would include the description of the nature of the Catalan–Balearic region, and since then Bofill had always emphasized the regional character of the museum. The Catholic and Conservative scientists group focused their research on describing the natural heritage of Catalonia; from their point of view, this was an essential step in the construction of the nation. Accordingly, most of the donations that they made to the museum contributed to the regional collections.

In addition, quite a number of the city's inhabitants, from a wide range of social backgrounds, came across objects during their daily activities that they considered marvels, or at least curiosities and therefore deemed them worthy of a place in the city's natural history museum. Due to their rarity, all kinds of teratological samples were sent to the museum. Antoni Porret, the museum's janitor, brought in a snail found in the Parc de la Ciutadella that carried a monstrous deformity on its shell;[32] a council vet donated a teratologic pig's trotter;[33] an inhabitant of La Barceloneta, the poor fishermen's neighbourhood in

30 'True dissent between faith and reason cannot exist'. The sentence was taken from the findings of *De Fide Catholica* proclaimed by the First Vatican Council (1870): Josep M. Camarasa, *Josep M. Bofill i Pichot, metge i naturalista. Semblança biogràfica* (Barcelona: Institut d'Estudis Catalans, 2009), p. 27. Some years later, when the Lliga achieved political control of local and provincial institutions, the ICHN was integrated into the Institut d'Estudis Catalans (IEC), in 1917, becoming an official institution.

31 He was an honorary member of the Sociedad Ibérica de Ciencias Naturales, created by the anti-Darwinist Jesuit Longinos Navás (1858–1938) in order to fight the pro-evolutionism of the Real Sociedad Española de Historia Natural. One of Bofill's collaborators and good friend was the priest Jaume Almera, who claimed to be interested in geology because it was the main battlefield in science for the defense of creationism. Jordi Bohigas, *'Per Déu i per la Ciència'. L'Església i la Ciència a la Catalunya de la restauració (1874–1923)* (PhD thesis, Universitat de Girona, 2011).

32 AMCNB, C17 E01; C28 E01.

33 AMCNB, C08 E27.

Barcelona, brought in a goat with two heads.[34] But maybe the most repeated quotidian marvel was found in the henhouses of the city and surrounding villages, and eggs that were considered rare because of their morphology, dimensions or properties were frequently sent to the museum.[35]

Certain places in the city were common sources of objects for the museum. Just outside the Martorell, in the Parc de la Ciutadella strollers and park workers found interesting birds, insects and even fungus.[36] The city also had its own fossil and mineral deposits, the most important being the mountain of Montjuïc, which was rich in Miocene pieces. These pieces could be found all over the place without much effort. Hence, this deposit was not only explored by practising palaeontologists, but also by ordinary people whose donations had to be classified in the museum because they were not familiar with classification procedures.[37] This was the case of Vicente Alba, a council guard who probably worked in the Montjuïc area and donated fossils to the museum on several occasions.[38] One of the biggest cemeteries in the city was, and still is, located on Montjuïc; up to 1903 the Council Cemeteries Board had collected 604 fossils from the mountain, at least 40 of which were given to the Martorell.[39] The Barcelona seaside and nearby villages also offered vacationers and sea workers natural curiosities that ended up in the museum: shellfish from the port, disoriented dolphins beached along the coast, a piece of seaworn slate and so on.[40]

Citizens were often not looking for anything, but simply came across their discoveries accidentally. The point is that these people knew that there was a home for exceptional natural objects in the city: the Martorell. One may argue that these examples show that the museum achieved visibility, centrality and legitimacy in a Barcelonese society that was heterogeneous and full of political tensions; however, this thesis cannot be verified until the relationships between the popular classes and the museum, that remain largely unknown, will be unveiled by further investigation.

In 1905, the Council had overcome a period of crises after an alliance had been established between republicans and Catalanist conservatives, the so-called Solidaritat Municipal. One of the main objectives of the Solidaritat was to promote culture in the city. In 1906 the Council created a new Junta de Ciències Naturals (Junta from now onwards) as an autonomous board of scientific experts to run the Martorell. Catholic and conservative scientists were strongly represented in the Junta. Its members included the botanist and teacher Joan

34 AMCNB, C01 E41; C09 E48.
35 AMCNB, C01 E30; C08 E86; C08 E84; C17 E01.
36 AMCNB, C09 E23; C17 E01; C28 E01; C01 E13; C01 E24; C08 E79.
37 AMCNB, C01 E44; C01 E04.
38 AMCNB, C02 E12; C08 E59; C02 E14; C08 E90.
39 AMCNB, C09 E19.
40 AMCNB, C01 E43; C09 E43; C09 E07; C17 E01; C22 E16.

Cadevall (1846–1921), who left his personal herbarium to the museum when he died; the above-mentioned priest Jaume Almera; Almera's disciple, the priest Norbert Font i Sagué (1874–1910), one of the most frequent donors to the museum until his premature death; or the doctor Josep M. Bofill i Pichot (1860–1938), who would play a pivotal role in linking the ICHN to the museum and the Junta. Therefore, through the board, important representatives of the conservative ideology shared by those who were the most frequent donors to the museum gained control over the museum itself. As we will see in the following sections, this control would become even tighter when several members of the Catholic conservative scientists group made their way into the museum and were professionalized through positions offered there.[41]

Catalanism and professionalization, 1907–1917

The Junta devised a plan for the Martorell in accordance with their values, which were Catalan nationalism, Catholicism and conservatism. This resulted in a transition from a project focusing on the region to a more explicitly political project focusing on the nation. This shift happened in parallel with the success of the political movement initiated with the foundation of the Lliga Regionalista in 1901, and which managed to take over the Council (1905) and the Provincial Government (1907) and finally to create the Mancomunitat (1914).[42] This rapid manoeuvre involved the implementation of a model of cultural and ideological intervention, which was presented as the only way for expressing the will to build the Catalan nation. The Lliga developed a propaganda apparatus in order to recruit members from all those associations and civic movements related to Catalanism that had been emerging from the 1880s. In the natural sciences, those who responded to the Lliga's call were excursionists, naturalists educated in religious schools and, above all, members of the ICHN. In this sense, it is quite symptomatic how regional donations took definitive pride of having a place in the Martorell over other kinds of acquisitions; if, as commented for the previous period, the ratio between regional and exotic donations was 50 per cent to 25 per cent, from 1906 to 1915 regional donations rose to about 70 per cent, with exotic ones falling to about 13 per cent.[43] This process culminated in 1917 with the incorporation of the ICHN collections into the museum, thus increasing and consolidating the importance of the Catholic conservative scientists group in the Martorell, both in terms of material reality and symbolism.

41 For a different context, Alberti has discussed amateur and professional identities: Samuel Alberti, 'Amateurs and professionals in one country: Biology and natural history in late Victorian Yorkshire', *Journal of the History of Biology*, 34 (2) (2001): 115–47.

42 The Junta would reflect this progressive growth of the political power of the Lliga and its allies in territorial scales, passing from exclusive control of the City Council to shared control with the Provincial Government and, after 1914, with the Mancomunitat.

43 The rest were 11 per cent from Spain and the Colonies, 4 per cent from Europe and some miscellanea or unspecified items.

Moreover, by this time members of this group were positioned inside the museum itself or on the board that regulated it. Therefore, these donors never lost control over the objects or collections that they contributed to. Apologists did not need to fear finding their donations in displays that could be considered evolutionist or not in full agreement with the official doctrine of the church. Accordingly, in addition to regular collaborators like the priests Almera and Font i Sagué, we also found contributions from Hermanos de las Escuelas Cristianas – or brothers of La Salle – from Escoles Pies de Sarrià, from the Jesuits and from Spanish rural priests. Members of the Catholic conservative scientists group who donated objects to the museum were not just donors, but also active builders of the museum, not only in a material sense, but also with regard to the ideas that the museum would transmit.

The Junta policies were very successful in terms of social mobilization. This period is characterized by a boom in donations to the museum. Indeed, if we compare it to the previous period, donations tripled in number and the number of donors doubled. It is worth mentioning that during this time the Junta managed to actively recruit several donors and even to pay a few collectors, who started contributing, albeit modestly, to the museum. Also, the Junta raised sufficient resources to buy certain collections and objects of interest, together with furnishings for old and, as we will see, new spaces for public exhibition and storage.

The Martorell was already considered too small immediately after becoming an exclusively natural history museum (see Figure 3.2). Almost 20 years later it was clearly insufficient to accommodate all the regional collections. In addition to this, the Junta was willing to introduce important novelties that also required space, like

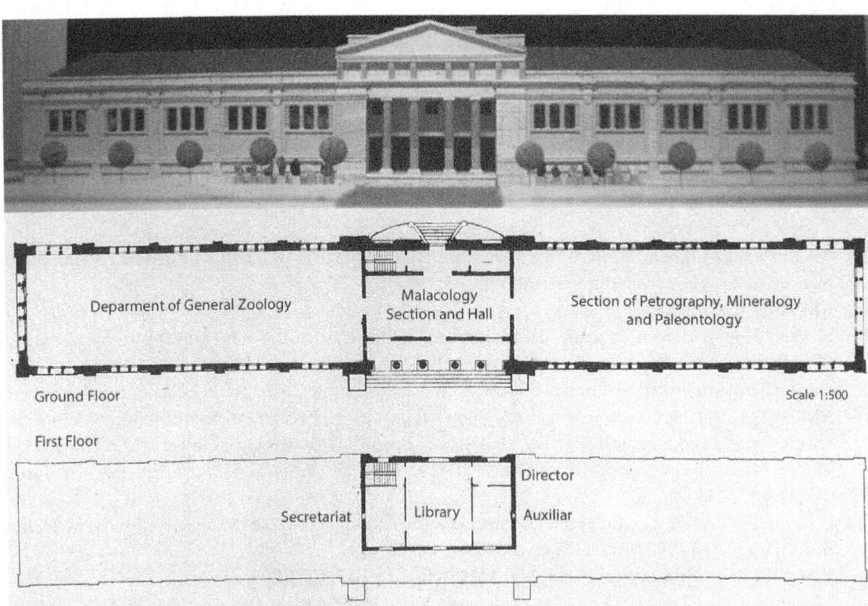

Figure 3.2 Ground and first floor plans of the Martorell Museum in 1916–17.

dioramas and large showcases representing scenes of animal life in a reliable 'natural' context.[44] The first one of these was installed in 1907 in the central showcase of the zoology room. It represented a group of six Pyrenean chamois on imitation rocks, 'as they are typically to be found in that region'.[45] Many more such dioramas were to come.[46] In 1909, thanks to the growing budget that marked these first years of the Junta's activity, a fair amount of renovation work to improve and maintain the building was done; for example, changing the parquet in the museum rooms that were open to the public, or building wood and glass cupboards for the herbarium.[47] In 1910, the part of the mineralogical collection on public display was reinstalled and the old cardboard boxes were substituted with what was described as 'an original system [which] we have no knowledge that it exists in other parts of the world; it consists in placing a sample in a metal clip, which is nailed to an oak base, which carries a plaque were the scientific determination of the mineral, its formulation, location, features, etc. are specified'.[48]

But perhaps the most important action that the Junta took during this period was to acquire management responsibilities for new spaces and buildings within the park: the greenhouse and the shade-house to accommodate new botanical and zoological species.[49] Indeed, having a botanical garden had been a long-standing aim of the Junta. Finally, after discarding a project for a radical modification of the Martorell building – which involved the construction of a first floor to cover the whole surface, the elevation of the central body and the enlargement of the lateral wards – [50] in May 1916 the Junta was granted permission to occupy the first floor of the Universal Exhibition's former restaurant, a modernist building known as Castell dels Tres Dragons, whose ground floor was used by the municipal school of music. The new space was soon baptized Museu de Catalunya.[51] Its ideological significance was clearly stated during the inauguration act by councillor Santiago Andreu, who acknowledged the Junta which he presided over as being part of the 'scientific flourishing or renaissance of our homeland'.[52] This comment, and others, demonstrated how the natural sciences had entered the political scenario of the Catalan national building

44 On this kind of dioramas, see Karen Wonders, *Habitat Dioramas: Illusions of Wilderness in Museums of Natural History* (London: Almqvist and Wiksell, 1993).

45 AMCNB, C21 E01.

46 In 1910, Soler had made so many 'naturalizations' of this sort that it was no longer possible to exhibit all of them in the public rooms of the Martorell. So they were stored in the 'annex laboratory' were the taxidermist worked and in the basement, 'waiting till more room could be found in the future': AMCNB, C25 E14. On these so-called biological groups see Nyhart, *Modern Nature*, pp. 235–6; and Santos Casado and Santiago Aragón, 'Vignettes of Spanish nature: Imagining a national fauna at the Museo Nacional de Ciencias Naturales, 1910–1936', *Historical Studies in the Natural Sciences*, 44 (2) (2014): 197–233.

47 It was made – with specific provisions for preventive conservation of the sheets – by the craftsman Antoni Vich: AMCNB, C21 E09, C23 E02.

48 Valls, 'El Museo de Ciencias Naturales'. AMCNB, C25 E14; C22 E05.

49 AMCNB, C20 E02.

50 AMCNB, C32 E17.

51 AMCNB, C32 E13.

52 Enric Aragonès, 'El mapa geològic i topogràfic de la província de Barcelona', *Treballs del Museu de Geologia de Barcelona*, 13 (2005): 115–280, p. 120.

process led by conservative Catalanism. From this perspective, the new Museu de Catalunya would bring to the public:

> the presentation of Catalan fauna in biological groups that are representative of the environment in which animals live in *our Nature*, being remarkable the fact that the will to imitate Nature was what caused these plants, herbs and stones that characterise it to be brought from their places of origin.[53]

The Museu de Catalunya also developed an oceanographic section, still in an embryonic stage, at the back of the first floor, and, importantly, 'laboratories' and 'classrooms' that were conceived and presented as cutting-edge spaces for the task of popularization that the Junta wanted to carry out (see Figure 3.3). The Junta itself explained the function of the laboratories within the museum in its guide from 1917:

> [...] museums would not serve their purpose if they were not accompanied by their corresponding working and research laboratories. The proper preparation of samples received, their study and classification, research about their organization and structure, as well as studying their constitutive elements, make the existence and operation of a series of installations which give origin to the Museum Laboratories essential.[54]

In 1917, yearly courses and lectures started in the classrooms of the Museu de Catalunya, given by the 'technical staff of the Junta', in line with the commitment to popularize natural sciences that the Junta had maintained from the beginning (see Figure 3.4). The courses were officially described as 'elementary, practical and about research on diverse subjects of natural sciences such as geology, mineralogy, entomology, zoology, micrography, and studies of nature and descriptive botany'. The courses were limited to 12 people, the museum provided all the work materials for both classes and also preparations and collections during excursions. Although on paper people from any social background could be admitted to these courses, as they were free, the actual definition of intended audiences reveals a rather narrow selection:

> These courses are especially recommended to student teachers, high-school students and students from the schools of Agriculture and Fine Arts and other people who want to be specialized in any branch of the natural sciences.[55]

No provision was made for the general public in this sense. Lectures and courses shared location with the brand-new annexes of the Museu de Catalunya, the previously mentioned laboratories, all together on the first floor of a beautiful modernist building at the entrance of the park. Underpinning all of

53 AMCNB, C32 Exp.23 (emphasis added).

54 *Guia d'Instal·lacions i Serveis de la Junta de Ciències Naturals* (Barcelona, 1917), pp. 19–21.

55 *Guia*, p. 131.

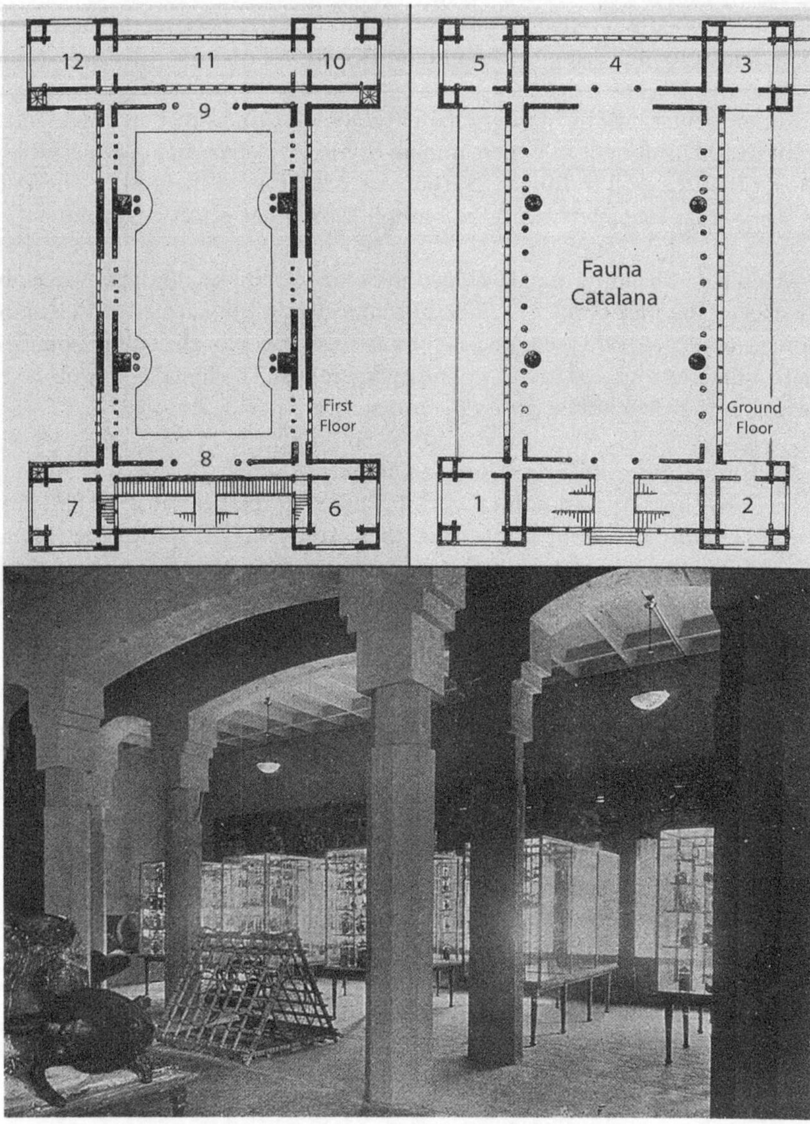

Figure 3.3 Plans of ground and first floor of the Museu de Catalunya and a view of the oceanographic section. 1. Marine biology and zoology laboratories; 2. Classroom; 3. Conversation and presidency rooms; 4. Oceanographic section; 5. Mineralogy laboratory; 6, 7. Entomology laboratories; 8, 9. Graphical representation of botanical and zoological dispersion; 10,12 Cryptogamy and phanerogamy laboratories; 11. Herbariums.

Figure 3.4 Classroom (above) and laboratory (below) in the new Museu de Catalunya, 1916–17.

this was a will for social visibility and a propagandistic purpose serving a series of essential aims that the Junta made explicit in their 1917 guide:

> Nowadays our Museum has the conditions to study everything that it receives, to create a nucleus of natural sciences amateurs, to resolve their doubts and guide them on the path that they have embarked upon, to raise the cultural level of the general public and to place before their eyes the marvel of Nature.[56]

56 Ibid., pp. 19–21.

Indeed, this decade is also characterized by the launch of the museum's and Junta's own publications, which contributed to shape the model of science popularization that the museum displays wanted to promote. The first thing to catch the eye about these publications is the institution's definitive adoption of Catalan as the main language for its public communications. The museum's section catalogues, compiled by its conservators and assistants were also published in Catalan. A second element worthy of note was the intense publication of booklets and short manuals to instruct amateur collectors. Thus the museum tried to help the donor to contribute effectively to the collections, simply by training him, or her, in practical methods of preparation, storage and classification of specimens and samples collected in the field or kept in a private collection.

Consolidation in a climate of growing social tension, 1918–1924

Officially renamed 'Museu de Ciències Naturals' in 1917 and operating within the framework of the Mancomunitat (1914–1925), the museum flourished not only occupying new spaces, but also obtaining all kinds of infrastructures, resources and staff. In this manner, as touched upon previously, outstanding members of the ICHN, followers of Artur Bofill and other Junta members were integrated as curators, managers and assistants within the museum. In November 1917, five technical posts were offered.[57] Two of these were taken up by professors at the Universitat de Barcelona;[58] another two by young naturalists, members of powerful families related to the Lliga and educated in Jesuit schools: Ignasi de Sagarra (1890–1940) and the industrial engineer Joan B. d'Aguilar-Amat (1882–1936); and the last one by the botanist and military pharmacist Pius Font i Quer (1888–1964), who would be the director of the museum from 1921, when Artur Bofill retired. In addition to this, a dozen official collectors were recruited and also several assistants for the laboratories and the museum sections. Also of note is the presence of several young women that started work at the museum; between 1917 and 1924, thirteen of these 'Museum's mademoiselles'[59] joined the staff, mainly as assistants to the curators in the collections, laboratories, library or administration.[60]

57 Julio Gómez-Alba, *El Museo de Ciencias Naturales y la Junta de Ciencias Naturales de Barcelona (1878–1941)* (unpublished typescript dated 1992), p. 248.

58 These were the professor of physical geography Maximino San Miguel (1887–1961) and professor of crystallography Francisco Pardillo (1884–1954). Unlike most of the newly incorporated personnel in the Martorell, these professors, who came from other Spanish regions were not especially interested in Catalan nationalism, if not directly against it.

59 As they were named in some contemporary texts as well as in the oral testimony given by Salvador Maluquer Jr. (*1926), recorded in 2003 by Francesc Uribe and Eulalia Garcia, curators at the MCNB, to whom we are grateful for granting access to this repository of museum memory created by their personal initiative.

60 Yet some of them managed to be named as authors in some of the publications printed by the Junta in these years. For example, Mercè Bohigas and Antonia Sánchez, authors of *Catàleg dels insectes del museu* (Barcelona, 1917) and Emilia Civit, author of *Catàleg de les Molses del Museu* (Barcelona, 1918).

The data on donors and donations in this consolidation period eloquently reflects the professionalizing push of the museum. In this period, official collectors contributed with virtually half of the entries to the museum. Among the donors there were more and more amateurs who, instead of occasionally participating with small donations, put together whole collections systematically and brought them to the museum. Several of them were ICHN members donating their personal collections. The kind of aristocratic colonial donor that characterized the initial period of the museum was reduced to a minimum; exotic donations fell to 6 per cent of the total, compared to the 82 per cent accounted for by regional ones.[61]

This evolution is also confirmed by the continuity of the popularization project that the Junta had initiated during the previous period. The Junta continued its publication effort and the use of Catalan as vehicular language. In this period, the publication of manuals for collectors, which to an extent had characterized the previous period, seems to be concluded.[62] The Junta continued to publish the museum's section catalogues and booklets about facilities in the park, albeit in a lighter mood. Finally, authors related to the Museum started publishing several works about Catalan flora, fauna and gea with other publishing houses. From 1906 to 1917, the Junta published about thirty works, and a similar number between 1918 and 1923, but this intensive production was stalled abruptly when the coup of Primo de Rivera led to the abolition of the Mancomunitat and the prohibition of Catalan in official documents and publications.

In reference to the other essential pillar of the museum's popularization project, namely the displays of materials in the spaces devoted to the general public, the problems of space continued to be the *leitmotiv* of the management's claims to the institutions. These claims were also present in the above-mentioned popularization publications and propaganda. In the Geology room, a project to close the corridor and use it as the herbarium of the Botany Department was approved; two sections of cupboards were constructed to keep the collections of dry plants.[63] The claims about the lack of space usually complained about the difficulties of ensuring the proper order and installation of collections, but also problems caused by the very cramped conditions in the spaces open to the general public:

> we need to fix with urgency [...] the part devoted to public exhibitions. Nowadays these have a considerably inferior value to the collections for study, and it is painful that our Museum, which despite its short history is so rich in research materials, can offer the non-specialized visitor no

61 As for the remaining ones, 7.2 per cent were from Spain and the colonies, 1 per cent from Europe and about 4 per cent were unspecified.

62 As far as we know, the last one in this period was Ignasi de Sagarra's, *Instruccions per als recol·lectors d'insectes* (Barcelona: Museu Martorell, 1918).

63 *Memòria de la Junta de Ciències Naturals 1919–1920* (Barcelona, 1920), pp. 33–4.

> more than the current inadequate facilities [...] we will need extraordinary resources to work towards a complete renovation of the public exhibits[64]

In reference to the Zoology room in the Martorell building, the way of exhibiting the samples in the Ornithology collection was reformed by replacing the old feet, made in different shapes and of different woods, with new feet, all with the same shape and material (oak wood). The changes in the bird collection forced the relocation of fish and reptiles to the showcases on the upper balcony, where they were much less accessible to visitors. It is interesting the way in which the director, Font i Quer, argued for this change and, at the same time, defended his preference for a kind of material – animals preserved in alcohol – that until that moment had been almost hidden from the public. Hence, the confinement of fish and reptiles to the balcony:

> is not only for the purposes of systematization, but also because these are considered insufficient in quantity and mostly not very worthy of occupying a prominent place in the Museum. On the contrary, the *alcoholarium* was hidden away at the end of the hall in the Museu de Catalunya. In order to render this collection more advantageous for the public and to turn it into a greater ornament for the room, we decided to place the five showcases more spaced out and in a prime position with good illumination.[65]

These museum reforms, designed to improve the conditions of the exhibits for the general public, also affected in this period other collections, namely entomology, malacology and birds and mammal furs.[66] However, this period of consolidation coincided with a period of very violent political and social tensions. Some contemporaries linked explicitly the problems that the museum was experiencing during these years to the 'interruptions' created by the political atmosphere and the working class strikes. During his last official appearance as director of the museum, in March 1920, Artur Bofill wrote in reference to the renovation of the geology room in 1919:

> This work had been interrupted owing to the recent social situation in Barcelona, [...] Due to the strikes, the windows that close the hall have only been finished recently, and the cupboards are still being made.[67]

The year after, the report by the newly appointed director Font i Quer, included in the Junta Report of 1920–1921 and signed on March 31, 1921, recalled that the publication of the Junta 'interrupted in recent times, has been resumed',

64 *Memòria 1919–1920*, p. 40.

65 *Memòria 1919–1920*, p. 44.

66 On tensions between the show collection and the study collection see Nyhart, *Modern Nature*, pp. 223–40.

67 *Memòria 1919–1920*, pp. 27–30, 33–4.

and justified the three-year delay in the publication of the 1918 annual report.[68] Later Font insisted: 'This year, in which the strikes have finished at the beginning of the current financial period, we have been able to finally finish the Geology room in the Martorell'.[69]

Conclusion

In the chapter we have studied the role of the Martorell as an arena where the local Barcelonese elites competed for symbolic power. This struggle not only occurred during the initial period of the Museum, when a new upper class, fusing old noble families with the upper-class industrial bourgeoisie, prevailed in the political arena of the Restoration. It continued after the colonial defeat of 1898 when society lost confidence in those political forces that had traditionally supported the monarchy and thus permitted the emergence of new political actors. In Catalonia, these new forces were, on the one hand, the progressive emergence of Catalan nationalism and, on the other, the activism of working class movements and republican radicalism.

The Museum was progressively colonized by a social elite with a clear commitment to Catalanism, Catholicism, and conservatism. We argue in favour of a model of the coproduction between science and the nation.[70] The case of the Martorell seems to confirm the ideas by Susan Sheets-Pyenson and Sophie Forgan about the symbolic power of the natural history museum.[71] The Martorell resembles, in certain aspects, the model of museum that Sheets-Pyenson explored in colonial contexts;[72] the kind of 'temple of science' that local bourgeois elites erected, not in a colonial, but in a provincial capital to show off their wealth and power.[73]

At the outset, a clique of members the upper class of Barcelona society, made up of colonial and industrial businessmen and aristocrats, had left the imprint of

68 *Memòria de la Junta de Ciències Naturals 1920–1921* (Barcelona, 1921), pp. 55–6.

69 *Memòria 1920–1921*, pp. 39–40.

70 Carol E. Harrison and Ann Johnson, 'National identity: The role of science and technology', *Osiris*, 24 (2009): 1–14; Maya Jasanoff, 'Collectors of empire: Objects, conquests and imperial self-fashioning', *Past & Present*, 184 (2004): 109–35. Interesting nuances about this issue have been introduced by Irina Podgorny and Miruna Achim, 'Introducción', in *Museos al detalle* (Buenos Aires: Prohistoria, 2014), pp. 15–26, esp. pp. 20–4.

71 Sheets-Pyenson, *Cathedrals of Science*; Forgan, 'Bricks and Bones'; Sophie Forgan, 'Building the museum: Knowledge, conflict, and the power of place', *Isis*, 96 (4) (2005): 572–85.

72 Susan Sheets-Pyenson, 'Civilizing by Nature's Example: the Development of Colonial Museums of Natural History, 1850–1900', in Nathan Reingold and Marc Rothenberg (eds), *Scientific Colonialism: A Cross-Cultural Comparison* (Washington: Smithsonian Institution, 1987), pp. 351–78. However, care may be taken to avoid adopting too rigid taxonomies of models of natural history museum as has been advised by Irina Podgorny, 'La mirada que pasa: museos, educación pública y visualización de la evidencia científica', *História, Ciências, Saúde. Manguinhos*, 12 (2005): 231–64.

73 In this sense, not unlike the cases studied in North England by Alison Kraft and Samuel Alberti, 'Equal though different: laboratories, museums and the institutional development of biology in late-Victorian Northern England', *Studies in History and Philosophy of Biological & Biomedical Sciences*, 34 (2) (2003): 203–36.

their eclectic, exotic and colonial former private collections on a museum that did not have a clear purpose. Donation was understood as an act of patriotism, initially aimed at regenerating the declining Spanish colonial empire. Later, after the rise of political Catalanism and its control over local institutions, it came to be regarded as a contribution to the building of the Catalan nation. Although regional objects had a strong presence since the museum was founded, it was not until its exclusive dedication to natural history that the emphasis shifted to regional collections. This coincided with the Artur Bofill's appointment as director of the museum.

In 1906, the City Council created the Junta de Ciències Naturals, which embarked on a period of extremely intense activity and designed a series of strategies to transform the whole Parc de la Ciutadella into a space for the confluence of natural sciences and entertainment. The strong presence of the Catholic conservative scientists in the Junta was a constant, at least until the end of the period studied in this chapter. The displays and activities that the Junta promoted were often rooted in the conservative Catalanist ethos, emphasizing amongst other things regional nature as the heritage of the Catalan nation, and showing the economic benefits of natural products.

After 1914, with the creation of the Mancomunitat and the strong presence of Catalanist forces in the Council, the Provincial Diputació de Barcelona and the government of Catalonia, the institutionalization of Catalan culture, including the natural sciences, reached its maximum. This resulted in an expansion of the museum that ended with the occupation of a new building in the Parc de la Ciutadella, the Castell dels Tres Dragons, renamed Museu de Catalunya. Between 1924 and 1930, Primo de Rivera's dictatorship brought to an abrupt halt the processes related with Catalan autonomy, and the museum had to modify, at least on the surface, its overtly Catalanist orientation.

A conflict that seems to be a general norm for natural history museums[74] is the conciliation between the need for proper dosages of illustration and entertainment for a general public and the need for training, education and assistance of a selection of specialized public. These two very different purposes forced the division of the spaces in a museum who already lamented the lack of space. A good balance between these two facets of the museum was not always struck, and, especially during the last decade analysed, the private part (laboratories, classrooms, full collections . . .), which were only open to a selected group of trainees and researchers, held the upper hand over exhibits for the general public. The reason for this most probably lay in the commitment of the Catholic conservative scientist group to consolidating a natural history tradition in Catalonia; thanks to the financial and institutional support of the Catalan institutions, they found in the museum a powerful tool to promote themselves and gather new recruits to their cause, first as volunteers and trainees, and finally as professionals holding positions within the Martorell. The museum, therefore,

74 Jim Bennett, 'Museums and the history of science. Practitioner's postscript', *Isis*, 96 (4) (2005): 602–8, here p. 606; Peter Aronsson and Gabriella Elgenius (eds), *Building National Museums in Europe 1750–2010* (Linköping: Eunamus, 2011), pp. 117, 531, *passim*.

played a key role in the professionalization of natural sciences in Catalonia. In Forgan's words, the natural history museums were 'architecturally prestigious and scientifically authoritative',[75] thus serving to show the values and success of the model of city, and society, that the bourgeois elite was erecting, and at the same time revealing the importance and prestige of a social group that was pushing for an important role within this model, namely professional scientists.[76]

In principle, the intended audience of the first public museum of Barcelona was the entire population, but in practice its policies were aimed at attracting the middle classes to the scientific and national project of those who ran the museum. It has been very difficult to find traces of popular classes participating in the museum and it seems that, despite the rhetoric of integration, the directors of the Martorell simply ignored them.[77] Nevertheless, the modesty of some of the donations and the information we have been able to gather about some contributors' neighbourhoods and professions give us some indication that these classes at least enjoyed a minority presence among the pool of donors.

The museum accepted almost all donations, even ones that were not considered very valuable. Was this just a way of capturing its audience through participation? This question is especially relevant when we consider the popular classes, whose relation with the museum is much less understood. While it is not hard to guess why the upper classes would contribute to a museum that fully represented their own values and expectations, the motivations that made members of the popular classes buy into this kind of project are much harder to fathom; and even more so if we consider the prevailing climate of social confrontation in the city during the period studied, in which anarchists, radical republicans and others organized political opposition, strikes and attacks against the establishment and the bourgeois elite. Some left leaning intellectuals related to the working class movement criticized static representations of nature, and the idea of the natural history museum itself.[78] Without forgetting that we will be investigating a specific but significant contribution, it is worth asking

75 Forgan, 'Bricks and Bones', p. 191.

76 This quest for authority by Catalan scientists, or more precisely the Catholic conservative group, can be interpreted as an example of a general phenomenon described by the sociologist Pierre Bourdieu as 'double determination of the ideology'. According to Bourdieu, intellectuals working to confer symbolic power upon a hegemonic model of society will produce doubly determined ideological forms; forms that will work to support this model and at the same time will bring prestige and authority to the intellectuals that produced them. Pierre Bourdieu, *Sobre el poder simbólico. Intelectuales, política y poder* (Buenos Aires: UBA-Eudeba, 2000), pp. 65–73.

77 Köstering finds the opposite for German natural history museums: Susanne Köstering, *Natur zum Anschauen. Das Naturkundemuseum des deutschen Kaiserreichs 1871–1914* (Köln: Böhlau, 2003), pp. 259–63.

78 Odón de Buen, whose position as professor of Natural History at the University granted him a place in the Junta from 1906 to 1910 as we mentioned earlier, never donated anything to the museum. In his memories, he wrote: 'I always wanted to study Nature in Nature. To know natural beings in collections, whatever the classification they are grouped under, dissected with more or less fidelity, is not to know Nature' (cited in Alberto Gomis, 'Odón de Buen. Cuarenta y cinco años de compromiso con la universidad', *Asclepio*, 63 (2) (2011): 405–30, p. 414).

ourselves how the museum achieved the status of exhibitor of uncommon objects in the collective imagination of Barcelona, and if during this process it competed with other non-formal spaces, like fairs or other kinds of private exhibitions. There are still many stories to unveil in order to understand the relationships that the Museum of Natural Sciences established with its donors and the citizens of Barcelona.

4 Laboratory medicine and surgical enterprise in the medical landscape of the Eixample district

Alfons Zarzoso and Àlvar Martínez-Vidal[1]

In October 1899 a lavishly illustrated 39-page printed advertising pamphlet appeared in Barcelona, describing in great detail the technical and architectural features of Dr Cardenal's 'casa de curación' ('healing house'), a surgical clinic founded in the year of the Universal Exhibition of 1888 and now, following the extension and refurbishment of the building, rejoicing in the name of 'Clinicum'.[2] The pamphlet explained that this little hospital was located in the Eixample district, the main urban extension of the city, very close to Provença Station on the Sarrià railway line. The building, a house with a garden in Passatge Mercader, had been the family home of Salvador Cardenal (1852–1927), a physician who, after graduating from the Barcelona Faculty of Medicine, pursued further training as a surgeon in several European cities and became one of the first proponents of Joseph Lister's antiseptic method in Spain.[3]

The pamphlet – clearly addressed to 'metges de capçalera' (general practitioners)[4] – gives a thorough description of all the clinic's services, with interior and exterior plans of the house showing all the facilities: the operating theatres and the annexed laboratories, including the 'x-ray chamber', as well as the bedrooms, dining rooms, bathrooms and so on. The final pages

1 We would like to thank Tatjana Buklijas and Thomas Schlich for their insightful and helpful comments of draft versions of this chapter. This research has been partly funded by the Spanish Ministerio de Economía y Competitividad in the frame of the project 'Hacia la consolidación de las especialidades médicas en Barcelona: unidad y división en el saber y la práctica de la medicina (1911–1936)' (HAR2012-34586).

2 *Clinicum. Casa de curación quirúrgica del Dr. Cardenal: noticia descriptiva y extracto del reglamento* (Barcelona, 1899).

3 Juan Riera, 'The dissemination of Lister's teaching in Spain', *Medical History*, 13 (2) (1969): 123–53, pp. 137–40; José Danon, 'De la antisepsia a la asepsia en la obra de Salvador Cardenal', *Medicina & Historia*, 61 (1996): 5–28.

4 Article 12 of the clinic's Regulations mentions this point explicitly: 'The general practitioner of every patient that is to undergo surgery at the Clinic may, by prior agreement with the Director, be present at and even take part in the operation and subsequently visit his patient, if both of them so wish'. *Clinicum*, 1899, p. 38. The aim of this article was to guarantee that general practitioners would not lose clients through competition from the clinic.

contain a summary of the clinic's regulations, detailing the board and lodging conditions and even the prices for each category (special, first and second category).

Undoubtedly, however, the most important novelty that this piece of persuasive publicity aimed to convey to the public was that the facilities had been adapted to practise surgery in strict aseptic conditions. Thus in addition to the 'antiseptic room', that is, the original operating theatre designed on Listerian principles, on the ground floor, the pamphlet presented a technological innovation: the 'aseptic operating room'. As we shall see later, this new theatre, located on the first floor of the building, was equipped to operate with a complete absence of microbes, avoiding fearsome post-operative infections.

The fact that the house had three floors (see Figure 4.1), plus a basement for the clinic's internal services, hampered movement within the building, and especially the transfer of patients to and from the theatres before and after operations. However, all these problems had been considered and solved, because, as the pamphlet pointed out, the clinic had a powerful electric traction lift (see Figure 4.2) which connected all the floors rapidly and efficiently:

> The [...] spacious lift, for passengers, is moved by a powerful machine with a 7-horsepower electric dynamo, which receives its power supply from the Barcelona Electricity Company (like all the House's power and lighting facilities).[5]

According to the pamphlet, the lift compartment, 2 m long by 1.2 m wide, was 'perfectly comfortable, like an elegant little chamber, or a kind of railway carriage',[6] enabling patients to be moved comfortably throughout the building and particularly between their own rooms and either of the two operating theatres (antiseptic or aseptic):

> [...] when the operation is over, the patient, still deep *in anaesthetic sleep*, is returned by the same procedure to his room and his bed, in which he wakes up without having been aware of having been moved or of any part of what has occurred.[7]

The remarkable changes in the nature of this 'healing house' between 1888 and 1899 and in the kind of health care services it offered, as well as its location at the very centre of the district known as Eixample, near a commuter

5 *Clinicum*, 1899, p. 24. From 1896 two large companies – the Compañía Barcelonesa de Electricidad (Barcelona Electricity Company) and the Catalana de Electricidad (Catalan Electricity Company) – competed in Barcelona to install direct current electric cabling inside and outside the ring roads where the city walls once stood. Joan Carles Alayo i Manubens, *Electricitat a Catalunya. De 1875 a 1935* (Lleida: Pagès editors, 2007), pp. 321 ff. See also Chapter 11 of this book.

6 *Clinicum*, 1899, pp. 24–5.

7 *Clinicum*, 1899, pp. 18–20. Our emphasis.

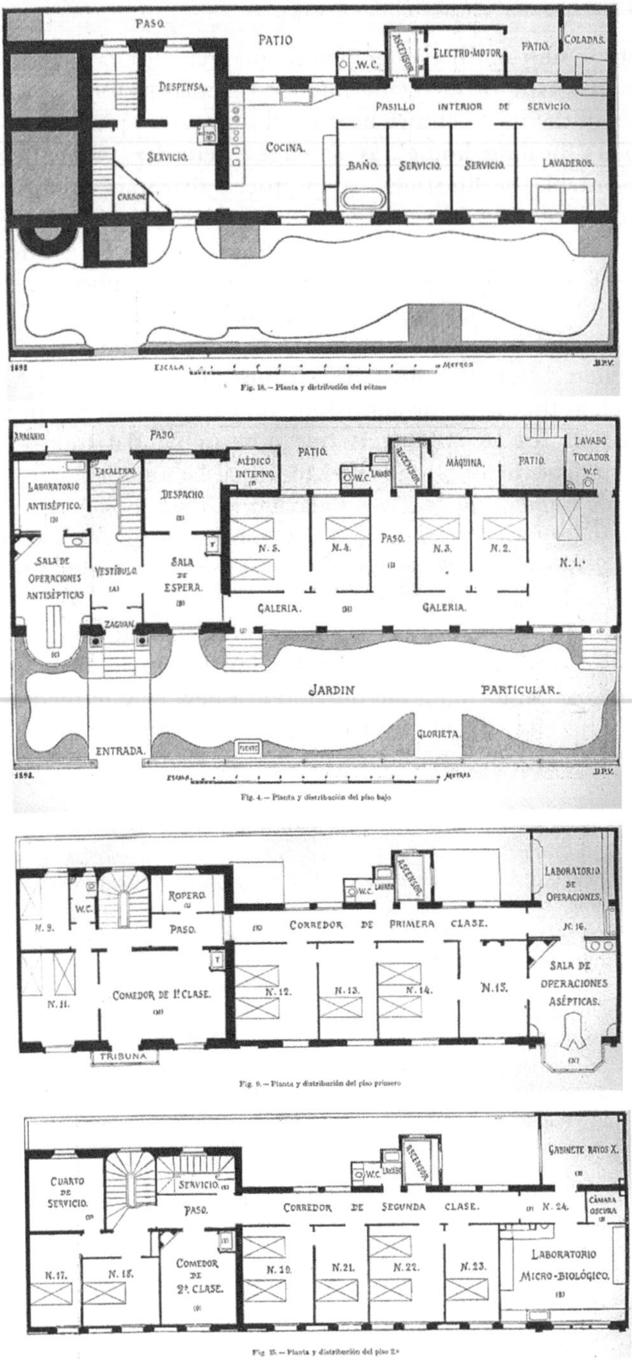

Figure 4.1 Floor plans of the Clinicum, Cardenal's 'Healing House', in 1899 (ground floor and two upper floors). The antiseptic operating room is placed on the ground floor, close to the entrance to the building, meanwhile the aseptic one is placed on the first floor.

railway station, sum up the geographical and architectural transformation that took place in a type of medicine intended primarily for the affluent. In the first decades of the next century new small surgical hospitals were built around Dr Cardenal's 'healing house', as Passatge Mercader, where it was located, became a veritable 'health street'.[8] In a process that started in the last third of the nineteenth century, laboratory medicine played a remarkable role in the emergence of new medical spaces and specialisms. One of the results was the abandonment of the old city and the establishment of innovative surgical sites in the new and prosperous extension of Barcelona.[9]

Laboratory medicine was established in the city within and also beyond the academic sphere. Slaughterhouses, dairy shops, pharmaceutical laboratories and so on, as well as the Academy and Laboratory of Medical Sciences,[10] a private scientific society run by volunteers, but quite influential, founded by a group of medical students in 1872, and the Municipal Laboratory[11] founded in 1885, were certainly burgeoning in Barcelona and its metropolitan area at that time. In 1891 the City Council undertook the modernization of subsurface drainage and sanitation according to the ambitious project submitted by the sanitary engineer Pere García Faria (1858–1927).[12]

In the mid-nineteenth century the location of the old General Hospital and the attached Faculty of Medicine, reinstated in 1843, in the Raval district played a crucial role in transforming the provision of new private medical services in the streets of the old city. But in the last decades of the century plans to move those institutions to the new bourgeois district resulted in a major change, as

8 *Medicine and Architecture in Barcelona: The Health Street: Passatge Mercader* [in Catalan in the original], http://www.medicinaiarquitecturabarcelona.cat/p_mercader.html (last accessed on 14 December 2014).

9 The expression 'Laboratory medicine' is used here as a historiographical category in the sense of a medical practise based in experimental scientific knowledge, mainly rooted in the microbiological paradigm but not exclusively. Laboratories (physical, chemical, microbiological and so on) could be spaces for the production of new medical knowledge as well as spaces used for routine diagnostic purposes. Deborah Brunton (ed.), *Medicine Transformed: Health, Disease and Society in Europe, 1800–1930* (Manchester: Manchester University Press/Open University, 2004), pp. 92–118. The introduction of new instruments of measure was a feature tied to the process of construction of a modern, scientific and technological medicine. In some contexts, such as London, those diagnostic novelties were plainly rejected by the gentlemen, well-accommodated and *élite* professionals that controlled medicine in the city. In front of them, young doctors, those advocating the use of instruments and trying to meet a demand through a different understanding of medicine. On this: Christopher Lawrence, 'Incommunicable knowledge: Science, technology and the clinical art in Britain, 1850–1914', *Journal of Contemporary History*, 20 (4) (1985): 503–20.

10 Josep M. Calbet i Camarasa, 'Els primers anys de l'Acadèmia de Ciències Mèdiques', *Gimbernat*, 28 (1997): 147–56.

11 Antoni Roca Rosell, *Història del Laboratori Municipal de Barcelona: de Ferran a Turró* (Barcelona: Ajuntament de Barcelona, 1988).

12 Xavier Irigoyen, 'Apropament al "Projecte de sanejament del subsòl de Barcelona" de Pere Garcia Faria'. *Espais*, 24 (1990): 49–56; Ferran Sabaté Casellas, 'Pere García Faria i els orígens de l'enginyeria sanitària', *Gimbernat*, 22 (1994): 227–36.

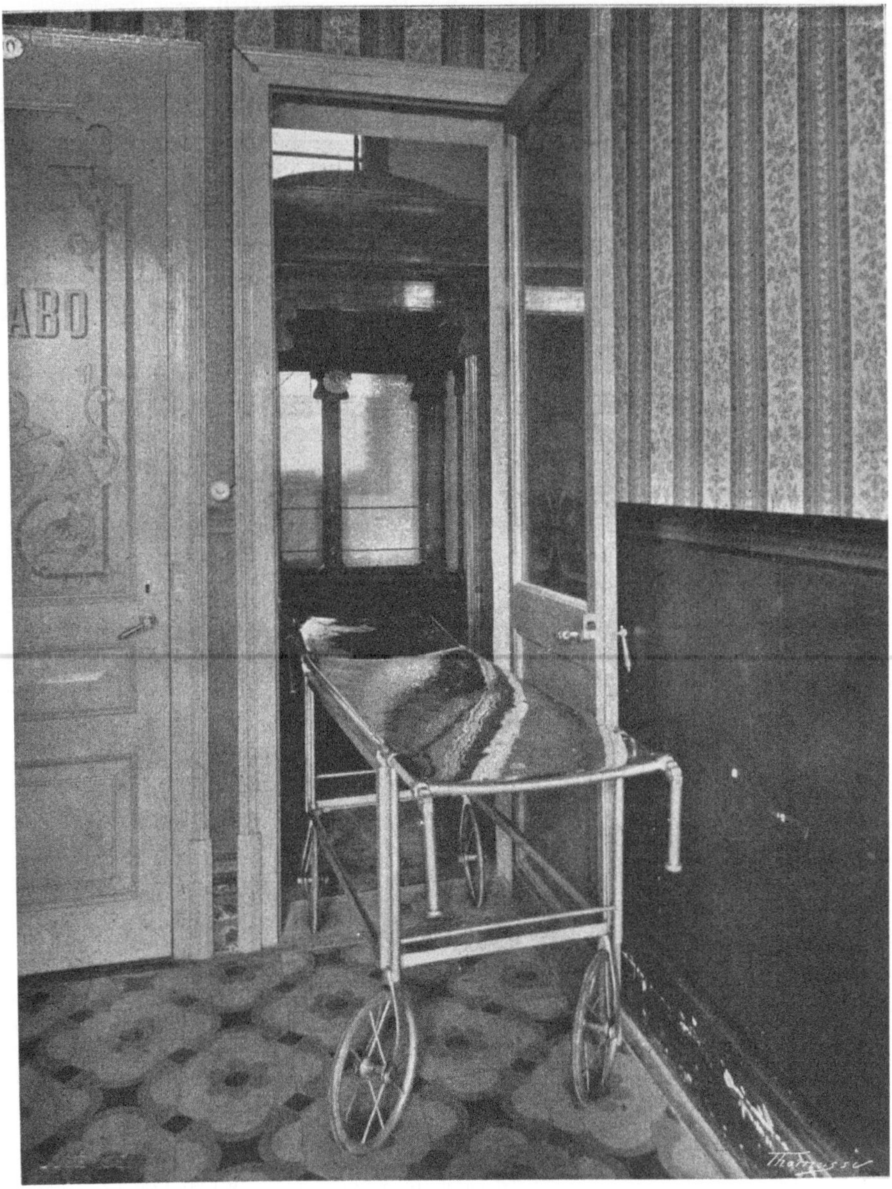

Figure 4.2 Photography of the trolley and lift at the Clinicum (1899).

small surgical clinics, and also large hospitals, were established in particular streets in the Eixample district. One of the most conspicuous consequences of this was the transformation of the medical landscape and architecture of the city at the end of the first third of the twentieth century in Barcelona.

Scientific spaces such as those created under the general heading of laboratory medicine contributed decisively to establishing a specific model of representation in the new urban planning model designed by and for the bourgeoisie of Barcelona.[13] Those spaces had a particular design that emphasized specific ways of doing science and practising medicine, centred on the specific locus of the operating theatre. Acceptance of the germ theory of disease was first expressed in the design and construction of the operating theatre and the bacteriological laboratory.[14] Though small in size, new surgery buildings, equipped with powerful lifts, electric lighting, x-ray chambers and chemical and microbiological laboratories shaped the identity of patients as well as that of doctors. In this chapter we highlight the physical role of these spaces, not only in accommodating the new requirements of laboratory medicine and new diagnostic and therapeutic tools, but also in terms of the image of the new buildings and their carefully planned location in specific areas of *fin-de-siècle* Barcelona. In doing so, we consider that the use of a rhetoric of modernity was implied in the new architecture of the clinics.[15]

This chapter will focus on the strategies developed by doctors – in particular, by two audacious surgeons – to create and meet a new demand primarily among the bourgeoisie by offering new health care services and modern medical spaces. It will concentrate on the emergence of modern surgery in Barcelona through a specific architecture based on new technologies and scientific innovation, as well as the means of communication used by the owners of the new health care spaces to reach new audiences through pamphlets and booklets, and even books and journals, advertising and illustrating those spaces.[16] In contrast to the holistic conception of the body underlying traditional medicine based on Galen, surgery, which involved a compartmentalized conception of the human body, strongly contributed to the process of medical specialization.

13 Andrew Cunningham and Perry Williams (eds), *The Laboratory Revolution in Medicine* (Cambridge: Cambridge University Press, 1992).

14 On the sense of progress through the use of germ theory, see Thomas Schlich, 'Farmer to industrialist: Lister's antisepsis and the making of modern surgery in Germany', *Notes and Records of the Royal Society*, 67 (3) (2013): 245–60. On the relationship between spatial divisions in the architecture of medical buildings and the discursive practices revealed in their interiors, see Lindsay Prior, 'The Architecture of the Hospital: A Study of Spatial Organization and Medical Knowledge', *The British Journal of Sociology*, 39 (1) (1988): 86–113.

15 On this notion of space as an active part of social life, see Erwin Goffman, *Asylums: Essays on the Social Situation of Mental Patients and Other Inmates* (New York: Doubleday, 1991, orig. 1961); Cameron Logan, Philip Goad and Julie Willis, 'Modern hospitals as historic places', *The Journal of Architecture*, 15 (5) (2010): 601–19.

16 A series of studies on the relationship between architecture and modern surgery in the early twentieth century from a perspective of engagement with the methods of material culture and the history of science and medicine can be found in John D. Thompson and Grace Goldin, *The Hospital. A Social and Architectural History* (New Haven: Yale University Press, 1975); Annmarie Adams and Thomas Schlich, 'Design for control: Surgery, science, and space at the Royal Victoria Hospital, Montreal, 1893–1956', *Medical History*, 50 (3) (2006): 303–24; Thomas Schlich, 'Surgery, science and modernity: Operating rooms and laboratories as spaces of control', *History of Science*, 45 (3) (2007): 231–56.

Throughout the nineteenth century surgeons in London enhanced their prestige and wealth and ascended the professional hierarchy by founding special small hospitals which appealed to the bourgeoisie and middle classes and 'special hospitals played their part in the transformation [...] of the care for the sick in the home to care in an institution from the doctor as peripheral in medical care of the majority to the doctor as central'. These special hospitals, in Lindsay Granshaw's words, 'led the way in Britain to the much greater medicalization of all hospitals'.[17] Obviously, Barcelona was not London, nor were the academic and professional contexts of Catalan and English surgeons similar. However, some of these trends could be detected in the private sector development of surgery in Barcelona.

This chapter seeks to identify and characterize the first spaces specifically related to the ideas and principles of laboratory medicine in Barcelona, focusing mainly on the kind of small hospitals that were flourishing in Eixample.[18] To explore the strategies implemented by entrepreneur surgeons, we will consider two models in which different types of surgical processes were practised. The first is Dr Cardenal's 'healing house', already mentioned, a small non-specialized surgical hospital built in 1888 and radically remodelled in 1899. The other, founded in 1891 by Ricard Botey (1855–1921) near the old General Hospital and transferred in 1898 to Eixample, not far from Cardenal's clinic, focused explicitly on the surgical treatment of ear, nose and throat (ENT) conditions, a significant step further in the emergence of medical specialization.[19] Both these 'healing houses', Dr Cardenal's and Dr Botey's, each with its own particular features and innovative resources against pain and infection, played a prominent part in spearheading laboratory medicine in nineteenth-century Barcelona.[20]

17 Lindsay Granshaw, '"Fame and Fortune by means of Bricks and Mortar": The Medical Profession and Specialist Hospitals, 1800–1948', in Lindsay Granshaw and Roy Porter (eds), *The Hospital in History* (London: Routledge, 1989), pp. 9–10, 199–220.

18 Hospitals, as large public health care institutions, have traditionally attracted the attention of historians. Recent research has also shown an interest in public and private hospitals: Guenter B. Risse, *Mending Bodies, Saving Souls: A History of Hospitals* (Oxford: Oxford University Press, 1999); John Henderson, Peregrine Horden and Alessandro Pastore (eds), *The Impact of Hospitals, 300–2000* (Oxford: Peter Lang, 2007); Laurinda Abreu and Sally Sheard (eds), *Hospital Life: Theory and Practice from the Medieval to the Modern* (Oxford: Peter Lang, 2013); Christopher Bonfield, Jonathan Reinarz and Teresa Huguet-Termes (eds), *Hospitals and Communities, 1100–1960* (Oxford: Peter Lang, 2013).

19 On medical specialization, see Alfons Zarzoso, Sara Fajula, Àlvar Martínez Vidal, Vicent Morera, Enrique Perdiguero and Judith Perona, 'Especialización y divulgación de la medicina en la ciudad de Barcelona (1868–1938)', in *Actas del XV Congreso de la SEHM. Transmisión del conocimiento médico e internacionalización de las prácticas sanitarias: una reflexión histórica* (Ciudad Real: SEHM-UCLM, 2011), 279–305.

20 Private clinics have been the object of interest among historians, especially those in Central Europe specializing in the treatment of psychiatric disorders: Edward Shorter, 'Private clinics in Central Europe, 1850–1933', *Social History of Medicine*, 3 (2) (1990): 159–95 and 'Women and Jews in a private nervous clinic in late nineteenth-century Vienna', *Medical History*, 33 (2) (1989): 149–83; Leslie Topp, 'An architecture for modern nerves: Josef Hoffmann's Purkersdorf sanatorium', *Journal of the Society of Architectural Historians*, 56 (4) (1997): 414–37; Andreas Killen, *Berlin Electropolis: Shock, Nerves*

The emergence of modern surgery between the old city and the new

Carrer de l'Hospital and Carrer del Carme were the streets that marked the geographical limits of the General Hospital of the Holy Cross (Hospital General de la Santa Creu), founded beyond the second wall of the medieval city in 1401, and the Faculty of Medicine. The overcrowded conditions, physical location and spatial limitations of these institutions were fiercely criticized by the bourgeoisie in the second half of the nineteenth century, especially after the launch of the Cerdà Plan for the extension – Eixample in Catalan – of Barcelona in the 1860s. Despite undergoing a process of radical restructuring and facing the prospect of being moved, the area surrounding the General Hospital and the Faculty of Medicine still exerted an influence and acted as a pole of attraction for a concentration of new private clinics. The vigorous growth of the new city and the pressing need to meet the demands of the bourgeoisie were to transform the location and structure of those small private hospitals.[21]

The search for light – and not just sunlight – and better hygienic conditions prompted some enterprising Barcelona surgeons, who very quickly saw the opportunity to achieve economic and social success by developing a new way of practising surgery based on the technical advances resulting from the campaign to combat pain (analgesia/anaesthesia) and infection (antisepsis/asepsis). In response to lack of privacy and fears of pain and infection, as well as the need to reduce the high mortality rate, some specialized practices, especially those related to surgery, gradually abandoned the General Hospital in the old city and moved to new premises in Eixample.

Some remarkable surgeons with little or no chance of making a career in academia or gaining a position in a provincial or municipal hospital chose to set up their own private surgical clinics. In such circumstances, the establishment of anaesthetic and analgesic as well as antiseptic and aseptic procedures as the basis for a new approach to surgery offered them an exceptional opportunity. This initiative involved a new medical culture which introduced spaces such as the operating room, where a new type of expert developed new techniques and created new medical knowledge.[22]

and German Modernity (Berkeley: University of California Press, 2006), pp. 48–80; Sabine Wieber, 'Sculpting the sanatorium: Nervous bodies and femmes fragiles in Vienna 1900', *Women in German Yearbook*, 27 (2011): 58–86. Also on sanatoria for the treatment of tuberculosis: Margaret Campbell, 'What tuberculosis did for modernism: The influence of a curative environment on Modernist design and architecture', *Medical History*, 49 (4) (2005): 463–88.

21 Eduard Aibar and Wiebe E. Bijker, 'Constructing a city: The Cerda Plan for the extension of Barcelona', *Science, Technology, & Human Values*, 22 (1) (1997): 3–30.

22 Some authors have seen the ways in which medical spaces such as the new operating theatres were structured and organized as material signs of the changes that were taking place in the status and self-image of surgeons: James T.H. Connor, 'Bigger than a bread box: Medical buildings as museum artefacts', *Caduceus*, 9 (2) (1993): 119–30. Allen M. Brandt and David C. Sloane, 'Of Beds and Benches: Building the Modern American Hospital', in Peter Galison and Emily A. Thompson (eds), *The Architecture of Science*, (Cambridge, Mass.: The MIT Press, 1999), 281–308. Thomas A. Markus, *Buildings and Power: Freedom and Control in the Origin of Modern Building Types* (London: Routledge, 1993).

The careers of two surgeons illustrate the tension between old and new medical spaces in the city at the end of the nineteenth century. In 1879 Salvador Cardenal joined the staff of the Hospital del Sagrat Cor (Sacred Heart Hospital),[23] a volunteer hospital founded the year before by a group of six charitable ladies and run by the Sisters of Charity of St.Vincent de Paul, and in 1883 he became its director. Nevertheless, he founded his own clinic, also in Eixample, in 1888. This initiative did not mean that he left his job at the Hospital del Sagrat Cor; he continued working in both institutions. Meanwhile, in 1891 Ricard Botey set up a clinic for ENT surgery in the old city, at 57–59 Carrer del Carme in the area around the General Hospital and the Faculty of Medicine. This was not an unusual choice of location for a private clinic, as can be seen from the prime example of the surgical clinic specializing in gynaecology established by Miquel A. Fargas on the first floor of 133 Carrer de l'Hospital in 1884. A few years later, following Cardenal's pioneering initiative, Fargas, Botey and other surgeons in Barcelona moved their clinics from the old city to Eixample.[24]

Here we are interested in two issues closely related to the material culture of surgery: scientific spaces and ways of understanding and performing surgery. A range of sources, such as the local press, medical periodicals, catalogues of medical instruments and so on, enable us to describe the physical features of the new spaces. On the other hand, communication of the new surgical procedures was crucial in order to publicize that surgery successfully, and books constituted an essential means to that end. The medical literature written by Cardenal and Botey provides us with a wealth of material that helps us to understand the ways in which the new surgery was introduced – travel, correspondence, books, translations, pictures, illustrations, and so on.

The two doctors we are studying in this chapter – Cardenal and Botey – share certain features that we consider vital to an understanding of the complexity of the process of training and defining experts.[25] Both of them harvested the first fruits of the knowledge derived from laboratory medicine, which gave rise to a new understanding of the microbiological world. In this sense, the surgery practised by these individuals was based on this new medicine and thus enabled them to make a local contribution to the transformation of surgery and the production of a new fund of knowledge and progressive specialization, and also to the introduction of radically new spaces for medical practice: clinics.[26] Studies of urban geography have detected a change in the model of health care – from the domestic, family or home visit model to the model of going to see the doctor or an association of doctors at their own premises, in polyclinic

23 This hospital moved to another site, not far from the first one, in 1883. See Joan Agustí i Peypoch, *Història de l'Hospital del Sagrat Cor (1879–1980)* (Barcelona: Fundació Uriach 1838, 1991), pp. 5–24.

24 *La Vanguardia*, 24 January 1893, 2 and 24 April 1898, 1.

25 George Weisz, *Divide and Conquer: A Comparative History of Medical Specialization* (Oxford: Oxford University Press, 2005). On the definition of who was considered a surgeon, see: Christopher Lawrence, 'Democratic, Divine and Heroic: the History and Historiography of Surgery', in Christopher Lawrence (ed.), *Medical Theory, Surgical Practise: Studies in the History of Surgery* (London: Routledge, 1992), 1–47.

26 Lawrence, 'Democratic, Divine and Heroic'.

buildings – which developed in parallel to the transformation of the modern city, characterized by planning and demographic changes and by changes in forms of transport and communication.[27] Their surgical practice in these spaces, together with their initiative and their ownership of the properties, gave them a role as entrepreneurs, taking the risk of investing and committing themselves to a particular idea of doing science based on a new form of medical architecture.

Another common feature connecting them is their training and their authority, acknowledged by their colleagues, in specific areas of surgery. As well as criticizing the Spanish system of teaching medicine, these surgeons went on scientific travels, either in the form of a 'grand tour' of European hospitals, clinics and university faculties or by systematically attending and participating in international conferences, to acquire training and observe both techniques and forms of procedure and the scientific rationalism of the new surgical spaces.[28] The creation of clinical spaces by these individuals is the basis of the ongoing production of new medical knowledge. Both of them underpinned their professional standing with intense promotional activity, whether by writing articles in the medical journals of the time or by creating specialized medical publications of their own. However, writing specialist treatises was the defining element that enabled them to establish themselves as part of a cohesive group. Their purpose was complex, defined in Ricard Botey's words as 'a mixture of science, display, professional activity and advertising'.[29]

The treatises not only present a thorough exposition of a surgical method and mode of practice. There is a clear desire to convince their fellow doctors, to win their trust, to convey and demonstrate the soundness of the new knowledge, to show the value of individual initiative in the context of a systematic critique of a centralized, speculative model of university training. The awareness of shaping a new activity, of creating new knowledge linked to new practices, techniques and spaces, led them to attach prime importance to the material culture of the new surgery and surgical specialisms. In this exercise of persuading others, the systematic use of various forms of visual representation of science becomes a central communicative strategy in both cases. The gradually increasing clientele of these surgeons' clinics, referred to them by urban doctors and to a lesser extent by rural physicians and other rural medical practitioners, had an explanation, in Botey's opinion: 'their behaviour has doubtless been influenced by the constant stream of scientific advertising imparted [...] by the monthly publications appearing in Barcelona'.[30] Advertising has to be

27 Paul Knox, James R. Bohland and Neil L. Shumsky, 'The urban transition and the evolution of the medical care delivery system in America', *Social Science and Medicine*, 17 (1) (1983): 37–43.

28 Australian hospital architects, administrators and politicians travelled widely in the first half of the twentieth century to study the latest developments in the field. Cameron Logan and Julie Willis, 'International travel as medical research: Architecture and the modern hospital', *Health History*, 12 (2) (2010): 116–33.

29 Ricard Botey, *Estudios clínicos sobre laringología, otología y rinología. Su práctica y enseñanza en Europa a fines del siglo XIX (1891–1900)* (Barcelona, 1903), p. 376.

30 Ibid., p. 333.

understood in this context not only as unethical and unproblematic, but as a paramount part of a visualizing argument in a strategy of persuasion developed exclusively in medical publications and just devoted to medical practitioners.[31]

A general surgical clinic: Cardenal's healing house

'Dr Cardenal's Surgical Healing House' was one of the first private clinics established in the Eixample district of Barcelona. Its founder, Salvador Cardenal (1852–1927), instituted it, with perspicacity, at number 13 Passatge Mercader, in January 1888 (see Figure 4.3), a few months before two medical science conferences – one on medicine[32] and the other on pharmacy – were held as part of the Barcelona Universal Exhibition.[33] It seems that at first the clinic was simply the private home of Salvador Cardenal himself and his family.[34] Very soon, however, alterations were made to the house to devote it primarily to health care purposes. Ten years later, in 1899, the original clinic had been remodelled, modernized and considerably enlarged, occupying number 15 Passatge Mercader as well, then an enclave surrounded by medical establishments.[35] The 'Surgical Healing House' had turned from 'a little town house with an almost square ground plan measuring 10 metres each side' into 'a building with a façade almost 30 metres long, three habitable floors and a basement for the service facilities, all of it south-facing and abundantly airy and sunny'.[36] To emphasize the architectural changes made, the establishment adopted the slightly pretentious title of 'Clinicum'. According to the publicity brochure published immediately after the opening of the new building, the refurbishment was motivated by 'the scientific requirements of modern surgery, the convenient and practical operation of a service for patients with a diverse range of surgical conditions, and the comfort, independence and material well-being of those patients and

31 Delia Gavrus, 'Dreams and men of action: Neurologists, neurosurgeons, and the performance of professional identity, 1920–1950', *Bulletin of the History of Medicine*, 85 (1) (2011): 57–92.

32 According to Riera, these conferences marked a turning-point in the introduction of antisepsis in Spain. Riera, 'The Dissemination of Lister's Teaching', pp. 129–30.

33 *Exposición Universal de Barcelona. Congresos de Ciencias Médicas celebrados del 9 al 15 de septiembre de 1888 por iniciativa y con subvención del Consejo General de la Exposición Universal de Barcelona. II. Congreso Médico* (Barcelona, 1889). Salvador Cardenal was a member of the organizing committee for the conference on medicine in his capacity as Director of the Hospital del Sagrat Cor.

34 Víctor Marí Balcells, 'Clíniques privades a Barcelona a les darreries del segle XIX', *Gimbernat*, 33 (2000): 163–81, pp. 171–4.

35 Work opportunities for doctors were plentiful in the new bourgeois city. Certain areas or streets, such as Harley Street in London or Mediziner-Viertel in Vienna, became veritable magnets for physicians, surgeons and specialists. Maria Rentetzi, 'The city as a context for scientific activity: Creating the Mediziner-Viertel in fin-de-siècle Vienna', *Endeavour*, 28 (1) (2004): 39–44; Charlotte Humphrey, 'Place, space and reputation: The changing role of Harley Street in English health care', *Social Theory and Health*, 2 (2) (2004): 153–69; Tatjana Buklijas, 'The Politics of fin-de-siècle Anatomy', in Mitchell G. Ash and Jan Surman (eds), *The Nationalisation of Scientific Knowledge in the Habsburg Empire (1848–1918)* (London: Palgrave Macmillan, 2012), 209–44.

36 *Clinicum*, 1899, p. 5.

Figure 4.3 Photograph of the facade of the Clinicum in 1899.

of the relatives or others accompanying them, insofar as this is compatible with those requirements'.[37] The brochure explained that the architect, Bonaventura Pollés i Vivó,[38] had succeeded in 'putting into practice all the scientific specifications and requirements of the project', so much so that all the previous surgical spaces had become obsolete, even those reproduced photographically in the latest edition of his *Practical Manual of Antiseptic Surgery*, published in 1894:

> In constructing these new operating rooms we have departed considerably from the system we adopted in the room we constructed in 1890, depicted in phototype in the third edition of my 'Practical Manual of Antiseptic Surgery'.[39]

The allusion to his *Manual* was by no means a trivial point. Cardenal had published three successively revised and expanded editions of a work devoted

37 Ibid., p. 6.

38 Bonaventura Pollés i Vivó (1856–1918) was a Catalan modernist architect who designed several buildings in the Eixample district in Barcelona, prominent among which was his own house, on the corner of the Rambla de Catalunya and the Carrer de València. He was municipal architect of the city of Vilanova i la Geltrú. Bonaventura Bassegoda, 'Nota necrológica: Buenaventura Pollés y Vivó', *Anuario de la Asociación de Arquitectos de Cataluña*, (1919): 153–6.

39 *Clinicum*, 1899, p. 15.

to antiseptic surgery.[40] The first, which appeared in 1880 under the title *Practical Guide for Healing Wounds and the Application of the Antiseptic Method in Surgery*,[41] was primarily concerned with the surgical training he had acquired abroad, and above all with his interest in disseminating the antiseptic method advocated by Joseph Lister (1827–1912), the scientific basis of which was not necessarily fixed or consistent with the various subsequent developments in microbe theories of infection.[42] Indeed, the book originally reproduced a series of four lectures that the author, Cardenal, had delivered to the members of the Academy and Laboratory of Medical Sciences of Catalonia. The second edition – considerably expanded, adapted and updated – came out in 1887 with the title *Practical Manual of Antiseptic Surgery*[43] and immediately sold out. The third edition, which occupies over 1,000 pages and is dated 1894, maintains the same title as the second.[44] It contains a collection of photographs of his clinic before the radical alterations carried out immediately afterwards, between 1897 and 1898. Needless to say, from the doctrinal point of view these two later editions of the work range beyond antiseptic surgery, and thus the successive developments in aseptic surgery, especially those propounded by German medicine of the time, are recreated, assimilated and updated in their pages.

As the author notes in the preface to the second edition (1887), the book's contents at that time reflected his reading of 'current surgical literature' and his own experience, derived from intensive professional practice in the city of Barcelona.[45] As a working surgeon practising in a 'populous city', with numerous calls on his time, Cardenal explains that it had taken him an enormous effort to write the book, to the point that in order to finish it, he confesses, he had had to go away: 'I left Barcelona, almost fleeing from the city, in search of repose in order to be able to complete these pages in a peaceful location'. Indeed, the book's preface is dated in the Swiss city of Geneva.[46]

In the preface to the third edition (1894) of his *Manual*, Cardenal presents himself as a firm advocate of aseptic surgery, which, according to him, did not represent in any sense a break with Lister's 'original' antiseptic method but rather its ultimate fulfilment. Even though the expression 'antiseptic surgery' was retained in the title, the text itself fully incorporated and assimilated the idea of the complete elimination of microbes in the operating field before a surgical act, by sterilizing all surgical equipment, from the facilities and furniture to the

40 Riera, 'The Dissemination of Lister's Teaching'; Juan Riera, *La introducción en España del método antiséptico de Lister* (Valladolid: Seminario de Historia de la Medicina, 1973).

41 Salvador Cardenal, *Guía práctico [sic] para la cura de las heridas y la aplicación del método antiséptico en cirugía: lecciones dadas en la Academia y Laboratorio de Ciencias Médicas de Cataluña por S. Cardenal* (Barcelona: Biblioteca Ilustrada de Espasa Hermanos, 1880).

42 Christopher Lawrence and Richard Dixey, 'Practising on Principle: Joseph Lister and the Germ Theories of Disease', in Christopher Lawrence (ed.), *Medical Theory, Surgical Practice. Studies in the History of Surgery* (London: Routledge, 1992), 153–215.

43 Salvador Cardenal, *Manual práctico de cirugía antiséptica* (Barcelona, 1887).

44 Salvador Cardenal, *Manual práctico* (Barcelona, 1894).

45 Cardenal, Prólogo, *Manual práctico* (Barcelona, 1887), pp. iv–v.

46 Ibid., p. vii.

instruments, including dressings, bandages, gauzes, suture thread and so on. In this context, the impossibility of sterilizing the surgeon's hands and those of his assistants was a hotly debated issue at the time:

> This is a matter that still seems trivial to many people, and yet it cannot be emphasised too strongly, for as copious recent research has shown, one of the greatest difficulties in achieving *Aseptic surgery* is the complete disinfection of the hands of the person operating and of those assisting him.[47]

In the 1890s a growing number of authors, especially in central Europe, were advocating the use of surgical gloves, whilst others resisted this innovation and firmly opposed it. Proponents of gloves considered them indispensable to ensure the asepsis of the operating field without having to subject the hands to the intense irritant action of disinfectant substances; their detractors, on the other hand, argued that with gloves the surgeon lost tactile control of the operation, to the detriment of his surgical skill. At the 27th Congress of the German Surgical Society, held in 1898, a whole session was devoted not only to the advisability of wearing or not wearing gloves when operating but also to their design, their size and above all the type of material – linen, leather, rubber and so on – employed in their manufacture.[48]

In any case, Cardenal considered that thanks to microbiology, surgical therapeutics had made more progress in the last few decades than in 'eighteen centuries of philosophical and doctrinal lucubrations'.[49] Assimilation of surgical asepsis required not only a grasp of the microbial world but also the introduction of measures to make it possible to operate, as far as possible, in absolutely sterile conditions.[50] It was therefore also necessary to have suitable spaces, architecturally adapted to the purpose, in order to be able to practise the new surgery.[51] The advertising pamphlet mentioned previously, published in 1899, shows that following its remodelling and expansion Dr Cardenal's clinic offered a double attraction: on the one hand, effective surgical procedures based on the most up-to-date surgical and technical advances: in other words, laboratory medicine; and on the other, a comfortable environment for patients and

47 Salvador Cardenal, *Manual práctico* (Barcelona, 1894, 3rd edition), p. 392. The preface, written by Cardenal, is dated 1 January 1895.

48 Thomas Schlich, 'Negotiating technologies in surgery: The controversy about surgical gloves in the 1890s', *Bulletin of the History of Medicine*, 87 (2) (2013): 170–97.

49 Cardenal, 1894, Preface.

50 Thomas Schlich, 'Asepsis and bacteriology: A realignment of surgery and laboratory science', *Medical History*, 56 (3) (2012): 308–43.

51 In addition to studying nomenclature (Weisz) and geographical location (Knox, Bohland, Shumsky), architectural historians have turned their attention to the architectural syntax of spaces for medical practice: doctors' surgeries, polyclinic buildings, clinics and hospitals. See Annmarie Adams, *Medicine by Design: The Architect and the Modern Hospital, 1893–1943* (Minneapolis, MN: University of Minnesota Press, 2008) and Annmarie Adams and Stacie Burke, 'A doctor in the house: The architecture of home-offices for physicians in Toronto, 1885–1930', *Medical History*, 52 (2) (2008): 163–94.

their families, similar to their own homes. The clinic had two separate sets of spaces, one for antiseptic surgery (theatre, laboratory) and the other reserved for aseptic surgery (theatre, microbiological laboratory, x-ray chamber, darkroom for developing). In addition, the new theatre was equipped for operating at any hour of the day or night, an undoubted advantage if the need arose, since 'at night the room is lit by incandescent lamps and reflectors with a total of 200 electric candlepower'.[52]

As for comfort, it must be said that Clinicum's advertising gave central prominence to this issue by offering cosy, appealing bedrooms (albeit of different categories: the one in the special category even had its own 'basin, dressing table and WC'), central heating, running water, basins, water closets, a reading room, electric light and so on. A complete strategy had been developed for attracting clients, avoiding the traditional hospital model:

> Since no operations, or even treatments of any real importance, are to be carried out in the bedrooms, we thought it entirely superfluous and impractical to give the rooms and furniture the cold, dry, polished character of the operating theatres, preferring instead to keep them, as far as possible, as comfortable, warm and cheerful as a room in a private house. Our personal experience has taught us that well-to-do patients find it abhorrent to leave the comfort and well-being of their own homes for an establishment with the cold appearance of a hospital, whatever pseudo-scientific explanations they may be given to justify it, and for this reason we have avoided anything that might remind them of a Nosocomium and have taken pains to give our bedrooms the appearance and conditions of domestic comfort of an English private hotel.[53]

A specialized surgery clinic: Dr Ricard Botey's healing house

As the nineteenth century drew to a close, Ricard Botey considered that the existence of certain surgical specialisms was an indisputable fact. By this time, one person could not cover every branch of surgery, and therefore it could not be taught as a single subject. The need to delimit practices and theoretical knowledge in the field of surgery was inescapable.[54]

Botey used the slogan 'electricity, cocaine and antisepsis', coined by the Parisian ear, nose and throat specialist Marcel Lermoyez, to highlight the three elements that made it possible to delimit the sphere of knowledge and practice

52 *Clinicum*, 1899, p. 15, n. 1.
53 *Clinicum*, 1899, pp. 26–7.
54 Ricardo Botey, *La especialización de los estudios laringológicos, rinológicos y otológicos y sus relaciones con la medicina y cirugía generales. Lección inaugural del curso clínico de laringología, rinología y otología de 1894–1895* (Barcelona: Tip. La Académica de Serra Hermano y Russell), 1896.

of this specialism.[55] Electricity had been responsible for introducing light into the cavities of the head and revealing a photographic view of them to the human eye, thus facilitating the creative task of expert knowledge in otolaryngology. The other two elements are well known for their role in transforming surgery as part of parallel processes aimed at overcoming two hitherto impassable obstacles: pain and infection. Anaesthetic cocaine had made it possible to use new instruments and new surgical techniques profitably, with no pain and no reflexes. And antiseptic methods had been the key to rendering such procedures harmless. Thus in this period at the end of the nineteenth century surgery opened the way to specialization in fields such as ENT, shaping the discipline, with its morbid or pathological entities, its instruments and its techniques, and an increasingly distinctive mode of working, in social terms.[56]

The possibility of performing major operations, such as tracheostomy, laryngofissure or atticoantrostomy, or dealing with chronic otorrhea, empyema of the sinuses and nasal and laryngeal tumours, required the creation of new work spaces: the clinic and, at its centre, the operating room. Botey considered that three influential factors determined the success of clinics, related to surgical practice, the well-being of patients and the administrative organization of the practitioner. The clinic was regarded as the most suitable place to perform complex interventions requiring difficult techniques 'with confidence of success', 'and without a clinic, the patient dies'. From the patient's point of view, 'artisans, clerks, industrialists and even moderately well-to-do members of the bourgeoisie are as well or better cared for in clinics than in their own homes'. Botey emphasized the need for clinics, 'given the deficiencies of our hospitals' and 'because the floating population and working-class people find care and shelter there in better material conditions than in their own homes and at a relatively fair price'. We shall focus again later on comfort and on the cost of the new surgery.[57]

A revealing aspect of Botey's administrative thinking lies in his obvious desire to save time by rationalizing the organization of work at the clinic.[58] A few

55 Ricardo Botey, *Estudios clínicos sobre laringología, otología y rinología. Su práctica y enseñanza en Europa a fines del siglo XIX (1891–1900)* (Madrid: Impr. y Libr. de Nicolás Moya, vol. 1; Barcelona, Tip. La Académica de Serra Hermanos y Russell, 1903, vol. 2). Botey refers to an article by Marcel Lermoyez published in *Le Jubilé des Annales*, January 1899. Botey, *Estudios clínicos*, vol. 2, p. 369. See François Legent, 'L'enseignement de l'ORL à Paris au XIXe siècle. Partie 1: la naissance de l'ORL et les obstacles universitaires', *Annales françaises d'oto-rhino-laryngologie et de pathologie cervico-faciale*, 127 (3) (2010): 103–8.

56 Botey systematically defines the content and limits of the specialism. See *Tratado de oto-rino-laringología para médicos y estudiantes* (Barcelona: Salvat y Cía, 1902), republished in 1908 and in 1918.

57 Botey, *Estudios clínicos*, vol. 2, p. 348.

58 This process, also known as the industrial organization of medicine, reveals the involvement of laboratory medicine in a social context guided by the values of corporate culture: efficiency, standardization, scientific organization of work. Steve Sturdy and Roger Cooter, 'Science, scientific management and the transformation of medicine in Britain c. 1870–1950', *History of Science*, 36 (4) (1998): 421–66.

examples will allow us to appreciate how this conformed to the values of a culture in the ascendant. Botey arranged visiting hours and days according to the specific nature of the pathologies treated, alternating between otology, rhinology and laryngology. He had two assistants in his practice: one took care of cleansing and dressing and the other was responsible for receiving patients, recording their details in the admissions book and filling in a card containing the patient's number corresponding to the clinical record book on the front and the diagnosis and prescribed treatments on the back. The patient took his or her record card to each consultation, 'thus avoiding questions and hesitations that would have wasted precious time'. The clinical record books were divided into two parts according to the pathology, one for otology 'with pictures of the ear-drum in dotted lines so as to be able to draw or paint the morbid process rapidly [...]' and another for rhinolaryngology, 'the pages of which contain images in dotted lines of the larynx and the base of the tongue and the nasopharynx with the choanae [...]'.[59]

Botey was also critical of doctors who set up clinics as a business, to make a profit, investing in luxuries to win clients who are 'attracted by novelty and comfort'. In his view, a doctor had to acquire practical experience, based on private practice in a flat and linked to sound professional conduct, building up his clientele, before opening a clinic. He defended his own case and put it forward as a model. A year after founding the journal *Archivos Internacionales de Laringología, Otología y Rinología* (1890), Botey established a polyclinic practice at no. 57, Carrer del Carme to provide free treatment exclusively for patients with ENT pathologies. The premises included a bed to hospitalize patients undergoing major surgery such as trepanning or tracheostomy. The increasing number of patients and of admissions for operations led him to set up a 'healing house' three years later at no. 59 in the same street, equipped with 'two small waiting rooms for patients, a sufficiently large consulting room, a small histo-chemical and bacteriological laboratory, a small operating room, a spacious, well ventilated bedroom furnished in some style to hospitalize a paying patient, and another more modest one, both with their respective beds, etc.'.[60] In order to define the quota of patients and to plan his income, Botey required payment of one peseta from any person who did not provide proof – in the form of a document issued by his or her general practitioner, local mayor or chaplain – of destitution or extreme poverty. Patients also had to pay for operations, complicated treatments and medicines, although they were only charged for the expenses incurred. And if they needed major surgery and hospitalization, the price rose to five pesetas a day [in 1903]. These measures were implemented by Botey gradually, as part of a definitive strategy for new patients. The money was collected in a small closed wooden box labelled 'for the expenses of the clinic'; according to Botey, 'this conveys to the public the idea that the small

59 Botey, *Estudios clínicos*, vol. 2, p. 503.
60 Ibid., p. 499.

contribution required of them is more in the nature of charity than of a payment equivalent to the services rendered, since they are thereby helping to maintain the benefit of such an institution'.[61] Botey had this rule and others written in a set of regulations and hung them in the waiting room at the clinic.

Why did Botey decide to abandon the old town in the Raval? His writings mention his fear of losing the patients he had in Carrer del Carme, given that the new location was a long way from the city centre. Eixample seemed, on the one hand, to be a geographical space where a 'real clinic' could be established with a genuinely new kind of medical architecture, and on the other, to be an opportunity to acquire a particular potential clientele: 'it was the focus of population growth; two trams and a railway line passed very nearby; it was very close to the new Faculty of Medicine and the Clinical Hospital; and it was surrounded, at no very great distance, by two hospitals and five or six clinics'.[62] The new clinic (see Figure 4.4), strategically located at no. 166, Carrer de Rosselló, next to Carrer d'Aribau, opened on 1 April 1898.[63] According to the data on new patients, consultations and personal visits published by Botey, the clinic's change of location slightly improved its results over the following five-year period of operation. Indeed, the specialized nature of the clinic and the relatively small number of patients gave him sufficient time to 'draw conclusions, study some cases at length and publish several scientific studies on the specialism'.[64]

Botey, like Cardenal, used his publications as a means of communication to establish the characteristics of the new form of surgery. The photographs of the facilities, rather than illustrating the text, reinforced the advertising message of this medical culture and showed the path to be followed in order to practice modern surgery on the basis of a new model of health care.[65] Botey's clinic occupied a four-storey town house, with a garden in which convalescent patients could stroll and where 'poultry and pigeons are raised for the upkeep of the clinic'. It was a place of work, not Botey's normal home. According to the advertising for his works, he had a practice at that time on the mezzanine floor of no. 4, Carrer de Vergara in Barcelona – the frontier between the old city and the new.[66] It is worth highlighting some of the architectural features of the clinic that will allow us to grasp the scope of the new medical culture prevailing in Barcelona.

In the basement there was a waiting room with seating for 70 people. The floor was of Portland cement and the walls and ceilings were plastered, all with a rounded finish and no sharp corners, to facilitate cleaning by flushing

61 Ibid., p. 502.
62 Ibid., p. 506.
63 *La Vanguardia*, 24 April 1898, 1.
64 Botey, *Estudios clínicos*, vol. 2, p. 530.
65 Sara Fajula, 'Las ilustraciones del cuerpo femenino en el *Tratado de ginecología* de Miquel A. Fargas Roca (1910)', *Dynamis*, 33 (1) (2013): 139–68.
66 *La Vanguardia*, 14 January 1893, 1.

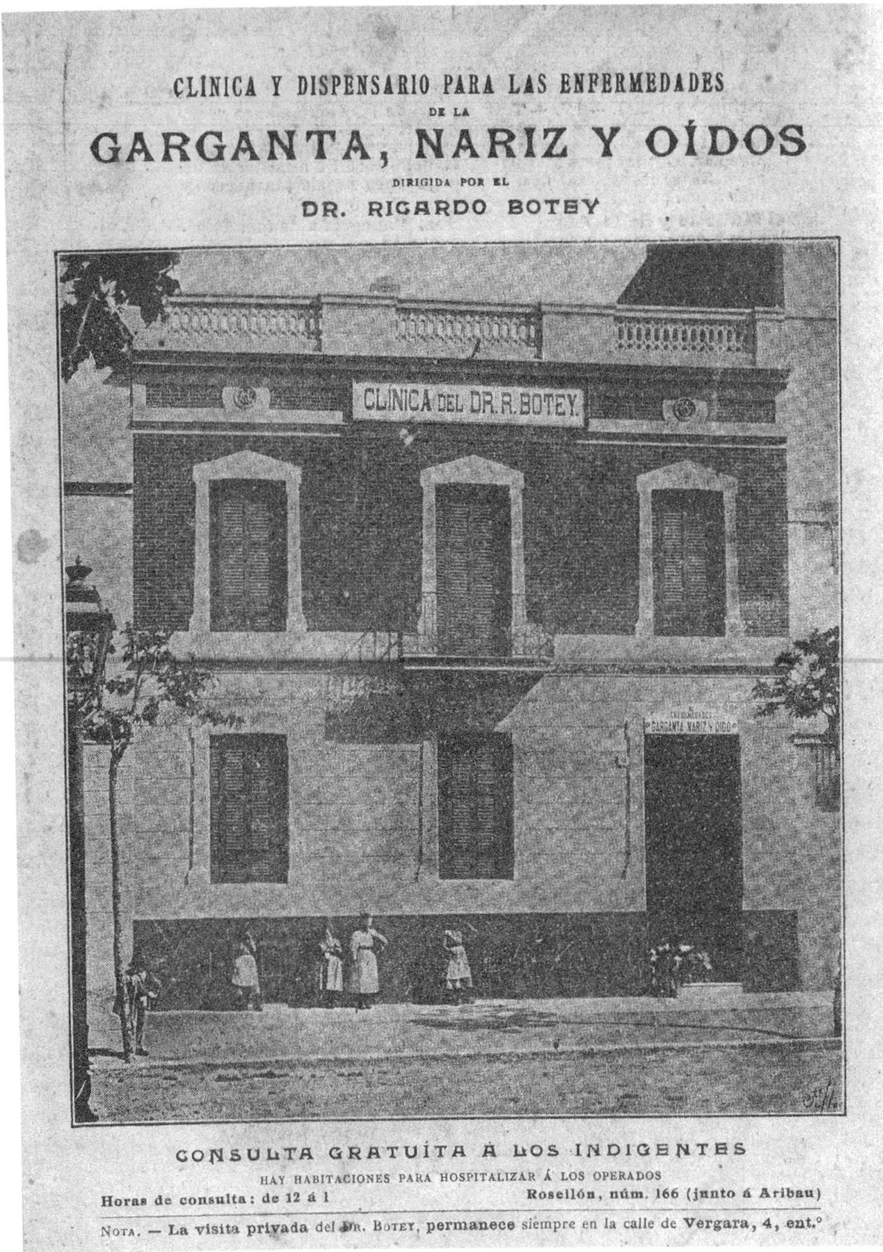

Figure 4.4 Advertisement of Botey's clinic published in Catalan medical press (1898).

with water using a Verdaguer trap that prevented the possible re-entry of foul air from the sewer system. The system of cleaning by irrigation was explained by the presence of patients with tuberculosis who expectorated and spat on the floor or in the spittoons. The presence of 'Verdaguer water closets' and of other foul water drainage systems indicates the availability of a sewer system in the street. Other services were not yet within reach in 1903. Thus, for example, when describing the consulting room, fitted out following the model of Dr Stanislav von Stein's clinic in Moscow,[67] with five exploration points equipped with their respective examination instruments and Auer lamps to reflect light from the mirror opposite, Botey complained that he would like to have installed 'a series of fixed-focus 50-candlepower Edison-Swan lamps [...], but I came up against the serious problem that the city cabling does not yet run along Carrer de Rosselló'.[68]

As well as the rooms for examinations, cleansing and dressing and pre- and post-operative treatments, which are described in great detail, showing the importance ascribed to material culture and to corporate culture as guiding principles of work in the new surgery, Botey highlighted other spaces – a lecture room, a library, a museum – that he considered crucial to his practical understanding of surgery. The little lecture room, with seating capacity for 20 students, 'has a blackboard for demonstrations, wall charts illustrating the anatomy of the ear, nose, etc.; a chair with a lectern; etc.'. In the mezzanine, the director of the clinic had an office for receiving 'wealthier patients who do not wish to go down to the consulting room to be seen with others'. Beside it there was a library, with international medical journals in the field (19 titles), general medical periodicals generated by exchanges with his journal (74 titles) and monographs and pamphlets in the field, published in the previous 20 years, in many languages (over 1,000 volumes). On this floor there was also the museum of pathological anatomy, with over 250 items, including 'many dry specimens and others in plaster, in wax, preserved in alcohol and in Le Prieur solution, of the ear, nostrils, pharynx and larynx, in normal and pathological states. It also has pathological specimens extracted in the operations performed in the consulting room and the operating room at the clinic, and finally a collection of normal and pathological microscopic slides of the hearing system, the speech organs and the nasal cavities'. On the second floor Botey had an additional space for clinical diagnosis, essential in that kind of laboratory-based surgery. In a large space, well lit by a big north-facing window, was the pathological, microbiological and chemical laboratory.[69]

67 Stanislav (or Stanislas) von Stein was director of the ENT clinic at the Imperial University of Moscow. Ricardo Botey, 'La clínica oto-rino-laringológica del Dr. S. von Stein'. *Archivos latinos de rinología, laringología, otología y de las enfermedades de las primeras vías respiratorias y digestivas*, 9 (1899): 33, 65, 97, 137.

68 Botey, *Estudios clínicos*, vol. 2, pp. 506–13.

69 Botey details the material available in the laboratory: a Zeiss microscope with ideal optical features for bacteriological examinations, a microscope for ordinary examinations, a Thomas-Jung

This way of understanding the specialism in terms of the structure and function of the architectural spaces of the clinic and the material objects contained in each space defines the new culture of surgery. Although the clinic's consulting room was an inexhaustible source of new knowledge, the operating room was its real nerve centre. The walls were painted with several coats of oil-based paint and varnished. The floor was mosaic, the angles and ceilings had no sharp corners and all the surfaces were smooth. The scientific principles of the concept of infection were clearly present. Another key element here, given that, as we have seen, the house did not yet have a supply of electricity, was light. It is difficult to imagine how an operating room could have been constructed in the narrow streets of the old city. Botey resolved the issue through architecture. He opened up the room to the light and let it shine in so as to be able to operate properly:

> the room is south-facing, with no walls or houses nearby, and more than half the operating table is lit from above [...] this sunlight, from the middle of the body upwards, is sufficient for all the clinic's needs. [...] this desideratum was achieved, without excessive expense, as follows: the whole expanse of the wall of this room that gave onto the garden, where there was a balcony, was demolished, and the part of the building immediately above was supported with two sturdy metal beams. An iron bay window was built there, with frosted glass both at the sides and in the ceiling; and in order to gain more space without modifying this gallery, the latter was built so that from a height of about 75 cm from the floor its far wall was set back 50 cm, thus forming a ledge or *étagère* of small glass tiles at this height, where the receptacles and other utensils required for the operation could be placed, and this made it possible to move the operating table up to the foot of this glass bay window, so that the patient's head and torso received an enormous amount of light.

The material contents of the operating room are a good demonstration of the soundness of the new surgical medical culture. The nickel and glass bed could be placed in various different positions and was equipped with a system for collecting fluids. The glass shelves, accessory tables, revolving stool and metal cupboards, the basin for disinfecting hands with two foot-operated containers, a marble sluice sink for depositing and cleaning instruments, with two taps providing hot and cold running water, sterilized using an autoclave with a siphon in the adjoining room. Botey reminds his readers that all the objects in the

microtome; wet chambers, syringes for injections, slides, cover slips, needles, pipettes, boxes for storing specimens, scales, chemical reagents and dyes, Pasteur and Ferran vials for cultures, a stove and an oven for sterilizing instruments, a Claude Bernard table for holding animals in experiments, scalpels, knives, saws and other instruments for autopsies and vivisection, equipment for fixing anatomical parts when preparing specimens, copious water, gas and so on. Botey, *Estudios clínicos*, vol. 2, pp. 524–5.

operating room were produced by a local manufacturer called Joaquim Sala in Carrer d'Estruch in Barcelona. In addition, on one of the walls there were three glass shelves bearing phenosalyl antiseptic solutions, inert solutions for aseptic cleansing, catgut, suture thread, drainage tubes and other surgical instruments.[70]

The clinic and the functions and design of its rooms enabled Botey to produce knowledge in a particular way. The detailed, illustrated clinical histories of his patients, organized and classified by pathology, together with the knowledge derived from the laboratory and the operating room or from discussions with craftsmen and instrument makers, facilitated the production of a vast body of writing on otolaryngology for publicity purposes: 'to sum up, I have written 4,571 printed pages of books and monographs and 1,335 pages of articles. Almost all these writings are on the specialism. To this activity must also be added the anatomical specimens and the drawings, numbering over 800, that I have produced, which represents additional work'.[71]

Conclusions

In the last decades of the nineteenth century, surgeons such as Salvador Cardenal and Ricard Botey established new clinics in the Eixample district. We have argued that this was the beginning of a new medical landscape based on a new approach to the spatial organization of health care architecture. These small hospitals came to represent a model for a new way of understanding and practising surgery. As we have seen, their owners consciously acted as medical entrepreneurs and medical writers, informing their medical colleagues through a range of media such as pamphlets, advertisements, medical journals and books. In line with these ideas, we have attempted to show how those media were strategically used, copiously illustrated with pictures and architectural plans, sections and designs, to establish a clear idea of how and what a modern private surgical clinic based on the laboratory sciences should be: a division of spaces according to the principles of the medical disciplines of the laboratory, alongside new rooms equipped with radically new diagnostic technology, where the new urban technologies of transport, lighting, heating and lifts played a crucial role in offering a new kind of surgery in a new medical setting, and comfortable, discreet, isolated and inviting rooms and spaces attracted potential patients from among the new citizens of Eixample.

Medical journals, urban guides and medical directories published in Barcelona in the first third of the twentieth century enable us to end this chapter by confirming that the private clinics studied here did indeed serve successfully as a model for the establishment of new clinics and medical practices in this area of the city in the 1920s and 1930s. More than thirty clinics, most of them

70 Botey describes some 70 ENT instruments he had designed himself, illustrating them with engravings and drawings, some of them by his own hand. Botey, *Estudios clínicos*, vol. 2, pp. 614–17 and 743.

71 Ibid., p. 614.

surgical, were created in the new district, competing with each other and with other medical institutions such as the great new hospitals of Barcelona.[72] These small surgical hospitals helped to consolidate specialties such as ENT, gynaecology, ophthalmology, urology and so on, all of them fuelled by a localized conception of diseases in the human body, facilitating a division of knowledge and practices to increase efficiency.

New forms of transport and communication clearly encouraged these changes, and work opportunities for doctors increased in the new city. Health care expanded as a business or trade in the industrial and liberal culture of the city of Barcelona. In this context, as we have shown, the new clinics made great use of medical advertising as one of the main ways of visualizing how a new medical paradigm was introduced in a very particular type of building and in a specific area of the city.

72 *Medicine and Architecture in Barcelona: Medicine in the Eixample district* [in Catalan in the original], http://www.medicinaiarquitecturabarcelona.cat/cliniques.html (last accessed 14 December 2014).

5 Technological fun

The politics and geographies of amusement parks

Jaume Sastre-Juan and Jaume Valentines-Álvarez[1]

On 23 January 1926 the readers of the conservative newspaper *La Vanguardia* could read an imagined description of the experiences of a peasant travelling to Barcelona. In an article entitled 'Mechanical Wonders', journalist José Escofet contrasted the quiet and slow-paced rural life to the overwhelming mechanical dynamism of 'modern' cities:

> The city starts dancing a marvelous dance around the foreigner. Twisted forms and fugitive colours everywhere, like in an urban landscape painted by a cubist painter. Tramways, cars, trucks, buses, they all rapidly run in opposite directions. [...] The peasant feels dizzy: he has seen the mechanical birds loudly flying near the roof terraces; he has witnessed the crash between a car and a tramway; he has gone for a sidecar ride; *he has visited an American amusement park.*[2]

The literary *topos* of the peasant visiting the city, scared and seduced at the same time, but in any case full of wonder, was hardly new, of course. This is but one of the countless examples in which it was used to build discourses on how 'modern' technology was changing the city.[3] What should be highlighted in José Escofet's description is the fact that amusement parks are included, alongside transport networks, as elements of the perceived experience of urban technification and transformation of public spaces. Addressed to different publics, the many amusement parks that flourished in Barcelona during the first decades of the twentieth century were part and parcel of its technological cityscape.[4]

1 We would like to thank Arwen Mohun, Carroll Pursell and the editors and contributors of this volume for their insightful and helpful comments of draft versions of this chapter.

2 José Escofet, 'Las maravillas mecánicas', *La Vanguardia*, 23 January 1926, 7. Emphasis added.

3 For a historiographical review of these discourses, see: Mikael Hård and Markus Stippak, 'Discourses on the Modern City and Urban Technology, 1850–2000: A Review of Recent Literature', in Mikael Hård and Thomas J. Misa (eds), *The Urban Machine: Recent Literature on European Cities in the 20th Century*, 'Tensions of Europe', electronic publication (www.tc.umn.edu/~misa/toc20/urban-machine), 35–56.

4 The publication of some popular books on amusement parks in Barcelona has unveiled a forgotten part of the city's past. Among them, the book by Ròmul Brotons has proven an invaluable starting point for our research, even though the historiographical approach differs very significantly from ours. Ròmul Brotons, *Parcs d'atraccions de Barcelona: de 1853 fins a l'actualitat* (Barcelona: Albertí, 2001).

The most relevant feature of these amusement parks – the American origin of which is stressed by Escofet – were the roller coasters, water chutes, bumper cars or witching waves that made Coney Island famous.[5] Amusement parks like Steeplechase, Dreamland or Luna Park became the model for an increasingly standardized cultural form which was exported and locally appropriated all over the globe, especially during the first three decades of the twentieth century.[6] Connected to the city centres by the new urban transport systems, these 'fun factories' – in Carroll Pursell's terms – offered new technology-intensive forms of mass leisure and new ways of obtaining pleasures through the commodification of risk, as historian Arwen Mohun pointed out.[7]

In this chapter, we argue that the emergence of this new kind of technological fun in Barcelona went hand in hand, in politically significant ways, with the dramatic urban transformations that the city experienced in the period 1888–1929. The chapter has two interrelated goals: on the one hand, to explore the politics of this standardized and mechanized fun; on the other hand, to explore the spatial patterns of the spread of amusement parks in Barcelona and how they reshaped the broader social geographies of fun.[8]

But isn't fun *just* fun? Can we trust the 'Just for Fun' slogan that greeted many amusement parks' visitors from America to Australia? Does technological fun have politics? If so, in what sense? Mechanical devices have generally been considered either as producers of purely kinesthetic fun, or as liberating ways of subverting both gravity and social norms.[9] This chapter will try to show how this new technologically-mediated fun was neither politically neutral nor

5 On Coney Island and American amusement parks, see: John Kasson, *Amusing the Millions: Coney Island at the Turn of the Century* (New York: Hill & Wang, 1978); Judith Adams, *The American Amusement Park Industry: A History of Technology and Thrills* (Boston: Twayne Publishers, 1991).

6 On the appropriation of amusement parks in Great Britain, see: Josephine Kane, *The Architecture of Pleasure: British Amusement Parks, 1900–1939* (Aldershot: Ashgate, 2013).

7 Amusement parks are only recently beginning to receive attention from the point of view of the history of technology. The scholars that are exploring their relevance beyond the case-study of Coney Island have focused on key issues such as the commodification of risk, the gendered nature of the rides, the dynamics of technology transfer, the standardization and exportation of American amusement parks, or the invention and mass-production of the rides. Arwen Mohun, 'Designed for thrills and safety: Amusement parks and the commodification of risk, 1880–1929', *Journal of Design History*, 14 (4) (2001): 291–306; Arwen Mohun, 'Amusement parks for the world: The export of American technology and know-how, 1900–1939', *ICON*, 19 (2013): 100–12; Carroll Pursell, 'Fun factories: Inventing American amusement parks', *ICON*, 19 (2013): 75–99. Pursell has recently explored the wider field of the intersections between technology and play in *From Playgrounds to PlayStation. The Interaction of Technology & Play* (Baltimore: Johns Hopkins University Press, 2015).

8 For a first approach to leisure in Barcelona around 1900, see: Teresa-M. Sala (ed.), *Pensar i interpretar l'oci: passatemps, entreteniments, aficions i addiccions a la Barcelona del 1900* (Barcelona: Publicacions de la Universitat de Barcelona, 2012).

9 John Kasson's book, for example, is a typical example of this approach, which considers that thrill rides 'plunged visitors into a powerful kinesthetic experience that [...] overturned conventional restraints, washed away everyday concerns, buoyed and buffeted participants as they submitted to its sway' and 'institutionalised the carnival spirit for a culture that lacked a carnival tradition' (cited in Pursell, 'Fun Factories', p. 76).

subversive by nature. On the contrary, we will argue that social control was also part of the story.

A useful theoretical framework for analysing the intersections between pleasure, technology and politics in amusement parks is the work of Tony Bennett, who has been a strong advocate for studies of popular culture from a Gramscian perspective since the 1980s.[10] In particular, in dealing with the politics of technological fun in Barcelona, we will explore what he calls 'regimes of pleasure'.[11] According to Bennett, the regimes of pleasure embedded in popular culture are 'not merely the forms of pleasure on offer but the system of signs and associated ideologies under which they are constructed and offered as pleasures'.[12]

In line with this perspective, by 'technological fun' we mean not only the pleasure obtained through mechanized thrill rides in amusement parks, but also its cultural articulation within specific discourses on technology and modernity. More specifically, we will explore amusement parks in Barcelona in parallel to hygienist thought, scientific management, regional planning and technocratic ideologies. In this sense, we argue that the changes in regimes of pleasure were related to the contemporary emergence of what Dominique Pestre calls new 'regimes of knowledge'.[13]

And, moving to the geographies of technological fun, what about the spaces in which it took place? Were they neutral containers or historical agents? How did the city shape the amusement parks and how did they shape the city? The majority of amusement parks in Barcelona were placed at the outskirts of the city. The design of their rides required large open spaces, which were not very common in Mediterranean city centres. But there were also symbolic and economic reasons for their peripheral location having to do with hygienist urbanization projects, tourism and land speculation.

In order to explore the urban geographies of technological fun, we will adopt a spatial narrative and focus on three key spaces of the city, which constituted

10 Tony Bennett, 'Introduction: Popular Culture and 'the Turn to Gramsci'', in Tony Bennett, Colin Mercer and Janet Woollacott (eds), *Popular Culture and Social Relations* (Philadelphia: Open University Press, 1986), pp. xi–xix. For a defence of the usefulness of Gramsci's thought for the history of science, see: Agustí Nieto-Galan, 'Antonio Gramsci revisited: Historians of science, intellectuals, and the struggle for hegemony', *History of Science*, 49 (4) (2011): 453–78.

11 In fact, Bennet uses the term in his analysis of the introduction of mechanical rides in Blackpool. Tony Bennett, 'Hegemony, Ideology, Pleasure: Blackpool', in Bennett, Mercer and Woollacott, *Popular Culture and Social Relations* (Philadelphia: Open University Press, 1986), 135–54.

12 Ibid., p. 136.

13 Dominique Pestre defines 'regimes of knowledge' as specific sets of institutions, instruments, spaces, theories, practices, values and so on, that coalesce at a specific time and the articulation of which defines the ways in which knowledge is socially produced. According to him, the regime of knowledge that crystallized in the period 1870–1930 involved the co-construction of a new industrial technoscience and a new technoscientific nation-state. Dominique Pestre, 'Regimes of knowledge production in society: Towards a more political and social reading', *Minerva*, 41 (3) (2003): 245–61; Dominique Pestre, *Science, argent et politique. Un essai d'interprétation* (Paris: Inra, 2003). On the rise of the engineering profession and its role in the emergence of a new regime of knowledge during the first decades of the twentieth century in Catalonia, see: Jaume Valentines-Álvarez, *Tecnocràcia i catalanisme tècnic a Catalunya als anys 1930* (PhD thesis, Universitat Autònoma de Barcelona, 2012).

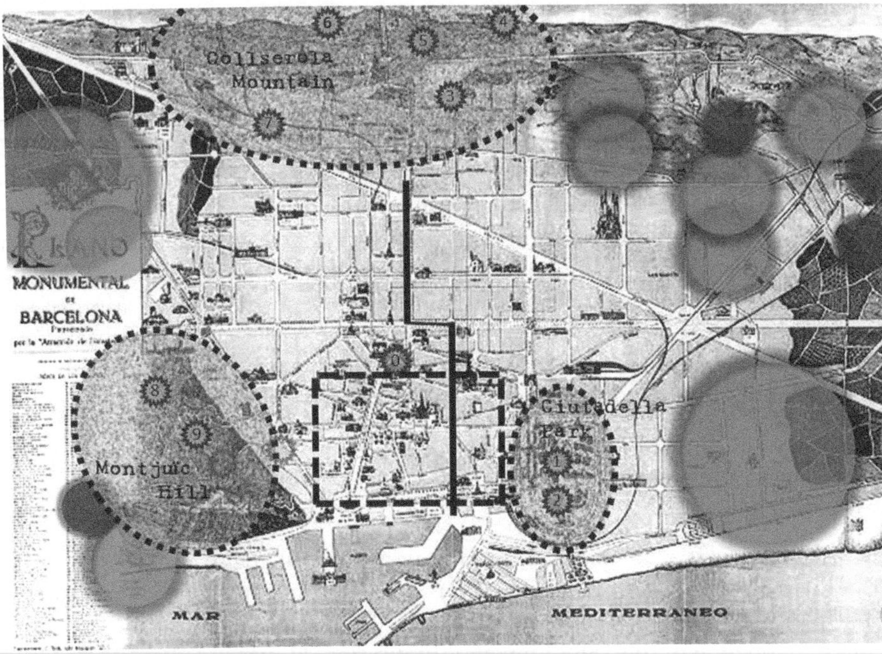

Figure 5.1 Touristic map of Barcelona, 1907. The ellipses represent the spaces of the 'technological fun belt', with numbers indicating amusement parks or sites in which mechanical rides were installed: 0. Plaça de Catalunya; 1. 1888 Universal Exhibition; 2. Saturno Park; 3. American Park; 4. Rabassada; 5. Tibidabo Park; 6. Lake Valley; 7. Turó Park; 8. Amusement Park at the 1929 International Exhibition; 9. Maricel Park. The shaded circles indicate the neighbourhoods forming the 'red-and-black belt'. The rectangle marks the old city centre and the two lines signal the Haussmannian Via Laietana (connecting the sea with the Eixample neighbourhood) and the Passeig de Gràcia (reaching towards the mountain).

what we call a 'technological fun belt': the Parc de la Ciutadella in the eastern part of the old city centre, the Collserola mountain in the northern part of the new expanded city, and finally the Montjuïc hill in the western part (see Figure 5.1). As we will see, this 'technological fun belt' was socially and politically entangled with the major urban transformations that Barcelona experienced during this period. In fact, we suggest that it should be read in parallel to the formation of the 'red-and-black belt', named after the colours of the main anarchist trade-union's flag. This is how urban historian José Luis Oyón called the new poor and segregated peripheries of Barcelona, where anarchist ideas were very prominent among the immigrants from other regions of Spain who mainly lived there.[14]

14 José Luis Oyón, 'The split of a working class city: Urban space, immigration and anarchism in inter-war Barcelona, 1914–1936', *Urban History*, 36 (1) (2009): 86–112.

We will deal with the following amusement parks that flourished in the spaces of the 'technological fun belt': Tibidabo Park (1901), American Park (1907), Lake Valley (1908), Saturno Park (1911), Rabassada (1911), Turó Park (1912) and Maricel Park (1930). Three main aspects will be considered: first, the publics and their experience of technological fun; second, the mechanical rides themselves – both as technical artefacts and signs; and third, the promoters and their ideologies. While all aspects will be present in the three sections, each one will emphasize one of them. The first section will highlight the different publics that enjoyed the amusement park in the Parc de la Ciutadella. The second section will emphasize the rides and their role in the urbanization of the Collserola mountain. And the third section on the Montjuïc hill will focus on the promoters and their ideology. The conclusions will bring together all these threads of the politics and geographies of technological fun in Barcelona.

Parc de la Ciutadella: Mechanical rides, civilizing spaces and interclass publics

Mechanical rides were brought to Barcelona – as in many other cities – by the big nineteenth-century celebrations of industrial capitalism: world fairs.[15] The first mechanical roller coaster to be installed in the city was the one in the amusement zone of the 1888 Universal Exhibition. However, when the exhibition closed, it was moved from the Parc de la Ciutadella to Plaça de Catalunya, by then a free-access open space where popular fair entertainments proliferated: cafés, circuses, panoramas, puppet shows, anatomical museums, animal shows, cinemas, freak shows and so on.[16] Many of these entertainments ended up some years later in the Paralelo Avenue, which became the new centre of popular entertainment. The new technological fun would have to wait for its return to the Parc de la Ciutadella until much later, when the Saturno Park opened its doors in 1911.

Right after the 1888 Exhibition, the plans for the Parc de la Ciutadella included quieter kinds of entertainment. The creation of the Parc de la Ciutadella in the late nineteenth century needs to be placed within the rational recreation movement. In order to morally uplift the population and foster social control, this movement opposed the wide spaces of public parks to the opacity of the narrow urban streets of the old city centre and the 'underworld'. It opposed the rational entertainment and disciplined behaviour of what Tony Bennett calls 'exhibitionary complex' institutions, such as museums and world fairs, to a more popular fun that could be found in the pubs, cabarets, music-halls and boxing arenas located in popular neighbourhoods.[17]

15 On the continuities between world fairs, amusement parks and other fairgrounds, see: Deborah Philips, *Fairground Attractions: A Genealogy of the Pleasure Ground* (New York: Bloomsbury, 2012).

16 Enric H. March, 'Ex-plaça de Catalunya', *Bereshit: reconstrucció de Barcelona i altres móns*, http://enarchenhologos.blogspot.com.es/2014/01/ex-placa-de-catalunya.html (last accessed 18 October 2014).

17 Tony Bennett, *The Birth of the Museum* (London: Routledge, 1995).

As Oliver Hochadel and Laura Valls analyse in Chapter 2 of this book, after the 1888 Exhibition the Parc de la Ciutadella became a site for science popularization, with the creation of the Museu Martorell, the zoological garden, the Umbracle, the Hivernacle, the mammoth sculpture and so on. The park's popularization programme was managed by the Junta de Ciències Naturals, which from 1907 on actively deployed a nationalist, Catholic and conservative agenda that conceived entertainment around natural sciences as a tool for moral uplift and religious apologetics, and as a way to enhance national identity.[18]

However, not all were mammoths and museums: other forms of leisure coexisted in the Parc de la Ciutadella. The Saturno Park, a big amusement park named after the famous Luna Park in Coney Island, was located in the heart of the Parc de la Ciutadella from 1911 until 1926. The businessman Ramon Bargués bought the main mechanical rides from the 1909 Valencia Regional Exposition and secured from the City Council the right to fence a big portion of the park and the licence to sell tickets at 10 cents to enter the perimeter.[19]

Now the Parc de la Ciutadella not only offered to its visitors the grass, the trees, the open spaces and the rational recreation, but also mass entertainment and the kinesthetic emotions from what the promoters called 'modern sports'. Among many other entertainments, the visitor could glide in the skating ring Patines Siglo XX, go round and round on an electric carousel, shoot in the shooting grounds, or dare to board the main thrill rides: the toboggan, the water chute and the roller coaster Los Urales. In the toboggan, the visitor could feel the acceleration of gravity in his or her own body while descending the spiral-shaped ramp. After ascending with the help of an elevator, he or she could experience the speed of the big roller coaster, which had an artificial lake at its centre. This was the lake in which the vehicles of the water chute were launched through a ramp, causing incessant screams. On windy days, the splashes must have bothered or refreshed the customers seated at the tables of the huge café, which was actually one of the most visited parts of the amusement park.[20]

Indeed, mechanical pleasures were not the only ingredients of Saturno Park's technological fun. Listening to music, eating, drinking, strolling around, and watching other people on the attractions, were also crucial components of the experience. All of them, however, took place in the context of a constant display of new technologies. Whether one was drinking lemonade with his fiancée or eating sweets with one's parents, there was a background noise, a mix between the roller coaster's mechanical clink-clonk and the technology-induced screams and laughs.

But who visited the amusement park? Although in the beginning it was rhetorically aimed at the bourgeoisie and, indeed, it hosted on many occasions

18 For an analysis of these Catalanist and Catholic scientists, see Chapter 3 of this book.

19 Brotons, *Parcs d'atraccions de Barcelona*, p. 67.

20 This description is based on several articles published in the weekly and daily newspapers *Papitu*, *Mirador*, *La Il·lustració Catalana* and *La Vanguardia*.

Figure 5.2 Postcards depicting mechanical rides at the Saturno Park, around 1913. *Left*: toboggan. *Right (top to bottom)*: witching waves, water chute and the roller coaster Los Urales.

aristocratic soirées, it also quickly attracted the artisans and workers of the nearby popular neighbourhoods, and thus ended up being an interclass space.[21] There is visual evidence that the publics of the Saturno Park not only wore bourgeois hats but also worker's caps (see Figure 5.2), and contemporary descriptions tell us about people picnicking on the grass and lovers immorally sheltered in the shadows at night.[22] The satirical weekly *Papitu* described the park as an interclass space two weeks after its inauguration:

> Now life comes from the Park [...]. There you can find men and women from the bourgeoisie, *cocottes*, troublemakers and children. If one rides the Urales, one might hear a woman complaining: 'I cannot stand so many slopes'. If one rides that exhilarating boat, one might hear a man equally complaining: 'I get too excited with such fast descents'. [...] There is only

21 Brotons, *Parcs d'atraccions de Barcelona*, p. 73.

22 'Some families bring their "dinner" with them and lie on the ground as if they were at the mountain. What a picture for foreigners!'. Juan Taixidó, 'Cultura y sport', *La Vanguardia*, 7 August 1911, 6.

> one problem: the Toboggan. Since everyone can see the soles of your shoes when you descend, you must make sure to wear the ones without holes, otherwise everyone will notice.[23]

The Saturno Park was a political battleground from the very beginning. The first years of the amusement park were a success in terms of audience, and new rides, like the Witching Waves or the Labyrinth, were incorporated every season. However, the privatization and commodification of the Parc de la Ciutadella led to protests from the political opposition in the City Council. Republicans and conservative nationalists accused the populist and anticlerical party in power to have perverted the public and moral nature of the park in order to benefit their own friends.[24]

In 1916, critiques grew significantly. Some of them included what they perceived as the aesthetic dissonance of the amusement park in the list of degradations and misbehaving that were sullying the Parc de la Ciutadella as a space for rational recreations. In the autumn of 1916, the pages of *La Vanguardia* described on more than one occasion a park 'turned into a centre of American attractions and particularly frequented by soldiers and maids'.[25] And, most interestingly, some explicitly contrasted the ugliness and immoral qualities of mechanical rides to the respectability of the exhibitionary institutions fostered by the Junta de Ciències Naturals. By that time, Francesc Darder, the director of the adjacent zoo, kept complaining to the mayor about the consequences of the noise both of the rides and visitors of the Saturno Park 'for the life of the poor animals', and asked for the removal of the amusement park.[26] And Carles Buigas, the engineer responsible for the Magic Fountain of the 1929 International Exhibition, wanted to 'take out from the floor in front of the museum [Museum of Fine Arts] [...] the banal and desecrating stalls that were part of the Saturno Park'. After all this criticism, when the concession expired in 1916, the Saturno Park was finally moved to another – less central – area of the park (near the Cascada), and the site it formerly occupied was converted into a rational garden by architect Nicolau Rubió.

So did the mechanical rides clash against rational recreations of the Parc de la Ciutadella and the civilizing project they embedded? The amusement park was clearly not part of the Catholic and nationalist popularization agenda of the members of the Junta de Ciències Naturals, who thought that the most morally uplifting kind of popular science was one that did not include the thrills and big iron structures of the mechanical rides. However, this should not lead us to think that we are dealing with the regimes of pleasure that the exhibitionary institutions opposed (popular fair entertainments). As we will see in the next

23 'La saturnal del Parc', *Papitu*, 7 June 1911, 400.

24 Brotons, *Parcs d'atraccions de Barcelona*, p. 69.

25 José Escofet, 'La ciudad y los parques', *La Vanguardia*, 28 October 1916, 8.

26 Rossend Casanova Mandrí, 'Francesc d'Assís Darder i l'origen del Parc Zoològic de Barcelona', *Revista de Catalunya Barcelona: Fundació Revista de Catalunya*, 142 (1999): 36–44, p. 38.

sections, amusement parks shared spaces, ideals and promoters with exhibitionary institutions.

In the present case, and despite the criticisms from the conservative ranks, the tensions with the Parc de la Ciutadella's authorities and the actual practices of its customers, the regime of pleasure on offer at the Saturno Park was clearly different from the one operating in the 1880s in Plaça de Catalunya (the space which had hosted the first mechanical roller coaster in the city, alongside other privately-owned little spectacles) or the one operating in the Paralelo Avenue (with its music halls and cabarets). The Saturno Park was a big-scale, regulated, gated and commodified space, with a unified decoration inspired in the style of the international exhibitions' amusement zones, and aimed at appealing with its technological fun to a broad audience ranging from bourgeois and middle classes to the 'respectable' part of the working class.

Collserola mountain: Urbanism, hygienism and upper-class leisure

While the Saturno Park, at the boundary of the overcrowded city centre, became an interclass space where working classes were 'getting some fun out of life' – as Billie Holliday sang – other spaces for a more exclusive leisure were opened far away from the Parc de la Ciutadella: a new segment of the 'technological fun belt' emerged in the Collserola mountain, where several amusement parks appeared during the first decades of the twentieth century: the Tibidabo Park (1901–present), the American Park (1907–1908), the Lake Valley (1908–1916), the Rabassada (1911–1934) and the Turó Park (1912–1927). This boom of amusement parks in Barcelona took place while major urban transformations and social struggles were changing the city. Both processes were entangled with land speculation, the promotion of tourism and hygienist and civilizing ideals.

Two great urban plans were at stake at the time: the construction of the Via Laietana Avenue and the 'making' of the Collserola mountain. On the one hand, the construction of the Haussmannian Via Laietana (1908–1913) split the maze-like urban structure of the old city centre with the aim of smashing the popular 'vices' – in hygienist terms – and better controlling the social upheavals which threatened the bourgeois hegemony.[27] The avenue stood for an icon of modernization as well as a self-affirmation of the ruling classes in a time of strikes and violent riots in a city known as the 'Rose of Fire' since the Tragic Week of 1909 (also known as the Glorious Week for some).[28]

The Via Laietana connected the sea and the rational grid-like Eixample district through the city centre. From the beginning of the century, the Passeig de Gràcia – the richest street of this district, with houses by Gaudí and other

27 Antoni Nicolau and Daniel Venteo (eds), *La construcció de la gran Barcelona: l'obertura de la Via Laietana, 1908–1958* (Barcelona: Ajuntament de Barcelona, 2001).

28 José Álvarez Junco et al., *Tragic, Red and Glorious: One Week in 1909* (Barcelona: Ajuntament de Barcelona, 2010).

Figure 5.3 Postcards from the Rabassada (top) and the Tibidabo Park (bottom), around 1915. The views of the healthy, planned and civilizing 'green and metal-coloured belt' are shown through the roller coaster at the Rabassada (top), the Tibidabo Funicular (bottom) and the Aerial Railway (bottom right).

prominent Art Nouveau architects – reached towards the mountain. The elite gradually looked for new, quiet, green and fresh spaces of residence and entertainment in the upper part of town. In this process, new kinds of commodified and technified risky pleasures were built there, far away from (and in opposition to) the perceived risks of the city centre in terms of safety and health.[29] The 'technological fun belt', in which the metal-coloured iron structures of the mechanical rides mixed with the green of trees, turned its back on the so-called 'red-and-black belt' (see Figure 5.3).[30]

The promoters of urban changes, land speculation, transport networks and amusement parks around the Collserolla area were the same people, all of them members of the upper bourgeoisie in Barcelona. In 1899, the S.A. Tibidabo Company was founded, among others, by one of the main property developers

29 For a spatial analysis of the 'territorialization' of health risks in Barcelona (in Yankel Fijalkow's terms), see: Celia Miralles, *La tuberculose dans l'espace social barcelonais, 1929–1936* (PhD Thesis, Universitat Politècnica de Catalunya, 2014).

30 The 1907–1916 period has been highlighted as the starting point in the making of the 'red-and-black belt', despite the fact that the major building boom took place after World War I (1917–1926). Glòria Andrés and José Luis Oyón, 'Las segundas periferias de Barcelona: vivienda y formas urbanas, 1917–1936', in José Luis Oyón and Juan José Gallardo (eds), *El cinturón rojinegro: radicalismo cenetista y obrerismo en la periferia de Barcelona, 1918–1939* (Barcelona: Carena, 2004), 17–43.

of the Eixample (Salvador Andreu) and the founder of the railway company Ferrocarriles de Montaña a Grandes Pendientes (Romà Macaya). Similarly, the company Rabassada Sociedad Anónima Inmobiliaria de Sports y Atracciones (1910) was directed by the businessman Josep Sabadell, the prominent leader of the main Spanish employers' organization, created to counteract the rising working-class empowerment.[31] The land of the Turó Park was owned by the Girona family, one of the most influential family of bankers in Barcelona.

All these projects, which included mechanical rides, were fiercely supported by the Sociedad de Atracción de Forasteros (Society for the Attraction of Foreigners; from now on, SAF), which was created in 1908. Its main goal was to develop private enterprises for promoting tourism.[32] The members of the SAF belonged to the financial, political and cultural elites of Barcelona, and some of them actively promoted the projects of amusement parks in Collserola. Engineer Marian Rubió (1862–1938), who will be a key character in this chapter, was at the same time the main manager of the Tibidabo Park (1904–1936) and the President of the SAF (1911–1936).

All amusement parks in Collserola were conceived by these social actors as a way to make money by selling healthy, moral and entertaining spaces to the visitors, who could delight in the panoramic views, rest in magnificent hotels and restaurants, and enjoy French cuisine and Tzigane, Catalan, Russian or American orchestras.[33] Or they could also go for a walk, ride a horse below the trees or drink the 'miraculous' waters of the Collserola wells. Turó Park was qualified by the press as 'a privileged site in the city, dominated by the fresh airs from the mountain', and the Rabassada simply as the 'Eden' (see Figure 5.3).[34] Tibidabo Park's main commercial hook was quite similar. And the Lake Valley, very influenced by the project in the Tibidabo, was also designed as a 'moral and hygienic' space.[35] Ironically, pure air was a commodity sold by the same bourgeoisie that owned the factories which were polluting the city. In July 1923, Salvador Andreu wrote to Marian Rubió, who was also a prominent figure in the popularization of hygienist practices:[36]

> My friend Mr. Rubió, I have had an idea. To publish on the front pages of the main newspapers in Barcelona a big-lettered advertisement: 'Let's go

31 Soledad Bengoechea, 'Los hombres de la patronal a principios del siglo XX: Luis Ferrer-Vidal, José Sabadell y Félix Graupera', *Historia Social*, 48 (2004): 69–85.

32 Albert Blasco Peris, *Barcelona Atracción (1910–1936). Una revista de la Sociedad de Atracción de Forasteros* (PhD Thesis, Universitat Pompeu Fabra, 2005).

33 Before the first mechanical rides were in operation, the S.A. Tibidabo Company promoted the glamorous and arabesque Gran Café Restaurant (1902). The Rabassada tourist resort was built around the luxurious Restaurant Hotel (1899).

34 'La Rabassada: El centro mundial de atracciones', *ABC*, 11 August 1911, 7; 'Inauguració del Turó Parch', *Ilustració Catalana*, 9 June 1912, 288.

35 'Una excursión al Tibidabo', *Tibidabo: revista mínima*, 2 (21) (1905), 3–5; 'Permiso para obras a D. Heriberto D. Alemany y Escardó', Folder 113–31 (1908), Arxiu Municipal del Districte de Sarrià-Sant Gervasi (AMDSG).

36 Rubió lobbied for hygienist policies in articles in *La Vanguardia*.

> breath in Tibidabo!'. Only this sentence, three or four consecutive days, in almost all newspapers. [...] You will see how each *peseta* coin will become five.[37]

However, besides enjoying the calm, amusement park visitors could also get their thrills by seeing a bullfight in the Turó Park, playing tennis or football in the Lake Valley, practising clay (or real) pigeon shooting in the American Park or betting in the Rabassada casino. And also by boarding the speedy and risky mechanical rides. Commodification of hygiene and commodification of risk were entangled issues in the Collserola mountain.

The 'conquest of the mountain' was a planned project conceived as a civilizing colonization of the area through the development of transport networks.[38] Despite the fact that Collserola was formerly a space for forest resources and hunting, popular picnics and family celebrations around wells, religious festivities, workers choir singing or even clandestine trade-union meetings, these sites were symbolically dressed with a virginal aureole.[39] In 1911, a newspaper described the Rabassada in the following terms: 'Man found it shining and fertile [...]. Its woods offered him rest and its shadow offered him health. Man arrived, and took possession of the place'.[40]

Technology was a key element in this conquest of nature for developing tourism, making money, urbanizing and keeping diseases at bay. In amusement parks, the role played by technology in these processes can be seen on three levels: first, the transport network surrounding the amusement parks; second, the display of technological innovations within them; and third, their very core: the mechanical rides. We will deal with them in turn.

As in the case of the 'trolley parks' in the United States, many amusement parks in Barcelona were intimately related to the spread of the transport and power networks.[41] Moreover, in many cases, the transport facilities were not only the mandatory means to arrive to the entertainment sites, but also true mechanical rides in themselves. The funicular railway in the Tibidabo (the first one in Spain and one of the first in the world) was inaugurated in 1901, and became the earliest technological thrill in the Collserola mountain (see

37 Quoted in: Antoni Lázaro, 'Marian Rubió i Bellvé i el Parc del Tibidabo', in *Els Rubió: una nissaga d'intel·lectuals* (Barcelona: Angle, 2003), pp. 51–6.

38 The idea of 'the conquest of the mountain' has been used to analyse the urbanization of the mountain by: Sergi Garriga Bosch, *El Tibidabo i la concepció metropolitana de Barcelona* (Master thesis, Universitat Politècnica de Catalunya, 2012), p. 83.

39 However, these activities kept taking place in Collserola after amusement parks appeared. See: Ferran Armengol Ferrer et al., *Un segle pujant al Tibidabo: història del Tramvia Blau i del Funicular del Tibidabo* (Barcelona: Ajuntament de Barcelona, 2002).

40 *Heraldo de Madrid*, 26 August 1911. Quoted in: Sergi Yanes and Turiscòpia, *La Rabassada: la utopia de l'oci burgés* (Sevilla: Punto Rojo, 2011), p. 82.

41 The S.A. Tibidabo Company was not only a property developer company and an amusement park company, it was also created as a transport company and an electric company.

Figure 5.3).[42] During the eight-minute funicular ride that ended at the Tibidabo Park, located at the top of the Collserola mountain, the passenger could admire the panoramic sights.[43] Before reaching the top, the funicular passed by the Fabra astronomical observatory (1904) and the Mentora Alsina experimental physics museum (1905). At the top, one could see the impressive Sagrat Cor Church, and, some years later, the antennas of Radio Barcelona.[44] Thus, before entering the amusement park, the visitor was not only amazed by the views but also greeted by some of the icons of science and religion that shaped the Barcelona skyline.

Likewise, the Lake Valley Park featured Mina Grott, the first electric passenger railway in Spain. The excitement of the journey through a 1300-metre-long ancient water tunnel connecting the two sides of the mountain with plenty of electric light bulbs and dioramas was the main technological ride of this park.[45] Moreover, even the access by car to the Rabassada through a dusty and winding road already meant a technological thrill for the richest elite.[46]

Amusement parks themselves were also spaces for displaying technology. In 1913, the first Automobile Fair in Barcelona was celebrated in the Turó Park, as a part of a wider Exhibition of Import and Export Products.[47] Sponsored by the Real Automóvil Club de Cataluña (Royal Automobile Club of Catalonia), the most famous international and national brands (for example, Ford and Hispano-Suiza) were shown to the public alongside the newest airplanes. During the fair, a French Borel monoplane flew over the Turó Park. Shows of acrobatic flights high above the amusement parks were usual tourist attractions in the early years of aviation.[48] Alongside bourgeois anti-TB charity galas, pet exhibitions, nationalist events, banquets, high-society meetings and elite sports championships, exhibitions of technology were very common. During World War I, the War Gallery (1916) in the Turó Park and the War Museum in the Tibidabo Park (1916–1940) – promoted and designed by Marian Rubió – displayed materials, habitat-like scenes of the daily life in the trenches and war technologies, including a reproduction of the famous Big Bertha howitzer.

This technological display was highlighted and enhanced by the display of 'non-technological' colonized societies. In 1913, the Tibidabo exhibited the

42 In 1901, a tramway climbing up the newly urbanized Tibidabo Avenue was also inaugurated. Both the tramway and the funicular were designed and built by the S.A. Tibidabo Company.

43 A 'virtual' visit to the park through images can be found in: Junceda, *Auca del Tibidabo* (Barcelona: David, 1930).

44 See Chapters 9 and 10 of this book.

45 Actually, it became the first and only ride of the park, since the initial project (with a big carousel, roller coasters and balloons) was never implemented due to conflicts over railway licences with other transport companies. Ricard Fernández Valentí, *El Mina-Grott: història d'un petit tren de Vallvidrera* (Barcelona: Ajuntament de Barcelona, 2004).

46 From 1914 on, the Tibidabo Coup, a pioneer car rally, was organized there.

47 'Exposición de productos de importación y exportación', *Industria e invenciones*, 31 August 1912, 88.

48 On the Tibidabo, see: 'Lo vol sobre'l Tibidabo', *Ilustració Catalana*, 19 March 1911, 150.

'primitive' forms of work and leisure of dozens of men, women and children of a 'Senegalese tribe', while the Turó Park exhibited an orientalist 'Far East Village' where the public could 'examine the rough customs of their inhabitants'.[49] These displays can be interpreted as having a dialectical relationship with the mechanical rides in amusement parks, as being a distorted mirror, an intended living proof of the superiority of industrial cultures over 'pre-scientific' peoples.[50]

Nevertheless, the most significant display of technology in amusement parks was that of their amusement devices. Some of them, like the electric or miniature steam-engine locomotives, seemed to be part of the transport network that had brought the visitor to the park. In the Tibidabo, some of the first amusements were 'optical trips', like seeing through a telescope (1904) that worked with a one cent coin, or using a mobile searchlight (1905). Even its three major mechanical rides can be regarded as technologies developed in order to enjoy more intensively the views of the mountain and the sea. In 1915, the Aerial Railway – a monorail inspired by the German Wuppertal Suspension Railway (1901) – allowed a burgeoning scenic view of the city over the trees and passed by several exotic dioramas representing the Alhambra palace in Granada and the Dragon Cave in Majorca (see Figure 5.3). In 1921, a huge 50-metre lever illuminated at night, the Talaia, allowed the users to observe Barcelona from an overall altitude of 550 metres. Finally, in 1928, a real-scale and self propelled replica of the first commercial plane that flew from Barcelona to Madrid in 1927 put the city under the visitor's feet in the only 'flight experience' many people in Barcelona had in their lives. These mechanical rides in the Tibidabo Park were designed by Marian Rubió himself and built by local-based enterprises under his supervision.[51]

The other amusement parks in Collserola, however, preferred to import rides, mainly from the United States.[52] In some cases, the importation was direct and complete. Roller coasters in the Rabassada (1911) and Turó Park (1914), for example, were authentic Scenic Railways.[53] These rides were patented by the most famous 'father of gravity', LaMarcus Thomson, and built by his own company, L.A. Scenic Railway Thompson Company, which exported its products all

49 *La Vanguardia*, 1 June 1913, p. 11.

50 For an analysis of the role of discourses on technology in the ideological construction of racism and imperialism, see: Michael Adas, *Machines as the Measure of Men: Science, Technology, and Ideologies of Western Dominance* (Ithaca: Cornell University Press, 1989). The literature on anthropological spectacles is huge. See, for example: Sadiah Qureshi, *Peoples on Parade: Exhibitions, Empire, and Anthropology in Nineteenth-Century Britain* (Chicago: The University of Chicago Press, 2011).

51 The enterprise in charge of the Talaia was Suministro de Construcciones Metálicas Hijo de Miguel Mateu, and the one that built the Plane was Talleres Estrada (from Sarrià, a town next to Collserola). The Aerial Railway was also locally manufactured.

52 The Electric Carousel in the Turó Park, however, was implemented by the Barcelona-based Torra Company.

53 In fact, the design of the Rabassada amusement park was assigned to John Calvin Brown, promoter of the White City Amusement Park in Manchester (1907). Yanes and Turiscòpia, *La Rabassada*, p. 93.

across the US and Europe.[54] As in many other parts of the world, many trendy rides for the 'factories of fun' in Collserola – electric carousels, witching waves, water chutes, electric cars, cakewalks, haunted houses – were an important part of the early globalization of products and technological know-how from the US.[55] But the circulation of technologies of fun from North America (sometimes indirectly, through other European countries like France) was not only a matter of production, but also a matter of commercial hook through the promotion of the so-called 'American way of life'.[56] Amusement parks were often referred to as 'American parks' or announced their ever new 'American rides'.[57] Indeed, the business of technological fun continuously produced innovations for attracting visitors, that is, for seducing consumers. From its inauguration, Turó Park promoted itself in the press as 'the best amusement centre in the world' and 'the trendiest park' with '40 modern rides'.

Furthermore, the mechanical rides helped spreading values and behaviours such as faith and confidence in technology. The physical integrity of the user boarding the risky rides completely depended on engineers and technicians, who worked hard to maximize both the safety measures and the illusion of risk through speed, acceleration or altitude.[58] Marian Rubió was fully aware of this key feature of amusement parks, and this awareness determined his technological choices. In a paper published in the journal *La Energía Eléctrica* in 1915, he assured that a major accident was 'absolutely impossible' in the Aerial Railway and that 'all the parts of the metallic construction have been designed to be totally safe'.[59] According to Rubió, everything was designed to 'feel the emotion of an aerial journey' and 'the beauties of that incomparable panorama'. However, the risk was real. During a stormy night in 1930, the Plane fell down.[60] No casualties were to be lamented, because the park was closed. However, other accidents (like the one that happened in the Rabassada just some days after its inauguration)[61] were a reminder that, even if commodified and controlled by engineers, risk was still risk.

54 From the 1910s, several international patents of mechanical rides systems were filed in Spain, as the journal *Industrias e Invenciones* pointed out.

55 Mohun, 'Amusement Parks for the World'.

56 On the late nineteenth-century and early twentieth-century roots of the globalization of American mass culture through amusement parks, among other popular entertainments, see: Robert Rydell and Rob Kroes, *Buffalo Bill in Bologna: The Americanization of the World, 1869–1922* (Chicago: The University of Chicago Press, 2005).

57 This influence is evident from their names. The amusement park called American Park was created in 1907. The following year, another amusement park with an English name was born next to the tiny Vallvidrera Dam: the Lake Valley Park. Moreover, most of the amusement parks were not called 'parc' or 'parque' (the words for 'park' in Catalan or Spanish), but 'park', with the English 'k'. On the other hand, the mechanical rides and entertainments often had English names.

58 Mohun, 'Designed for Thrills and Safety'.

59 Marian Rubió, 'Ferrocarril aéreo de la cumbre del Tibidabo (Barcelona)', *La Energía Eléctrica*, XVII:12 (1915): 177–80.

60 Lázaro Díaz, *El Tibidabo desde dentro: historia de un parque de atracciones* (Barcelona: Artual, 2004).

61 Juan Baguñó, 'Sr. Director de . . .', *La Vanguardia*, 5 August 1911, 4–5.

Montjuïc hill: Engineers, ideologies and entertaiment

While engineer Marian Rubió was actively participating in the 'conquest' of Collserola and the design of the Tibidabo amusement park, he had another huge project in mind: the 1929 Barcelona International Exhibition and the urbanization of the Montjuïc hill.[62] This project was initially conceived as an international exhibition of electric industries in 1914 – the year in which Rubió himself designed the first mechanical ride in the Tibidabo Park – but it was finally postponed because of the outbreak of the Great War.

After Collserola, Montjuïc was the next space to be urbanized and 'engineered for fun' in the city. The Montjuïc hill was close to the city centre and flanking the very popular Poble Sec neighbourhood. Throughout the nineteenth century and the first decades of the twentieth century, this hill became a preferred site for expressions of self-managed fun by the working class, such as picnics by the many wells (known as *fontades*) or popular dancing (see Figure 5.4). However, from 1914 on, the hill was transformed by engineers and architects in order to host new ways of entertainment that overlapped with the former ones. During the 1920s, the public works to transform the hill attracted a great mass of migrants from other regions of Spain. They mainly settled in the areas forming the 'red-and-black belt', while at the same time they were building what would become the last part of the 'technological fun belt'.

The figure of Marian Rubió, who had a powerful influence in this process, embodies the links between the new regimes of pleasure and the new regimes of knowledge. Marian Rubió had already been involved in the 1888 Universal Exhibition, where he was appointed representative of the Military Engineering Corps (by then he was a young and promising military engineer in Minorca).[63] From then on his career took off, as we have seen in the preceding section. His leading role in the Tibidabo Park project, his presidency of the SAF, his ties with the right-wing and nationalist party La Lliga,[64] and his pro-Allied writings in *La Vanguardia* during World War I made him probably the best known engineer in Catalonia.

This trajectory led him to be appointed the chief engineer and director of the next Barcelona International Exhibition, a position he even maintained during General Primo de Rivera's Dictatorship (1923–1930) due to his high military rank, regardless of the change in political regime. But the patriarch of the family was not the only Rubió involved in the plans to urbanize Monjuïc: he was joined by two of his sons, the architect Nicolau Rubió and the industrial engineer Santiago Rubió, who some years earlier had already helped their

62 On the fact that the International Exhibition was conceived as a way to reorganize the city in capitalist terms, see: Ignasi Solà-Morales, 'L'exposició internacional de Barcelona (1914–1929) com a instrument de política urbana', *Recerques: història, economia, cultura*, 6 (1976): 137–45.

63 *Els Rubió*, p. 42.

64 Blasco, *Barcelona Atracción (1910–1936)*, p. 49.

Figure 5.4 The urbanization of the Montjuïc hill, from the autonomous fun of the *fontades* and other popular activities (*top left*: people climbing up the hill, 1910s) to the civilizing promenades in well-planned gardens (*bottom left*: Laribal garden) and the technological fun of amusement parks (*right*: promotional image of the Maricel Park, around 1930).

father in the design of transport systems, architectural spaces and amusement devices in the Tibidabo Park.[65]

The 1929 Exhibition was celebrated as a professional triumph by the Asociación de Ingenieros Industriales de Barcelona (the Catalan association of industrial engineers) and it marked a turning point of their social and political rise, which was strengthened with the proclamation of the Second Spanish Republic (1931–1939) and especially with the administrative and economic reforms fostered by the Generalitat de Catalunya (the Catalan government) following its creation in 1932.[66] A visit to the exhibition will highlight the links of the Rubió family and the engineering professionals with the urbanization of the Montjuïc hill and the ideological features of the site in which the

65 At the same time, Marian Rubió's brother, the Gaudinian architect Joan Rubió, was the master mind behind some of the most gorgeous houses in the Tibidabo Avenue: Casa Roviralta (1903), Casa Fornells (1903) and Casa Casacuberta (1907). Joan Rubió was also an influential member of La Lliga and became district councillor in the Barcelona local government in 1905.

66 Valentines-Álvarez, *Tecnocràcia i catalanisme tècnic*.

mechanical rides of the exhibition's amusement park and the Maricel Park were to be located.

After taking the tram line number 31 from Plaça Catalunya to Plaça Espanya, the exhibition's starting point, the passenger could ascend to the National Palace using a funicular railway designed by Santiago Rubió, who already was a distinguished expert in transport engineering (particularly for mountain sites devoted both to Catholic worship and tourism).[67] From the National Palace esplanade, the visitor could admire the monumental Maria Cristina Avenue with the eclectic architecture of its pavilions, the many illuminated glass obelisks and the light spectacles of the 'Magic Fountains' that Marian Rubió had dreamt of and Carles Buigas had finally made real.[68] Standing there, and looking ahead into the horizon, the observer could see the entire city with the Tibidabo's Talaia at the top. Using the new escalators, the visitor could explore the main pavilions of the fair, such as, for example, the Palace of Metallurgy and the Palace of Work. The first one displayed an authentic old Catalan forge as an icon of the 'technological soul' of Catalonia. The display was carried out by Santiago Rubió and Antoni Gallardo, pioneer industrial archaeologists that looked for this kind of 'lost' national technologies in the Pyrenees from the 1910s.[69] The Palace of Metallurgy (first called Palace of Pedagogy, Hygiene and Social Institutions) was aimed at popularizing and spreading the Taylorist principles,[70] just as Marian Rubió had done in *El trabajo humano* (1916), one of the first (and few) books in Spain dealing with the 'human motor', scientific management and psychotechnics.[71]

After visiting the pavilions, one could go for a walk at the Laribal and Font del Gat gardens, with their tidy Mediterranean vegetation, their evocations of classical and Arabic culture, and their shaded promenades and fountains (see Figure 5.4). All these gardens were designed by the main disciple of the French landscape architect Jean Claude Nicolas Forestier in Barcelona: the architect Nicolau Rubió.[72] He was the main responsible of the 'ordering of nature' in the Montjuïc hill, which implied replacing the self-constructed houses and shanty towns of gypsies and migrants that occupied the slopes of the hill with

67 Santiago Rubió had already designed the funicular of Sant Joan in the mystical Montserrat mountain (1916–1918) and the funicular of the Núria Sanctuary in the Pyrenees (1917, not implemented). He also designed the Gelida funicular (1922–1924) and had been the director of the construction of the first line of the Barcelona subway (1919–1922).

68 Marian Rubió, *Ensayo de fuentes monumentales y juegos de agua en la exposición de Barcelona, Ibérica*, 249 (1918): 250–2. See also Chapter 11 of this book.

69 Valentines-Álvarez, *Tecnocràcia i catalanisme tècnic*, pp. 125–9.

70 The journal of the SAF pointed out the urgent need for the nation to popularize the Taylorist principles beyond technical journals. 'Exposición de Barcelona de 1929', *Barcelona Atracción*, 204, June (1928): 163–8, pp. 166–7.

71 Marian Rubió, *El trabajo humano* (Barcelona: Publicaciones de la Cámara Oficial de Industria de Barcelona, 1916).

72 Nicolau Rubió was director of the Barcelona Town Hall's Parks and Gardens section and designed the spaces of the Parc de la Ciutadella and the Turó Park when their amusement parks closed in 1926 and 1927.

a rational urbanism of planned gardens and exhibition pavilions.[73] He was one of the main promoters of the 'garden city' idea in Catalonia and from the 1920s on he successfully lobbied the Catalan government to reorder the whole nation through regional planning.[74]

Finally, the visit would not be complete unless the visitor enjoyed himself or herself at the big amusement park that was installed in an ancient quarry. Once again, as in the case of the Parc de la Ciutadella, we find amusement parks going hand in hand with world fairs, and being part and parcel of exhibitionary complex spaces. The amusement park of the 1929 Exhibition had several mechanical rides, like a great roller coaster, the thrilling Devil Ball, laughter barrels and a miniature passenger steam-engine railway. At night, it was illuminated by hundreds of light bulbs, thus joining the many light spectacles of the exhibition. In line with the contents of the exhibition, a mechanized and civilized fun was colonizing the Montjuïc hill.

Initially, technology was to play a major role in the museum projects that had to consolidate the civic use of the hill after 1929. Once the exhibition closed, the head of the Rubió clan lobbied to devote the central space of the exhibition to build what he portrayed as the biggest technical museum in the world: the 'Technotheque'.[75] However, the attempt was not successful, and the spaces around the National Palace were finally the site for other more traditional museum projects such as the Art Museum of Catalonia (1934) and the Archaeological Museum of Catalonia (1935).

Nonetheless, technology remained a relevant feature in the Montjuïc hill. Several mechanical rides were moved from the initial site to a nearby location, where a new amusement park called Maricel Park opened its doors in 1930. In its promotional advertisements, the park claimed that 'thrill and happiness' were guaranteed and offered 'the most modern' and 'the most thrilling rides in the most delightful and fresh place of the city'.[76] It also offered 'the latest in American parks', like bumper cars and the Tumble Bug, a very recent invention by the leading roller coaster designer and founder of the Traver Engineering Company, Harry Traver (1877–1961).[77] Chronicles written shortly after the

73 Mercè Tatjer and Cristina Larrea (eds), *Barraques. La ciutat informal del segle XX* (Barcelona: Ajuntament de Barcelona, 2010).

74 The Regional Planning of Catalonia was promoted by himself several years before, when he was Secretary of the Societat Cívica 'La Ciutat Jardí' (Garden City). The project was not developed until the creation of the Catalan government in 1932. Nicolau M. Rubió Tudurí and Santiago Rubió Tudurí, *El pla de distribució en zones del territori català. Regional Planning* (Barcelona: Impremta de la Casa de Caritat, 1932), p. 8.

75 Jaume Sastre-Juan and Jaume Valentines-Álvarez, 'Envisioning a Technotheque for Barcelona. Some thoughts on the technical museum as a cultural form (1914–1939)', unpublished paper, presented at the 8th STEP Meeting, Corfu (Greece), 21–24 June 2012.

76 As stated in several advertisements in the press (particularly in *La Vanguardia* and *Mirador*).

77 David Pescovitz, 'Roller coaster biographies. Harry Traver', *Britannica online* (Project 'Roller coasters. Inventing the scream machine', *Encyclopædia Britannica*), 1998–1999, http://kids.britannica.com/coasters/i_traver.html (last accessed 3 May 2014).

inauguration remarked that 'the roller coaster [...] confers a great atmosphere of modernity to the side of the hill dominated by these impressive iron structures' where 'thousands of light bulbs draw the [amusement] park's skyline'.[78] The sounds of American jazz, as well as the 'unavoidable screams of women', filled the air, and the central restaurant offered 'American dinners' under the lighting designed by Carles Buigas, the designer of the famous light-and-colour fountains of the exhibition (see Figure 5.4).

Everything seemed to be modern, middle class, clean, civic, streamlined, magic. Ticket office clerks were dressed in a red jacket and white trousers. Everybody, everything, was in red and white: saleswomen, the skating court, logos, shops, panels, mechanical rides and so on. While the red and the white of the last space of the 'technological fun belt' shined, the inhabitants of the shanties that previously occupied that same Montjuïc area were being relocated into a different belt, red and black, poor and radical, which emerged in other outskirts, for example just right on the other side of the hill, in the peripheral neighbourhoods of Can Tunis and Eduardo Aunós' 'Casas Baratas'.[79]

Conclusions: New regimes of pleasure and knowledge

In 1913, the journal *Industria e invenciones* published an article written by the engineer Marcel Hegelbacher on the newest innovations in mechanical rides, which was entitled 'Science in the service of pleasure'.[80] By asking which were the urban geographies and the politics of what we call 'technological fun' in Barcelona, we have explored the complex meanings behind Hegelbacher's title and we have argued that amusement parks were spaces in which both 'science' and 'pleasure' were being redefined in relation to one another during the first decades of the twentieth century.

Amusement parks were businesses with both local and international dimensions, and their emergence has to be understood within the context of the growing leisure industry and the commodification of entertainment. At the same time, most amusement parks were part and parcel of broader urbanization projects linked to property speculation and the development of tourism industry. But money is not the whole story. These urban projects were politically inscribed within hygienist and paternalist programmes. We have seen how amusement parks were promoted not only by the most powerful elites in the city, but many times by the very same individuals that were fostering hygienist policies, scientific management, regional planning and technocratic ideologies, like the Rubió family. The regime of pleasure of the new mechanized and standardized technological fun was closely tied to the rise of new technical elites and in tune with the new regimes of knowledge.

78 Manuel Sarret, 'Unes hores a Maricel-Park', *Mirador*, 24 July 1930; *La Vanguardia*, 14 July 1930.

79 Teresa García Castro, 'Barrios barceloneses de la dictadura de Primo de Rivera', *Revista de geografia*, 8 (1–2) (1974): 77–97; Tatjer and Larrea, *Barraques*.

80 Marcel Hegelbacher, 'La ciencia al servicio del placer', *Industria e invenciones*, 60 (117) (1913): 117–8.

The analysis of the geographies of technological fun in Barcelona also points in a similar direction. The urbanization processes in the Collserola mountain and Montjuïc hill went hand in hand with a change in their use for entertainment purposes. Amusement parks were important agents in the partial transformation of former peripheral spaces of autonomous and self-managed fun into urban spaces offering a new kind of entertainment, also including other exhibitionary institutions, such as industrial museums, lighting displays and world fairs. Parallel to the formation of a 'red-and-black belt' in Barcelona, a 'technological fun belt' was formed in green spaces that were being 'conquered' by the city's bourgeoisie through the technical elites and urban planners.

How were the new rides experienced by its several publics? This is always the most difficult question. On the one hand, we suggest that the symbolic construction of technological fun mattered. But on the other hand, it would be misleading to think that these discourses on technology and the ideologies of the promoters completely determined the reading of the experience with mechanical rides, which was unpredictable and multi-sided.[81] In any case, amusement parks became an important part of the technological cityscapes, and were spaces where the population became acquainted with new technology beyond labour and experienced new pleasures that meant a subversion of everyday behaviour in terms of gravity and even of social norms – for example the gendered codes of proximity between bodies – but not of the dominant symbolic order. The loops, twists and crashes of the water chute at Saturno Park, the roller coaster at the Rabassada, or the bumper cars at the Maricel Park took place in a carefully engineered, commodified and civilizing environment that was insistently portrayed as 'modern' and 'international' by its promoters.

In conclusion, the ludic experience of visiting an amusement park in Barcelona in the beginning of the century was mediated by discourses on technology and society – hygienism, scientific management and ideologies of progress – that were embodied in the spaces around – and within – the amusement parks, their rides and their displays. In this sense, technological fun was not just fun, and amusement parks were not only entertainment sites, but also an arena in which certain ways of looking at technology and the social order tried to seduce hearts and minds.

81 Bennett, 'Hegemony, Ideology, Pleasure: Blackpool', pp. 135–54.

Part II

Resistance – counter-hegemonies

6 The Rose of Fire

Anarchist culture, urban spaces and management of scientific knowledge in a divided city

Álvaro Girón Sierra and Jorge Molero-Mesa[1]

Recent memories of the 1888 Universal Exhibition could not conceal the poor living conditions and social strife that plagued the city's working class in November 1889.[2] That month Barcelona hosted the Segundo Certamen Socialista (The Second Socialist Contest), a competition organized by various anarchist groups and labour societies. At the Palacio de Bellas Artes (Palace of Fine Arts), one of the emblematic buildings erected for the exhibition, and before an audience of over 20,000, contestants read a long series of prize-winning speeches. Some delivered high praise for scientists (for example, Lamarck, Spencer, Darwin, Tyndall and the like) or for science in general. The undisguised aim was to revitalize anarchism while simultaneously demonstrating the working class's competence on literary, sociological and scientific issues.[3]

Forty years later, in 1929, Barcelona was not only a great modern industrial power, it had also become one of the most important centres of anarchist activity – an exception in a world which, in general, had turned its back on the movement. These changes took place while Barcelona was transformed into something very different from the ideal of the functional and socially coherent *new city* envisioned by urban planners. Far from any sort of rational planning, its spectacular urban and demographic expansion was chaotically moulded by market forces that imposed a deep socio-spatial segregation.[4] A veritable abyss had opened up between the proletarian neighbourhoods and upper-middle-class districts. The divide was reflected in lifestyles and dress codes, but even

1 The critical remarks made by the referees, José Álvarez-Junco and Chris Ealham, have been instrumental for improving this chapter. The research made has been financed by the Spanish Ministerio de Economía y Competitividad through the research projects, 'Science and the city. Natural history, biology and biopolitics in the divided city: Barcelona and Buenos Aires (1868–1936)' (HAR2013-48065-C2-1-P) and 'Health, illness and social order in Spain. The social inclusion-exclusion dynamics based on class and profession through medical discourses and practices in the 20th century' (HAR2009-13389-C03-01).

2 Angel Smith (ed.), *Red Barcelona. Social Protest and Labour Mobilization in the Twentieth Century* (London/New York: Routledge, 2002).

3 Josep Termes, *Historia del anarquismo en España (1870–1980)* (Barcelona: RBA, 2011), pp. 113–18.

4 Eduard Masjuan, *La ecología humana en el anarquismo ibérico. Urbanismo 'orgánico', neomaltusianismo y naturismo social* (Barcelona: Icaria, 2001), pp. 57–60.

more so in the human body itself. Hygienists and workers alike were driven by the consequences of pauperism, disease and physical deformities to remark upon the 'differentiating characteristics [...] of the proletarian race'.[5]

This chapter attempts to explain how this process affected anarchist efforts to *manage* scientific knowledge by using a panoramic view. A brief sketch of the political and spatial *geography* of libertarian scientific communication in the city is provided to highlight how social polarization reinforced another ongoing process: the widening gulf between the anarchist cultural universe and that of middle-class radicals. As interclass bridges became increasingly untenable, not only intellectuals but even scientists and technicians began to arouse suspicion amongst the anarchist ranks. After 1910 the cultural fabric of anarchist–syndicalism grew into a network oriented towards the autonomous management of scientific knowledge.[6] It is not easy to bring the experience of those efforts to modern readers. The relationship between anarchism and science, on a global scale, has been largely overlooked by historians of science, Piotr Kropotkin's evolutionary thought being the sole exception.[7] In the case of Spain and Barcelona, the situation is only slightly better. Stimulated by previous contributions made by historians of political and cultural thought,[8] there has been growing interest recently in the relationship between Darwinism and the libertarian movement,[9] the connections between New-Malthusianism, anarchist groups, contraception and eugenics,[10] and some important, though incipient, contributions on medicine and anarchism.[11] However, it can be said that we are just now beginning to scratch the surface. Small wonder no serious attempt has been made to link history of science, anarchism and urban development in Barcelona.

This is an oversight that appears paradoxical in light of the exceptional importance given by the movement to scientific culture. Anarchists drew from established Enlightenment, radical and socialist traditions in which confidence in the transformational possibilities of rationality and science was a defining

5 Joaquín Romero Maura, *'La rosa de fuego'. El obrerismo barcelonés de 1899 a 1909* (Madrid: Alianza 1989), pp. 140–1.

6 This could also be tied to the articulation in Barcelona of an urban public sphere of proletarian provenance. See Chris Ealham, *La lucha por Barcelona. Clase, cultura y conflicto* (Madrid: Alianza 2005), pp. 78–104.

7 Daniel P. Todes, *Darwin without Malthus. The Struggle for Existence in Russian Evolutionary Thought* (New York/Oxford: Oxford University Press, 1989), pp. 123–42.

8 José Álvarez Junco, *La ideología política del anarquismo español (1868–1910)* (Madrid: Siglo XXI, 1991), pp. 65–92.

9 Álvaro Girón Sierra, *En la mesa con Darwin. Evolución y revolución en el movimiento libertario español (1869–1914)* (Madrid: CSIC, 2005).

10 Eduard Masjuan, 'Procreación consciente y discurso ambientalista: anarquismo y neomalthusianismo en España e Italia, 1900–1936', *Ayer*, 46 (2002): 63–92; Richard Cleminson, *Anarchism, Science and Sex: Eugenics in Eastern Spain, 1900* (Oxford: Peter Lang, 2000).

11 Jorge Molero-Mesa and Isabel Jiménez-Lucena, '"Otra manera de ver las cosas". Microbios, eugenesia y ambientalismo radical en el anarquismo español del siglo XX', in Marisa Miranda and Gustavo Vallejo (eds), *Darwinismo social y eugenesia. Derivas de Darwin: cultura y política en clave biológica* (Buenos Aires: Siglo XXI, 2010), 143–64.

characteristic. The promise of social harmony resided in revolution, but also in science. A supreme guarantor of material progress, it seemed to provide support to a materialist world view that left little room for one of anarchists' *bête noirs*: organized religion. But the utility of scientific knowledge was not limited to its hypothetical potential in eroding the Catholic Church's authority. One of the fundamental objectives of the libertarian movement was the education of the working class. Social revolution would remain incomplete without a revolution of minds.[12] This would be achieved not only by introducing new ideas, but by completely overcoming atavistic traditions, customs and habits. As in the case of republicans and socialists, the aim was to disseminate *rationality* in daily life.[13] Science was to spearhead this process of emancipation. In short, the *management* of scientific knowledge was a central aspect of an alternative political subculture which saw itself as completely *rational*, and in this way different from both bourgeois culture and, not without tensions, prevalent patterns of popular culture.

It is difficult to ascertain the causes of this paltry historiographical presence. In the present case, however, there is a legitimate suspicion that the popular image of anarchism, which is still associated with nineteenth-century terrorism, raises doubts about whether science and anarchism actually have anything to do with one another. Nonetheless, and despite the fact that some libertarians supported terrorism, the vast majority of them had little or nothing to do with the logistical aspects of such attacks and many openly criticized such a form of political violence.[14] On the other hand, some historians have portrayed anarchism as a form of modern millenarianism in which eschatological elements played a prominent role. They also linked its alleged *primitive* features with industrial underdevelopment. However, this is hardly consistent with the popularity of the libertarian movement in a big industrial city as Barcelona. Moreover, other historians have highlighted the social and historical rationality of the Spanish anarchist movement's actions.[15] That said, the relative weakening of the Catholic Church's influence among workers did not eliminate certain emotional needs although the daily praxis of the libertarian militants might be described as *rational*. The redemptive role assigned to science is far from incompatible with utopian expectations.

Anarchism is a historiographical object that resists definition. It was a polymorphous movement with a wide range of tendencies.[16] However, it is noteworthy that the vast majority of Barcelona's libertarians shared communal, if

12 Jaume Navarro, 'Los educadores del pueblo y la *revolución interior*. La cultura anarquista en España' in Julián Casanova (ed.), *Tierra y libertad. Cien años de anarquismo en España* (Barcelona: Crítica, 2012), 191–217, p. 192.

13 José Álvarez Junco, *Alejandro Lerroux. El Emperador del Paralelo* (Madrid: Síntesis, 2005), p. 169.

14 Rafael Núñez Florencio, 'El terrorismo', in Julián Casanova (ed.), *Tierra y libertad. Cien años de anarquismo en España* (Barcelona: Crítica, 2012), 61–87.

15 Girón, *En la mesa con Darwin*, p. 21.

16 Susana Tavera, 'La historia del anarquismo español: una encrucijada interpretativa nueva', *Ayer*, 45 (2002): 13–37.

not collectivist, ideals, which took the principle of solidarity as a fundamental axiom. They were quite far from the extreme-right movements that define themselves as libertarians in the USA nowadays.[17] Nonetheless, the social philosophy of the libertarian movement does not fit well within any of historiography's traditional classifications. Born from the cleft between supporters of Karl Marx and those of Mikhail Bakunin at the First International, anarchism is usually identified as an unorthodox branch of socialism. However, origins notwithstanding, historical research made it clear that Spanish anarchism and liberalism do share a number of significant similarities. In the words of J. Álvarez Junco, anarchist ideology fell within the intellectual framework of liberal rationalism, faith in progress, the goodness of mankind, and harmony between nature and society.[18] This is consistent with one of the libertarian's most cherished beliefs: science was supposed to reveal the justice and harmony in nature, one of the great enlightenment myths that anarchists readily adopted.[19]

Historians often claim that the libertarian movement in Catalonia must be understood within the coordinates of the profound unity of the left's political culture, dominated by the values and principles of republicanism, which survived right up to the Spanish Civil War.[20] But care should be taken not to interpret the political subcultures of anarchism, socialism, and republicanism as one indistinguishable force. The libertarian notion of revolution, for example, included the immediate objectives of socializing the means of production and complete abolition of the state. That was far from the wildest dreams of the most radical branches of republicanism and affected decisively libertarian views on science. While science itself was rarely questioned, scientists were not so warmly embraced. Reticence towards them was based both on their alleged bourgeois standing and anarchists' vision of modern history. Libertarians acknowledged that the liberal bourgeoisie had been a progressive factor in the emancipation of mankind. It was not the case in the last decades of the nineteenth century: the upper-middle class had already fulfilled its historical mission. The bourgeoisie only devoted its efforts to defend class privileges.[21] It was the *people* that now had to shoulder the redemptive mantle. The *virgin* intelligence of those with no privileges to defend, that is, the workers, made it possible for them to completely accept the self-liberating implications of scientific knowledge. That could not be expected of the bourgeoisie anymore, scientists included. On their

17 José Álvarez Junco, 'La filosofía política del anarquismo español', in Julián Casanova (ed.), *Tierra y libertad. Cien años de anarquismo en España* (Barcelona: Crítica, 2012), 11–31, pp. 25–6.

18 Álvarez Junco, 'La filosofía política del anarquismo español', pp. 16–17.

19 José Álvarez Junco, 'La subcultura anarquista en España: racionalismo y populismo' in Yves-René Fonquerne and Alfonso Esteban (eds), *Culturas populares. Diferencias, divergencias, conflictos* (Madrid: Universidad Complutense, 1986), 197–208, p. 198.

20 Pere Gabriel, 'Republicanismo popular, socialismo, anarquismo y cultura política obrera en España (1860–1914)', in Javier Paniagua et al., *Cultura social y política en el mundo del trabajo* (Valencia: Fundación Instituto de Historia Social/UNED, 1999), 211–22, p. 220.

21 Álvarez Junco, *La ideología política del anarquismo español*, p. 203.

view, social status does not so much affect the making knowledge process as it does the *moral* capacity of different social classes to accept the implications revealed by new scientific truths.[22] Thus, the opposition between *bourgeoisie science* and *working class science* was often depicted as the confrontation between a science consciously limited by class interests – and thereby compromised, incomplete and truncated – and one guided by a proletariat whose only real interest is truth.[23]

On top of this, science was also conceived as collective capital, a sort of knowledge bank accumulated by mankind from generation to generation. On that understanding, anarchists saw this *universal heritage* as having been kidnapped by one class and, in a sense, *enclosed* within institutions dominated by the sphere of *privilege*, such as universities and academies. From the libertarian perspective, it was more important to disseminate knowledge that had already been discovered than promote new discoveries. The aim was for science to become 'the life of all', and to cease to be an unattainable luxury for workers. That is why science occupied such an important role in anarchists' editorial activities. However, at first glance, what was edited and distributed shows that the literary universe of Barcelona's anarchist elites was not all that different from that of the middle class, which paid significant attention to the works of Spencer, Haeckel, Darwin and Lombroso. Were the anarchists, as some historians claim,[24] mimicking the dominant models of *high culture*? No, they were using *high culture*. It may be obvious that Ernst Haeckel, Herbert Spencer, Ludwig Büchner, Enrico Ferri, Théodule Ribot and Lombroso were read and disseminated, but all readings modify their object and the established order of the original text.[25] The manifest resignification of terms like 'survival of the fittest', 'normality', 'degeneration', 'disease', 'environment', indicates a *use* of scientific knowledge far beyond the fiction of a mere pyramidal diffusion of knowledge.[26] As Anselmo Lorenzo said, 'from bourgeois science we will take the truth, and we will do away with the sophisms that serve as the basis of privilege [...]'.[27]

An urban geography of anarchist science

Theoretical digressions aside, it is important to understand the geography of scientific communication within the city. It is by no means easy to draw a map of anarchist cultural institutions in Barcelona, which, to a greater or lesser extent,

22 Romero Maura, *'La rosa de fuego'*, p. 193.

23 Anselmo Lorenzo, 'Incapacidad progresiva de la burguesía', *La Revista Blanca*, 145 (1904): 1–7, p. 2.

24 Carlos Serrano, 'Cultura popular/cultura obrera en España alrededor de 1900', *Historia social*, 4 (1989): 21–31.

25 Michel de Certeau, *L'invention du quotidien. I. Arts de faire* (Paris: Gallimard, 1990), pp. 244–5.

26 Carlos Tabernero, Isabel Jiménez-Lucena and Jorge Molero-Mesa, 'Libertarian movement and self-management of knowledge in the Spain of the first third of the 20th century: "Questions and answers" section (1930–1937) of the journal *Estudios*', *Dynamis*, 33 (1) (2013): 43–67.

27 Anselmo Lorenzo, 'Ciencia burguesa y ciencia obrera', *Acracia*, 22 (1887): 354–9, p. 355.

were involved in the debate of scientific ideas. In the case of late nineteenth-century Paris, we know that the areas in which activists were recruited – the mostly proletarian north-eastern part of the city – do not coincide with the areas of greatest anarchist activity (Montmartre).[28] Could the same be said of Barcelona, or are these two unique cases? What we know for sure is that in the 1880s and 1890s most libertarian activity was concentrated in city's old sector, particularly the run-down areas today known as Raval. This was a traditional stronghold of artisans and workers. At the same time, the role of artisans and labourers from nearby villages (for example, Sant Martí de Provençals, Sants, and Gràcia) that had just been incorporated into Barcelona in 1897 should not be neglected.

Yet anarchists also inherited a federalist tradition which inclined them to decentralization. Anarchist cultural life could have been geographically complex and not limited to workers' official organizations: cafés played also an important role. Libertarian cafés explicitly distanced themselves from popular taverns: they provided board games, but also books, newspapers and pamphlets. They organized collective readings and even presented theatrical works. The mobility and decentralization of this extensive network further increased in 1888 when the syndicalist organization, Federación de Trabajadores de la Región Española (FTRE), was officially dismantled. It dissolved into affinity groups of between five and ten members – 'grupos de afinidad' – which became the nuclear elements of libertarian action until the emergence of anarchist–syndicalism in the 1910s.[29] Instability was also inherent in the anarchist movement. Frequent cycles of repression meant the constant appearance and disappearance of groups, networks and their respective media mouthpieces. Even more decisive was the international character of the libertarian movement. Indeed, there was such a broadly-flung network of communication that applying the concept of *local context* requires extreme caution.

Anarchist readings of Darwinism, for instance, had little or nothing to do with the debates that were being generated on the topic in Barcelona and throughout Spain.[30] Instead they owed more to the thought of the Russian anarchist Piotr Kropotkin, a theorist who greatly influenced the libertarian movement on a global scale. Such interpretations were also coloured by the recognition, evident after the mid-1880s, that some venerated figures in the *Darwinist* field, Ernst Haeckel[31] and Herbert Spencer,[32] were attempting to discredit

28 Richard D. Sonn, *Anarchism & Cultural Politics in Fin de Siècle France* (Lincoln/London: University of Nebraska Press, 1989), pp. 51–2.

29 George Esenwein, *Anarchist Ideology and the Working-Class Movement in Spain, 1868–1898* (Berkeley: University of Califonia Press, 1989), pp. 130–5.

30 Agustí Camós, 'Darwin in Catalunya: From Catholic Intransigence to the Marketing of Darwin', in Eve-Marie Engels and Thomas Glick (eds), *The Reception of Charles Darwin in Europe* (London/New York: Continuum, 2008), pp. 400–12.

31 Robert J. Richards, *The Tragic Sense of Life. Ernst Haeckel and the Struggle over Evolutionary Thought* (Chicago: The University of Chicago Press, 2008), pp. 312–18.

32 Daniel Becquemont, *Darwin, darwinisme, évolutionnisme* (Paris: Editions Kimé, 1992), pp. 144–9.

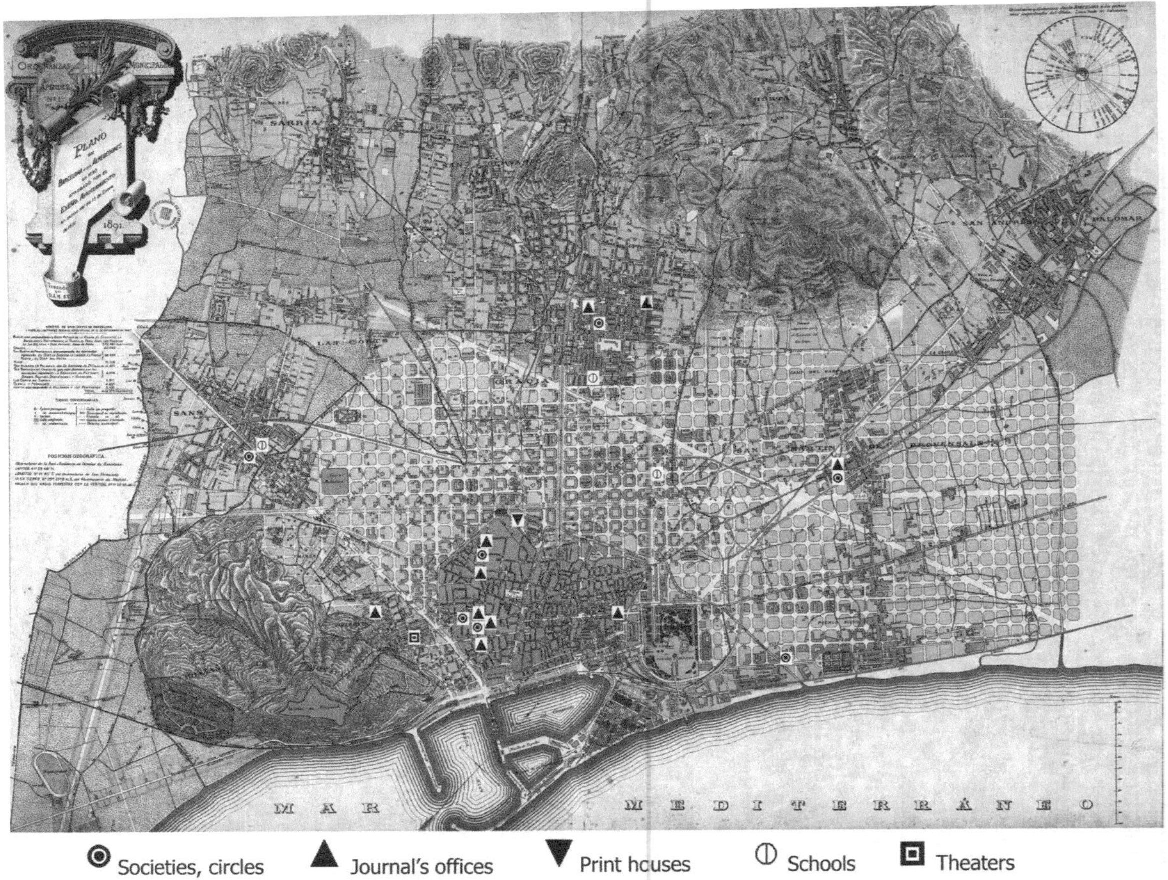

Figure 6.1 The geography of libertarian cultural activity in Barcelona and its hinterland (1888–1896).

socialism. In this respect, the central problem for libertarians was the dominant interpretation of the Darwinian metaphor of struggle for life. Presenting nature as a battlefield challenged the libertarian belief in nature as a harmonious order.[33] In the face of this challenge, anarchists from Barcelona attempted to remove the Malthusian sting from Darwinism by denying the existence of an imbalance between population and (food) resources. Above all, particularly during the 1880s, they supported what Antonello Lavergata identifies as one of the most common forms of softening the edge of literal interpretations of the struggle for life. Libertarians began to support the claim that, while the most ferocious forms of struggle could have played a significant role during the early stages of evolution, as humanity progresses, the struggle is moderated,[34] or even transferred to less bloody arenas, such as that of the intellect. Although wars of mutual destruction occurred along the way, solidarity was seen as the final stage of this process.[35] This was hardly original. The Parisian publication *Le Révolté*, of enormous influence in the libertarian movement on a global scale, followed the same path.[36]

Presenting things this way had its advantages: modern society's ferocious struggle for life was seen as the pathological preservation of behaviours typical of primitive evolutionary stages. Actually, this argument became a rhetorical device frequently used to condemn capitalism. It also, however, had equally obvious disadvantages: interpreting the universe of the living as a great calamity is contradictory to visions of a harmonious nature. In 1890–1896, Piotr Kropotkin provided the way out in his celebrated series of articles on mutual aid and Darwinism published in *The Nineteenth Century*, one of the most prestigious English monthly reviews. He inverted the terms: solidarity is the predominant factor in nature, not a Malthusian struggle between individuals for space and resources. Mutual aid, on the contrary, is the main factor explaining progressive evolution. Social life and ethical norms are not the result of an extended evolutionary process that takes us increasingly apart from our *natural* past. On the contrary, morality is based on social instincts that we inherited from our animal ancestors.[37] After 1892, parts of Kropotkin's articles were translated and published in the influential Barcelona weekly, *El Productor*. Kropotkins' views on Darwinism were widely used for internal debates, specifically, during the first years of the twentieth century, when the conflict between those anarchists who

33 Diego Núñez, *El darwinismo en España* (Madrid: Castalia, 1977), p. 56.

34 Antonello Lavergata, 'Les bases biologiques de la solidarité', in Patrick Tort (ed.), *Darwinisme et société* (Paris: Presses Universitaires de France, 1992), 55–87.

35 Álvaro Girón, 'The Moral Economy of Nature. Darwinism and the Struggle for Life in Spanish Anarchism (1882–1914)', in Thomas F. Glick et al. (eds), *The Reception of Darwinism in the Iberian World* (Dordrecht: Kluwer, 2001), 189–203.

36 'La loi de la force et le concert pour l'existence', *Le Révolté* 9 (7–13 May 1887): 1.

37 Álvaro Girón Sierra, '¿Anarquía y darwinismo? Piotr Kropotkin en España (1882–1914)', in Gustavo Vallejo and Marisa Miranda (eds), *Políticas del cuerpo. Estrategias modernas de normalización del individuo y la sociedad* (Buenos Aires: Siglo XXI, 2007), 171–98, pp. 178–81 and p. 186.

supported *kropotkinian* communal ideals (the majority) and radical individualists inspired by Stirner and Nietzsche arose.[38] It is clear that Barcelona's libertarian elite managed scientific knowledge – *Darwinism* – to make it fit with their deep beliefs and the peculiarities of local agendas. But this *management* should not be understood without a selective appropriation of what had been said initially on that topic by the French libertarian press and Kropotkin's reading of Darwinism.

Now, it is the press that occupies the leading role in interaction between the local and the global. It is through the press that Barcelona became a node in a network of libertarian communication. How was it *read*? In a city where not all workers were literate, collective reading certainly had a role, the exact relevance of which remains to be determined.[39] It is not a minor point: the reading aloud of flyers had already been instrumental in the creation of a public sphere in France since the eighteenth century.[40] In the case of Barcelona's anarchists its utility was clear: as in Buenos Aires it could help the workers get to know great works of thought and it would facilitate the recruitment of new supporters.[41] Obviously the nature of available sources makes it difficult to find concrete examples linked with science. However, there exists written evidence of one particularly significant collective reading a few years before the Universal Exhibition of 1888. In April 1885, at the headquarters of the Barcelona Society of Printing Press Workers, and before an audience comprising 'the members and their families', a direct translation of some articles published in *La revue socialiste* under the title 'Transformisme et socialisme' was read out loud.[42] Significantly, the goal of his author, the French socialist and free thinker Louis Dramard, was to challenge Ernst Haeckel's attempts to disassociate Darwinism and socialism.

The foregoing considerations do not exhaust the question: the political geography of the city is equally decisive. From a politico-cultural perspective, one can describe the history of libertarian cultural institutions as a process of generating their own independent spaces. Their progress reflected, in a sense, the growing isolation of the working class in the city. That was a long process, not exempt from setbacks. Already during the *Sexenio Democrático* (Democratic Six Years) (1868–1874), the internationalists of the Spanish Regional Federation (FRE) of the International Association of Workers (AIT) tried to drive a wedge between the workers and the middle class, aiming to establish thereby their hegemony within labour organizations. Towards the 1880s and 1890s things began to change. Anarchists created their own institutions – casinos, cafés, publications, secular schools and so on – that tended in principle to reinforce class

38 Ibid., pp. 191–8.

39 On collective reading among workers, see: Jean-François Botrel, 'Teoría y práctica de la lectura en el siglo XIX: el arte de leer', *Bulletin Hispanique*, 100 (2) (1998): 577–90.

40 Roger Chartier, *El mundo como representación* (Barcelona: Gedisa, 1996), p. 138.

41 Juan Suriano, *Anarquistas. Cultura política libertaria en Buenos Aires, 1890–1910* (Buenos Aires: Manantial, 2001), p. 117.

42 Louis Dramard, 'Transformisme et socialisme', *La Revue Socialiste*, 1 (1885): 24–39.

divisions.[43] They were relatively complex organizations that sponsored cultural activities, like soirees or conferences, and even began to edit periodicals. In the 1880s the social circle of workers known as 'La Regeneración' (located at 2 Sant Olegari Street) stood out for its influence in the city as the driving force behind two of the most important publications in the history Spanish anarchism: *El Productor* and *Acracia*. In the latter, the relationship between socialism and Darwinism was discussed at depth.

However, Barcelona was not yet a completely divided city. There were cultural spaces for interclass collaboration between libertarians and radicalized middle-class groups, especially the republican federals by the late 1880s. Anticlericalism, a constant feature of the Leftist movements of southern Europe throughout the nineteenth century and up to the Second Vatican Council, cemented the collaboration between republicans, anarchists and freemasons. The context was favourable to it as well. There was a feeling that it was possible to actively compete with the power of the church. Moreover, anticlericalism was known to be a powerful mechanism of popular mobilization.[44] Such informal alliances were solidified, at the same time, through the freethinking movement, one of the most important political and cultural currents in Western Europe.[45] The movement developed extensively in Barcelona from the last decades of the nineteenth century. Though it was presented as an interclass movement of heterogeneous social composition, the free thinkers of Barcelona came to defend political principles that belonged to the political culture of the working-class movement.[46] Freethinkers undertook the secularization of civil life through the dissemination of scientific rationalism, the popularization of non-Catholic social rites (for example, civil weddings and funerals) and, above all, the creation of secular schools. From this perspective, science occupied an essential place in the symbolic realm. Already in 1869, the preamble to the provisional regulations of the Free-thinking Society of Barcelona, led by the anarchist doctor Gaspar Sentiñón, stated: ' [...] the good cannot exist outside of the truth, and [...] there is no other truth than that provided by science'. [47]

Actually, freethinkers devoted a good part of their time to scientific communication. The press organ of the Free-thinking Society of Barcelona, *La Humanidad*, was instrumental in the propagation of a markedly materialist reading of Darwinism, in which the influence of Ernst Haeckel was noteworthy. From 1885, and until the beginning of the 1890s, it was replaced by an association of free thinkers known as 'La Luz'. This nucleus of activity combined republican leaders like Cristobal Litrán, a large faction of notable anarchists

43 Esenwein, *Anarchist Ideology*, p. 124.

44 Álvarez Junco, *Alejandro Lerroux*, pp. 337–42.

45 Jacqueline Lalouette, *La libre pensée en France, 1848–1940* (Paris: Albin Michel, 2001).

46 Albert Palá Moncusí, 'Sociabilité et libre-pensée en Catalogne (1860–1909) in *Séminaires du CRIDAF 2004–2005: La sociabilité dans tous ses états* (2004), p. 2 and pp. 12–13, http://www.univ-paris13.fr/CRIDAF/TEXTES/ PalaLibPens.pdf (last accessed May 2014).

47 *La Federación*, 14 November 1869, 2.

(with the exception of the anarchist–communists), and a few high-ranking members of the Freemasonry, like its organizer Rossend Arús.[48] Defined by its radical atheism, 'La Luz' collided with the spiritists. That might have caused a bit of internal debate: some spiritists were politically radicalized and collaborated actively in free-thinking activities.[49] The anarchist and engineer Fernando Tarrida del Mármol defended the uncompromising materialist position of 'La Luz', basing it on scientific authority. He made liberal use of his particular reading of the political and religious implications of thermodynamics, then a matter of debate both within and outside of Spain.[50] On top of this, he provided a sketch of Darwin's theory, which helped him to conclude that there was a logical explanation of the universe and man that dispensed 'entirely the hypothesis of the divine spirit'.[51] On the other hand, there were other spaces in which collaboration was part of daily life. That was the case of a celebre printing house 'La Academia', publisher of an important number of scientific books. Founded in 1877, it was owned by the republican federal Evarist Ullastres, giving jobs to some 60 workers. A number of them were anarchists who frequented the meetings of 'La Luz'. Its disappearance, in the 1890s, represented a severe setback for the libertarian movement as a whole.

Things got worse when the cycle of terrorism/repression gained momentum in the last years of the nineteenth century. The bridges between *moderate* anarchists and republicans collapsed. The indiscriminate repression put an end to the freethinking movement itself. On top of it, reductionist conflation of anarchism with terrorism aggravated the situation. This conflation took as its scientific basis the extremely popular criminal anthropology of Cesare Lombroso. The anarchists – many of whom were 'insane', according to the new criminal taxonomy – were turned into the example of choice for the positivist criminology avowedly indebted to Darwin and Spencer. But the Barcelona libertarians were extremely critically aware in their own right. They did not limit themselves to discrediting the abusive identification of 'anarchist' with 'degenerate' but went further to attack the very foundations of Lombrosian thought, especially his theories about the biological nature of crime, synthesized in the emblematic figure of the natural-born criminal.[52]

In the early years of the twentieth century, the old alliance between of free thinkers, radical republicans and some leading anarchists somewhat revived. In

48 Álvaro Girón Sierra, 'Una historia contada de otra manera: librepensamiento y 'darwinismos' anarquistas en Barcelona, 1869–1910', in Clara Lida and Pablo Yankelevich (eds), *Cultura y política del anarquismo en España e Iberoamérica* (México: El Colegio de México, 2012), 95–133, pp.117–19.

49 Gerard Horta, *Cos i revolució. L'espiritisme català o les paradoxes de la modernitat* (Barcelona: Edicions de 1984, 2004). On spiritism in Barcelona also see Chapter 7 in this book.

50 Stefan Pohl-Valero and Favio Cala Vitery, 'Energía, entropía y religión. Un repaso histórico', *Revista de la Academia Colombiana de ciencias exactas, físicas y naturales*, 34 (130) (2010): 37–52.

51 Fernando Tarrida del Mármol, 'El espiritismo', *La Luz*, 21 (1886): 4–5.

52 Álvaro Girón, 'Los anarquistas españoles y la criminología de Cesare Lombroso (1890–1914)', *Frenia* 2 (2) (2002): 81–108.

Figure 6.2 Anarchism and scientific education: young students at Francesc Ferrer's Modern School.

1901 Francesc Ferrer Guardia founded La Escuela Moderna (The Modern School) a universal symbol of the fight between free thought and inquisitorial Catholicism.[53] Ferrer's goal was to create a both a model of *rationalist* education and an emancipatory tool for workers. It was based on strong scientism and uncritical faith on positivism. The influence of Ernst Haeckel's materialist evolutionism permeated the entire intellectual fabric of The Modern School. In its daily activity, the collaboration of libertarians with figures of scientific respectability – such as the naturalist, republican and free thinker Odón de Buen[54] – was not infrequent at all. The history of this pedagogical experiment ended in the worst possible manner. In 1909, Barcelona experienced one of the most important popular uprisings of its entire history: the Setmana Tràgica (the Tragic Week). Ferrer was falsely accused of promoting the riots and was executed by a firing squad. The question at hand is whether, at its base, this type of interclass collaboration amongst the *liberal elements*, typical of the nineteenth century, had already become an anachronism in Barcelona by the onset of the new century. Already in the 1890s the wave of terrorism made widespread the dystopian bourgeois nightmare of an uncontrollable, violent city. The strike of 1902, and the Tragic Week, augmented an existing upper-middle class's distrust and many fled to the outskirts of the city.[55] Collaboration became both spatially and symbolically difficult.

53 See: Juan Avilés Farré, 'Republicanismo, librepensamiento y revolución: la ideología de Francisco Ferrer Guardia', *Ayer*, 49 (2003): 249–70.

54 Agustí Nieto-Galan, 'A republican natural history in Spain around 1900: Odón de Buen (1843–1945) and his audiences', *Historical Studies in the Natural Sciences* 42 (3) (2012): 158–89, pp. 174–9.

55 Ealham, *La lucha por Barcelona*, p. 45.

In 1910 a new actor entered the scene causing important changes: the anarcho-syndicalist Confederación Nacional del Trabajo (National Confederation of Labour) or CNT. Within a few decades it became the most important trade union in Spain. The CNT was born in a city characterized by an accelerating – and disorganized – expansion. Thousands of new immigrants coming from southeast Spain settled in a secondary periphery deeply segregated – both socially and politically – from the rest of the city. Libertarian cultural life, adapted to this socio-spatial changes. From the end of World War I until the 1930s, libertarian cultural associations known then as ateneos racionalistas (rationalist athenaeums), began to be active in the working-class suburbs where the anarchists had traditionally found the most support: such was the more or less intermittent operation of the rationalist athenaeums of Sants (1917–1923), Gràcia (1918), and Poble Nou (1918). By the 1930s these sorts of initiatives were extended to the new secondary peripheries and adjacent municipalities in accordance with the changes in the political layout of the city. The old working-class suburbs – and the densely populated centre (El Raval) – had ceased to be areas of undisputed anarchist hegemony. The secondary peripheries had been converted into the new breadbasket of libertarian adherents. [56]

The consolidation of the anarchist–syndicalism in the first decades of the twentieth century created new fault lines that somewhat reinforced the effects of existing socio-spatial segregation. This unavoidably affected everything related to the management of scientific knowledge. If, until the turn of the century, the organic collaboration between the unions of Barcelona and technicians and intellectuals was permitted and even sought out, the anarchist–syndicalism that emerged after the conference of the Regional Confederation of Societies of Worker Resistance and Solidarity (Barcelona 1908) left out of its corporate structure all those who were not manual labourers. At the same time, anarchist–syndicalism distanced itself from any intellectual, or elitist, anarchism.[57] What were the reasons behind this? Some of those intellectuals, like Anselmo Lorenzo, played key roles both in the cultural life of the movement and in science communication. But not all of them fitted well within anarchist ranks. That was the case of some of the young middle-class intellectuals interested in anarchism during the 1890–1910. Close to modernist literature, bearers of a Nietzschean language condemning the philistinism of the bourgeoisie, these young intellectuals were rejected because of their episodic and superficial adherence to anarchism.[58] There were also criticisms of the sterility of *intellectual*

56 José Luis Oyón, *La quiebra de la ciudad popular. Espacio urbano, inmigración y anarquismo en la Barcelona de entreguerras, 1914–1936* (Barcelona: Ediciones del Serbal, 2008), p. 444.

57 Antonio Bar, *La CNT en los años rojos: del sindicalismo revolucionario al anarcosindicalismo, 1910–1926* (Barcelona: Akal, 1981), pp. 102–5 and pp. 128–9. On relations between the CNT and technicians and scientists, see: Jorge Molero-Mesa and Isabel Jiménez-Lucena, '"Arm and brain": Inclusion-exclusion dynamics related to medical professional within Spanish anarcho-syndicalism in the first third of the 20th century', *Dynamis*, 33 (1) (2013): 19–41.

58 This occurred in Barcelona, but something very similar happened in Buenos Aires and Paris: Suriano, *Anarquistas*, p. 133.

Figure 6.3 Anarchist's decentralized cultural network reaching the old industrial municipalities annexed in 1897. El Clot's Libertarian Athenaeum.

anarchism that had settled down in Barcelona since the mid-1890s: the utility of seemingly abstruse debates was no longer evident for manual workers.[59] Another prevailing argument for the exclusion of intellectuals, and with them technicians and scientists, was rooted in the distrust of workers for a sector of society from which the politicians and 'parasites of all kinds' were recruited.[60] On the other hand, anti-dirigisme guided the new union. Workers' emancipation 'must not be sought in the will of others but in their own personal and collective efforts'.[61]

It is clear that the resolutions of the CNT congress had considerable effects on the cultural life of anarchism in the city, in part because unions, rather than

59 Romero Maura, *'La rosa de fuego'. El obrerismo barcelonés de 1899 a 1909*, pp. 240–1.
60 Bar, *La CNT en los años rojos*, p. 547.
61 Congreso de Constitución de la CNT (Barcelona 30 October–1 November 1910), http://archivo.cnt.es/Documentos/congresosCNT/CONGRESO_CONSTITUCION_CNT.htm (last accessed October 2014).

small like-minded groups of libertarians, had become the hallmark of anarchist activity. In fact, the congress forced anarchist activity bifurcate into two separate spheres of action. One was the labour union membership, constituted by manual labourers and serving purely to press demands and form protests. The other sphere was made up of libertarian athenaeums and other such cultural organizations where manual and intellectual workers could collaborate to achieve the social revolution. Such organizations, along with the rationalist schools, were necessary for the syndicate's objective of 'promoting the intellectual culture of the worker' and to achieve their maximum revolutionary autonomy. The impact of this strategy was also noticeable in the anarchist media. The bulletin of the CNT, *Solidaridad Obrera*, was used almost exclusively for revolutionary agitation and the organizational or strategic aspects thereof. Articles of science popularization or cultural interest appeared only sporadically in its pages. On the other hand, some journals and magazines intentionally incorporated such educational content (*Salud y Fuerza, Generación Consciente*). Other journals aimed to influence the strategic orientation of the syndicate with doctrinal articles. That was the case of *La Revista Blanca*, one the most important anarchist cultural journals at the time.

Anarchist science without scientists

The consolidation of CNT in the first decades of the twentieth century brought about many changes. Propaganda and the dissemination of all kinds of knowledge reached a high point of sophistication through associative networks, which were maintained by the massive use of all forms of media and communication.[62] They touched on every branch of the sciences, but obviously those having to do with subjects related to the life (and death) of workers took precedence in an organization that strived to maintain its class-based profile. As such, problems related to phenomena associated with health and sickness came to be a kind of 'thematic' specialty within these networks.[63] The magazines *Salud y Fuerza, Generación Consciente* and *Estudios* gained prominence.[64] However, as aforementioned, the relationship between anarchist–syndicalist members and experts in various branches of scientific knowledge was founded on suspicion and mistrust, basically because they were considered but another piece of the social machinery sustaining the capitalist system. The scepticism was extended to the written works of these experts. As Felipe Alaiz (1887–1959) put it, the scientific readings available at the time were 'characterized by undiluted erudition'. That is to say, such works

62 Javier Navarro Navarro, *A la revolución por la cultura: prácticas culturales y sociabilidad libertarias en el País Valenciano (1931–1939)* (Valencia: Universitat de València, 2004).

63 Josep Lluís Barona, 'Ciencia y revolución en la España de Martí Ibáñez', in José Vicente Martí and Antonio Rey (eds), *Actas del I Simposium Félix Martí Ibáñez: medicina, historia e ideología* (Valencia: Generalitat Valenciana, 2004), 17–38.

64 Francisco J. Navarro, *El paraíso de la razón. La revista Estudios (1928–1937) y el mundo cultural anarquista* (Valencia: Edicions Alfons el Magnànim, 1997).

contained, 'excessive allusions unintelligible to the majority of readers'. This style of writing was not an innocent professional convention; it had a cryptographic aim that seemed to endow the technician with a kind of sorcery. The technicians 'who had degrees and titles [...] instead of socializing their knowledge, instead of making it accessible, they put it in a sack [...] available only to their familial professional dynasty'.[65] What was required, was to wrest away from technicians and intellectuals the management of knowledge: a network of specialized teaching societies that, from militant eclecticism, would use knowledge retooled for the social transformation they aspired to induce.[66] This network would not be limited to the reading of texts but would treat scientific knowledge like all other knowledge, always at the service of the social cause. Let us analyse some of the consequences of this approach and its significance for libertarian movement.

Self-teaching and self-management of knowledge brought into existence groups of libertarian workers with a great understanding of the topics that concerned the proletariat or had a strong social impact through their connection to health and sickness.[67] Consequently, it could often be the case that the topics of some conferences were already well known by the audiences of libertarian associations. That was usually the case when the invited speaker just summarized contents of books widely available in working class associations' libraries. Many of the reviews of these conferences, drafted by workers for the anarchist–syndicalist press, included after their description of the address a critique of its content or a section clearly pointing out what the speaker should have said before the audience. In October of 1935, Martín Zabiña commented in *Solidaridad Obrera* on the seminar delivered by Doctor Ramón Torra Bassols at the Fraternal Centre of Palafrugell under the title, 'Should the people worry about the problem of venereal diseases?' After briefly summarizing the doctor's procedure, he continued: 'And now, with all the respect I owe the kind lecturer, let me to direct a few comments and objections to his address'. Among these objections he emphasized the 'deficient anatomical explanation' as well as the doctor's failure to explain 'the prophylactic means to avoid infections'. Additionally, he recalled that the products recommended for the treatment of venereal diseases (mercury and silver nitrate) 'do not cease to produce other disasters in the body that, sometimes, are worse than the disease itself'. The reviewer justified his critique in didactic terms: the lecturer was informed that,

65 [Felipe Aláiz], 'La escuela y el libro. [Libros de técnica y técnica de libros]', *Solidaridad Obrera*, 29 July 1933.

66 Isabel Jiménez-Lucena and Jorge Molero Mesa, 'Problematizando el proceso de (des)medicalizació. Mecanismos de sometimiento/autogestión del cuerpo en los medios libertarios españoles del primer tercio del siglo XX', in Marisa Miranda and Álvaro Girón Sierra (eds), *Cuerpo, biopolítica y control social: América Latina y Europa en los siglos XIX y XX* (Buenos Aires: Siglo XXI, 2009), 69–93; Isabel Jiménez-Lucena and Jorge Molero-Mesa, 'Good birth and good living. The (de)medicalising key to sexual reform in the anarchist media of inter-war Spain', *International Journal of Iberian Studies*, 24 (3) (2011): 219–41.

67 Social medicine was among the topics most frequently debated in libertarian athenaeums: Anna Monjo, *Militants. Participació i democracia a la CNT als anys trenta* (Barcelona: Laertes, 2003), p. 380.

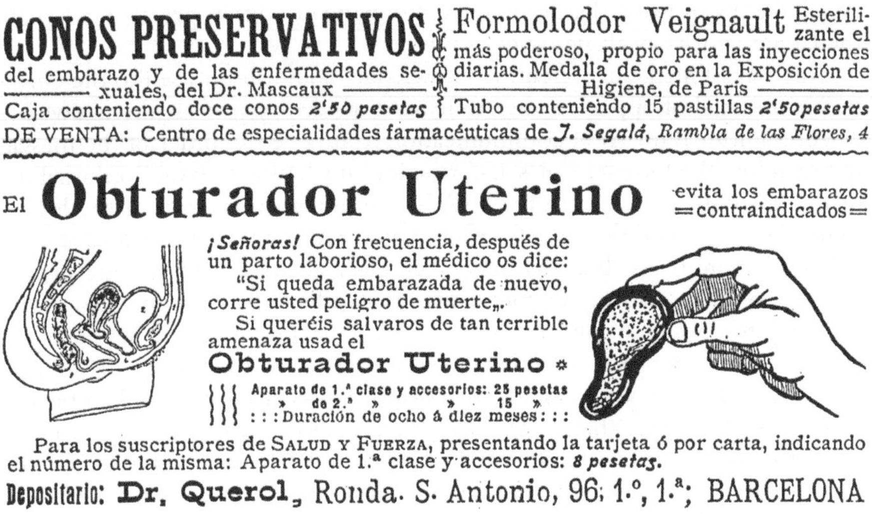

Figure 6.4 Revolution, technology and sexual reform in Barcelona. Contraceptive device's advert (1905).

if having the opportunity to address them again, anarchists would expect to 'listen for those aforementioned parts that we deem essential to the character of such seminars'.[68]

As we can see, anarchist–syndicalists had their own ideas about what was 'proper' to disseminate given the 'character' of the seminars. This shows that the libertarian audiences were not prone to passively accepting hegemonic knowledge. They rather had a truly different interpretation of concepts that, in many cases, had undergone a process of appropriation and development independent of the 'official' theories. Anarchist–syndicalists also maintained discrepancies with intellectuals belonging to other subaltern groups if the positions they held did not fit with libertarian interests. That was the case with naturism, widely popular among subaltern groups of various ideological persuasions at that time. In 1935 José Castro (1890–1980) offered a seminar at the *Casa del Poble* of La Cenia (in Tarragona) under the title 'What and How should Humanity Eat?' The reviewer, J. Balada, affirmed in the *Solidaridad Obrera* that, Castro, 'fully convinced that he is doing good work, *works by means of speaking and writing*, in favour of trophology naturism'. But he criticized the fact that Castro had declared liberty to be a branch of naturism, when, in reality, liberty was, 'the base of all things'. Balada recognized the existence of a multitude of naturist groups and accepted certain coincidences with the ideas they defended but

68 Martín Zabiña, 'Una conferencia. Objeciones', *Solidaridad Obrera*, 1 October 1935.

concluded by insisting that, 'our concept of naturism differs greatly from that of the large majority of those other groups'.[69]

This way of managing scientific knowledge did not limit itself to the correction or rebuttal of opposing ideas to reassert its ideology, but it also allowed the use of traditional forms of communication to reverse the order of the message, transcending the functions of 'self-formation' and 'cultural extension' attributed to the anarchist cultural centres.[70] We are thinking on the use of the lecturer himself as a mean of propaganda. In 1916, José Salvat wrote in *Solidaridad Obrera* about a lecture on syphilis that had taken place at the Labour Centre in Mercaders Street, Barcelona. He complained that it had 'made an impression only on the conscious men', already aware of the social problem represented by this disease. However, Salvat underlined that the event had not been in vain. The lecturer, Dr Marsal, was satisfied with how well received his talk had been by a 'nucleus of workers that, with outstanding consideration, were able to show respect to people who, uninterestedly, helped instruct the working class'. The recipient of the message were not the workers coming to listen or learn, but the lecturer himself, in this case a doctor who had unintentionally turned into an instrument of communication supporting the interests of anarchism. This was the core idea behind Salvat's proposal to organize this type of events *en masse*:

> [...] this will result into more lecturers and a more solid relation with doctors, lawyers and other men representing different fields of knowledge. *Studying our way of acting and thinking*, they will approve with their presence the cultural work accomplished in the Labour and Social Studies Centres, refuting the bureaucracy whose only tendency is the devaluation of our work through defamation, portraying us to uneducated and unaware people as criminals who shall be fought.[71]

This use of scientific knowledge supports the image shown by Felipe Alaiz in 1933 in *Solidaridad Obrera*. According to it, 'the majority of technicians, professors, journalists, engineers and doctors' did not know the full works of anarchist thinkers because they lived in the 'world of the books', whose academic contents portrayed anarchism as a utopia, and considered the 'free association and assistance outside the State absurd'. However, it was contradictory, according to Alaiz, that the workers looked at holders of university degrees with fear, even when 'he had read less than them'.[72] The practices and strategies of communication about the spread of scientific knowledge, in the cases that we have presented, would act like mechanisms of social inclusion/exclusion according

69 J. Balada, 'Propaganda naturista', *Solidaridad Obrera*, 29 March 1935. The italics are ours.

70 Javier Navarro Navarro, 'El papel de los ateneos en la cultura y la sociabilidad libertarias (1931–1939): algunas reflexiones', *Cercles*, 8 (2005): 64–103, p. 95.

71 José Salvat, 'El Dr. Marsal en el Centro obrero', *Solidaridad Obrera*, 27 December 1916. The italics are ours.

72 [Felipe Aláiz], 'La cuestión de los intelectuales', *Solidaridad Obrera*, 2 November 1933.

to the interests of the parties involved.[73] For Alaiz, this contradiction could be explained by the fact that 'a graduate is enclosed in a system'. This 'system', as clearly stated in his article, would be a mechanism of the state to legitimize the expert's superiority regardless of their knowledge. It would coincide with the 'bureaucracy' previously denounced by Salvat, contributing to the spread of a false image of the anarchists.

Holding a university degree legitimized the holder with scientific authority against the body of knowledge that a manual self-taught worker could have accumulated. Denouncements of this situation were employed in anarchist media to highlight the contradictions of the capitalist system, especially when libertarian cultural centres were shut down, accused of training violent revolutionaries. In 1933, in the midst of political repression against the libertarian cultural centres, *Solidaridad Obrera* claimed that, 'the social level of the proletariat was the result of the work of syndicates and athenaeums, not of the State', and that it was frequently ignored that, 'every one of the men of the athenaeums knows more about hygiene, physiculture and heliotherapy than ninety per cent of rural teachers'.[74] In fact, individuals educated and trained for the good of the community, possessing a cultural level similar or superior to the graduates, were usually represented as criminals. Workers conscious of this problem felt rage and indignation. In *Solidaridad Obrera* it was related how a mailman asked an inspector at 'Patria', a benefit society for the sick, how he could be compensated for the sickness and death of his wife. The doctor of the society had concluded that the wife had died of a stomach ulcer and that, since it was a chronic illness, it did not merit any economic compensation. The mailman answered: 'I assure you, Sir, that no sickness can be chronic that did not exist before she fell sick on this occasion, and it is demonstrable that in one month she was both healthy and dead! Is a sickness "chronic" that lasts for one month? That's what the doctors and the Rulebook will say, but I say otherwise'. Faced with this question, the inspector, 'with a strong voice and a scornful expression, at the same time leaving', answered: 'You're a mailman! You have no right to discuss these things!' The denouncement concluded: 'What was this "inspector" trying to say with those intentionally hurled provocations? [...] I tell you from here that I am a "mailman" and I consider myself at least as dignified as him, or more.'[75]

The dignity acquired by manual labour was strengthened and justified, in anarchist ideology, by the cultural dignity procured by the acquisition of technical knowledge. This helps us understand the impact of popular science magazines on the working class. It also explains the need to seek out forms of direct communication, such as the 'Questions and Answers' sections, a format used to elaborate the daily schedule of plenary sessions and congresses of the unions

73 Jorge Molero-Mesa and Isabel Jiménez-Lucena, '(De)legitimizing social, professional and cognitive hierarchies. Scientific knowledge and practice in inclusion-exclusion processes', *Dynamis* 33 (1) (2013): 13–17.

74 'Otra vez los bárbaros. Peligran las escuelas racionalistas', *Solidaridad Obrera*, 30 June 1934.

75 Juan Catalán, 'Patria', *Solidaridad Obrera*, 8 March 1917.

of the CNT.[76] But the recuperation of the real experiences of the various audiences is still a challenge.[77] It is certain that anarcho-syndicalism put questions directly to the working class, and that direct channels of communication were launched. But did these communication practices go beyond the limits of self-taught workers or politically informed militants? Tensions might have been created when anarchists tried to popularize readings of what is 'scientific', 'rational' or 'healthy'.[78] The anarchist propagandists did not work upon a *tabula rasa*, but on individuals and groups with their own cultural traditions. It could not be taken for granted that anarchist's alternative readings of scientific theories would find, always and everywhere, a fertile reception among the fragmented and divided working class of Barcelona.

Conclusion

In 1929 it was clear that the libertarian management of science was deeply affected by the creation of a social abyss within Barcelona. The old alliance between anarchists, federal republicans and freemasons was no longer operational. The geography of communication was modified by the explosive growth of the city and libertarians created a network adapted to the new realities. In 1888 suspicion about the bourgeois character of the science produced by scientists and technicians was emerging. In the 1920s it was almost systemic, reflecting both the class-based orientation of anarchist–syndicalism and the tough reality of a divided city. But this is not the only consideration we should take into account: imaginary spaces are important too. Anarchists sought to mobilize a working class that lived, almost without exception, in a miserable situation. Segregated from the rest of the city, the workers of this 'modern' Barcelona had few reasonable expectations of ascending the social hierarchy. General statistics reflect a partial improvement over time (a decrease in the mortality rate, for example) but the vast majority of immigrants who arrived in the city in the 1920s had, as the sole horizon of their lives, improvised neighbourhoods with horrifying living conditions. Part of the attraction of anarchism resided precisely in its utopian character: it enabled thinking about the possibility of a different world, new and radically distinct from the devastating visible reality. What's more, anarchism puts 'on trial that which currently exists'; it creates a 'contingency of order'.[79] If the real Barcelona was a hostile space, a true prison for workers, anarchism provided an escape route.

76 Tabernero, Jiménez-Lucena and Molero-Mesa, 'Libertarian movement and self-management', p. 51.

77 See: Jonathan Topham, 'Beyond the 'Common Context': The production and reading of the Bridgewater Treatises', *Isis* 89 (2) (1998): 233–62; Jonathan Topham, 'Rethinking the History of Science Popularization/Popular Science', in Faidra Papanelopoulou, Agustí Nieto-Galan, and Enrique Perdiguero (eds), *Popularizing Science and Technology in the European Periphery, 1800–2000* (Aldershot: Ashgate, 2009), 1–10.

78 On social-democratic readings of Darwinism, see: Kelly, *The Descent of Darwin*, pp. 135–41.

79 Paul Ricoeur, *Ideología y utopía* (Barcelona: Gedisa, 1989), pp. 315–6.

The anarchist utopia, like many other utopias, tried to be *rational*. Consequently, science acquired a prominent role. However, the following non-trivial fact must not be forgotten: the recreational character of science. In the science sections of anarchist newspapers and magazines of Barcelona, topics as disparate as malaria, the investigations of Marie Curie and aerial navigation were discussed. Many times they were treated as apparently aseptic objects of science popularization, far from any *immediate* political utility. But they were about something else: apart from contributing to the formation of the self-taught worker, the reading of scientific texts became the ideal type of entertainment. It allegedly took workers away from the stupefying effects of gambling or drinking and was different from other highly questionable *regimes of pleasure*.[80] Not all is about *honest* entertainment. The anarchists spoke of the *right to life*, not only in the sense of a right to the basic means of subsistence but also meaning that society was obliged to offer satisfaction to *all* vital necessities. Scientific popularization is not only a way of returning a good expropriated by the bourgeoisie (that is, science); it is also an essential element for the development of a truly full life. Put that way, popular science usually evoked spaces quite distinct from the gloomy alleys of the run-down areas of the old city, the cotton factories or the neighbourhoods presided over by the ominous shadow of factory chimneys. The reading of scientific texts opened escape routes through imaginary spaces: the infinite cosmos, microscopic organisms and other marvels and mysteries of the universe.

80 Amparo Sánchez Cobos, 'Sociabilidad anarquista y configuración de la identidad obrera en Cuba tras la independencia', in Clara Lida and Pablo Yankelevich (eds), *Cultura y política del anarquismo en España e Iberoamérica* (México: El Colegio de México, 2012), 219–58, p. 230.

7 The city of spirits

Spiritism, feminism and the secularization of urban spaces

Mònica Balltondre and Andrea Graus[1]

In September 1888, more than 2,000 spiritists from around the world gathered for the first time at an International Spiritist Congress. Although Paris was the cradle of spiritism, Barcelona held that first international meeting. The meeting took place in the dance hall of a theatre that had been turned into a conference room, with spiritist mottos, garlands of flowers and international flags hanging from the ceiling. The press was divided between those who reported the event as an intellectual conference and those who considered it a meeting of members of a sect. The local organizers were extremely pleased with the variety of social strata among those present, along with the presence of many women. Both the congress participants and audience illustrate the plural nature of the movement in Barcelona and, as we will argue, the degree to which it engaged the attention of the Barcelonese.

Along with spiritualism, spiritism (in Spanish: espiritismo) refers to the revival in the belief in spirits and the spread of mediumistic activities that took place across much of Europe and the United States, starting from around 1850. In the French and Spanish contexts, spiritism was linked to the influence of the French intellectual and pedagogue Allan Kardec (1804–1869). In the 1860s, Kardec transformed communication with spirits into a more or less well-defined school. This Kardecism was a set of alleged scientific principles that proved the eternal survival of the human spirit through reincarnation. Its moral code combined utopian socialism and eastern philosophical traditions with a new understanding of Christian values based on fraternity and love. Spiritism became highly popular in France, Spain, Italy and some Latin American countries such as Brazil.

Some spiritist practices, such as table-turning or automatic writing, were mostly carried out in a rural setting; but spiritism, as an organized movement, was primarily an urban phenomenon. As Lynn Sharp pointed out, spiritism exemplified 'a mentality that adapted traditional beliefs to a modern, secularizing

1 We would like to thank Andreas Sommer, Ángel González de Pablo and Annette Mülberger for their valuable comments on an earlier draft. We are also grateful to the editors of this volume, Oliver Hochadel and Agustí Nieto-Galan for their helpful suggestions.

age' based in cities.[2] The movement resonated within the urban population to the extent that it could be labelled an 'urban spirituality'. One of its most appealing aspects was that it presented itself as a scientific religion. Followers thought they were researching spiritual realities and they believed their corpus of natural, occult and moral ideas to be grounded on positivistic principles. According to Kardec, since spirits were considered natural entities, spirit communication through a medium was itself a natural phenomenon. Mediumistic accounts of séances could be used to legitimize and validate knowledge on life after death. In spiritist circles, séances had the status of empirical evidence. In Spain, as in France, constructing spiritism as a 'scientific' religious alternative to Roman Catholicism was crucial for the influence and spread of the movement.

There has been much historiographical research into spiritism in France; less on spiritism in Spain. Most work on Spanish spiritism either consists of general overviews or is by scholars who focus on the writing of spiritists.[3] For instance, much work concerns a leader of the movement: Amalia Domingo Soler, but only as a spiritist author.[4] Two exceptions to this trend are the work of Gerard Horta and that of María Dolores Ramos, focused mainly on spiritism in Barcelona.[5] Both scholars situate the movement within its local context, contextualizing their ideas. Our approach here is an attempt to move further in this direction and analyse the movement directly as an urban social actor.

We understand the spiritist movement in Barcelona as a civil association that stands between the state and the family. Treating spiritism as a civil association means identifying it as part of civil society; considering that those involved

2 Lynn L. Sharp, 'Fighting for the afterlife: Spiritists, Catholics, and popular religion in nineteenth-century France', *The Journal of Religious History*, 23 (3) (1999): 282–95, p. 295.

3 For a very good general picture of the movement in Spain see: Lisa Abend, 'Specters of the secular: Spiritism in nineteenth-century Spain', *European History Quarterly*, 34 (4) (2004): 507–34. For the reaction of Spanish physicians to spiritism see: Ángel Gónzalez de Pablo, 'Sobre los inicios del espiritismo en España: La epidemia psíquica de las mesas giratorias de 1853 en la prensa médica', *Asclepio*, 58 (2) (2006): 63–96. For connections between spiritism and psychical research (called 'metapsychics' in Spain), see: Annette Mülberger and Mònica Balltondre, 'Metapsychics in Spain: Acknowledging or questioning the marvellous?', *History of the Human Sciences*, 25 (2) (2012): 108–30.

4 Among others, see: Alicia Cerezo Paredes, '"Cada espíritu es un libro": Spiritualism in turn-of-the-century Spain', *Decimonónica*, 10 (1) (2013): 1–16; Christine Arkinstall, *Spanish Female Writers and the Freethinking Press 1879–1926* (Toronto: University of Toronto Press, 2014), pp. 23–59; Amelina Correa Ramón, 'Librepensamiento y espiritismo en Amalia Domingo Soler, escritora sevillana del siglo XIX', *Archivo Hispalense*, 83 (254) (2000): 75–102.

5 Gerard Horta, *De la mística a les barricades* (Barcelona: Proa, 2001) and *Cos i revolució. L'espiritisme català o les paradoxes de la modernitat* (Barcelona: Edicions de 1984, 2004). Mª Dolores Ramos, 'Heterodoxias religiosas, familias espiritistas y apóstolas laicas a finales del siglo XIX: Amalia Domingo Soler, Belén de Sárraga Hernández', *Historia Social*, 53 (2005): 65–84 and 'La república de las librepensadoras (1890–1914): laicismo, emancipismo, anticlericalismo', *Ayer*, 60 (4) (2005): 45–74. Horta offers a big picture of spiritist practices in Catalonia in relation to other heterodox groups, such as freethinkers, the freemasons and anarchists. Although his work is one of the most complete on Catalan spiritism to date, his approach is more anthropological than historical. Ramos is interested in the feminine side of spiritism.

in the movement contributed their knowledge and actions to the shaping of urban spaces.[6] If we were only to historicize the general ideas and doctrines of spiritists, we would miss the local importance of spiritism, which explains its contributions to and success within local contexts. Furthermore, we would miss the participation of spiritism in the social and political agenda of its spaces of influence.

Our goal is therefore to map spiritists within their urban landscape by looking at the associations they financed and the activities they organized, taking into account that those activities were directed at all the citizens of Barcelona, whether they were spiritists or not. Through their activities, spiritists helped to create a heterodox network of practices and knowledge. In this way, spiritism came to represent a secular alternative for the city of Barcelona. They advocated for a secular state and society, pursuing religious freedom.

Compared to the situation in other cities such as Paris or London, spiritists had a stronger political presence in Barcelona. That was not because they formed a political movement *per se*, but because the urban environment of Barcelona was highly politicized. The spiritist movement in Barcelona found its ultimate meaning and place in the polarized anticlerical battle in *fin-de-siècle* Spain.

The local geography of the movement

Spiritism started as a clandestine movement in Spain, but its supporters were able to organize themselves quickly. Spiritism could be openly defended under the first republican Spanish government in 1873 and throughout the following period known as the Bourbon Restoration (1875–1931). In that parliamentary monarchy, conservative and liberal political parties alternated in government. By then, religion divided Spanish society. Each government encountered difficulties satisfying both the Roman Catholic Church's requests and secular demands. Some liberal and republican politicians and intellectuals made efforts to secularize society, while those defending the Church's interests maintained that Roman Catholicism should continue to regulate civil society.[7] Fundamentalist Catholics defended the notion that family, authority and property were divine rights, not civil rights. In cities such as Barcelona, anticlericalism challenged the supremacy of the Church and its control over the social system and institutions. Within this anticlerical and mainly republican environment, spiritism found allies in collectives such as anarchists, the freemasons and freethinkers (see Chapter 6).

In Catalonia, many spiritist federations came into existence in the last quarter of the nineteenth century. Barcelona became one of the focal points of

6 For discussion of the use of 'civil societies' in the history of science see: Thomas H. Broman, 'Some preliminary considerations on science and civil society', *Osiris*, 17 (2002): 1–21.

7 William James Callahan, *Church, Politics, and Society in Spain, 1750–1874* (Harvard: Harvard University Press, 1984), pp. 248–78.

the movement, especially after a public burning of spiritist books in 1861.[8] Spiritist activity in Barcelona was mostly concentrated in the outlying towns of Gràcia and Sants – both integrated into the city in 1897 – and the part of the old town known today as the Raval. This last area was home to one of the most prominent and active spiritist centres: the Centro Barcelonés de Estudios Psicológicos (Barcelona Centre of Psychological Studies). The Raval had been the first industrialized area of the city, and also embraced traditional manufacturing and artisans. Spiritism was also quite popular in the industrial towns surrounding Barcelona with a politically committed working class, such as Sabadell and Terrassa. Thus, in Catalonia, spiritism became most popular in industrial areas where social networks, working-class communities, and trades unions were already established.

Unlike the scenario in Madrid, where military personnel and low ranks of the nobility were involved, the spiritist movement in Barcelona was led by white-collar workers, such as teachers, notaries and printers, along with the petit bourgeoisie of tradespeople and manufacturers. It also attracted followers from among the more skilled sectors of the working class.[9]

Putting Barcelona on the map of spiritism: The first spiritist congress

As we already mentioned, Barcelona hosted the first international spiritist meeting. In this section, we describe that meeting in order to reconstruct the primary concerns of the local spiritist leaders at the time and discuss the consequences of the congress for the movement. We also depict the audience and the public image of the meeting.

Since Barcelona was celebrating the Universal Exhibition in 1888, the city offered an excellent location to hold an event which many foreigners could attend and which would garner international attention. For the local organizers, it was a great opportunity to make spiritism visible in the city and eventually to turn Barcelona into a reference point for the international spiritist movement.[10] To maximize visibility, they planned conferences open to the public along with private working sessions.

The congress took place between September 8 and 11. With expectations running high, the local committee saw the meeting as an excellent occasion to

8 In 1861, the Bishop of Barcelona Antonio Palau (1806–1862), confiscated more than 300 spiritist books and pamphlets that had been illegally imported by the French librarian Maurice Lachârte (1814–1900), living in exile in Barcelona for political reasons since 1856. The bishop considered the books harmful for Catholicism. On 9 October 1861, the spiritist works were burnt on the Ciutadella esplanade. See Florentino Barrera, *El Auto de fe de Barcelona* (Buenos Aires: Vida infinita, 1980); Horta, *Cos i revolució*, pp. 310–5.

9 For a mapping of skilled and unskilled working-class neighbourhoods, see José Luis Oyón, 'The split of a working-class city: urban space, immigration and anarchism in interwar Barcelona, 1914–1936', *Urban History*, 36 (1) (2009): 86–112.

10 'Acta de la sesión del congreso preparatorio', *La Luz del Porvenir*, 9 (45) (1888): 1–4, p. 2.

promote the doctrine in 'solemn' terms.[11] They insisted that their doctrine and morals were based on scientific principles and that they were not an obscure group playing evil tricks or trying to communicate with spirits for fun. The organizers expected great men of science to attend, such as the astronomer and spiritist Camille Flammarion (1842–1925) and the chemist and psychical researcher William Crookes (1832–1919). Those men would surely have helped to popularize and lend credibility to spiritism, but they did not attend. Nevertheless, spiritist leaders from France and Italy, such as Kardec's successor Pierre-Gaëtan Leymarie (1827–1901), joined the executive committee and gave talks during the congress. Foreign spiritist delegates and adherents represented more than 10 countries, from Cuba to Russia.[12] The prominent attendance of Latin American delegates led some conservative newspapers ironically to point out that spiritists finally had managed to communicate with inhabitants of the 'other world'.[13]

The idea of celebrating a spiritist congress while an International Exhibition was taking place had been entertained by the Centro Barcelonés for a long time. In 1875, its members tried to convince American spiritualists to incorporate a spiritist meeting into the International Exhibition in Philadelphia. According to the viscount Antonio Torres-Solanot (1840–1902), then president of the Spanish Spiritist Society, spiritism should be placed in the tenth category of the exhibition, which contained 'objects illustrating the efforts made to ameliorate the physical, intellectual, and moral condition of the human being'.[14] His proposal illustrates the spiritists' conviction that their ideas would help the social and moral progress of humanity. It also shows that, in comparison to spiritualism, spiritism was more concerned with transforming societies. The initiative did not succeed; either in Philadelphia or in Barcelona, where the congress did take place during the 1888 Universal Exhibition but not *within* its programme.

The Barcelona spiritists had difficulties finding a place to hold the congress; the city was full of events due to the Universal Exhibition. Finally, they managed to find a suitable venue in the Salón Eslava: one of the city's quite large theatres. It held over 2,000 people and was located close to the grounds of the Universal Exhibition. Thus the spiritists certainly hoped to attract a large public.[15] The theatre was a functioning dance hall and, each afternoon, the organizers had to quickly transform it into a conference room. They wanted to make the place look solemn but also maintain enthusiasm for the event. Therefore,

11 Antonio Torres-Solanot in *Primer Congreso Internacional Espiritista. Representaciones, adhesiones, sesiones públicas, sesiones privadas, conclusiones, documentos, etc. Reseña completa* (Barcelona: Imprenta Daniel Cortezo y Cª, 1888), p. 80.

12 Those countries were France, Italy, Belgium, Russia, Rumania, the United States of America, Chile, Peru, Mexico, Argentina, Cuba, Venezuela and Puerto Rico. *Primer Congreso Internacional Espiritista*, pp. 65–71.

13 *La Dinastía*, 12 August 1888, 1.

14 Torres-Solanot in *Primer Congreso Internacional Espiritista*, p. 51. Italics in the original.

15 *Primer Congreso Internacional Espiritista*, p. 58.

Figure 7.1 First International Spiritist Congress (1888).

it was decided the decoration would include all kinds of spiritist propaganda. Placed on the main platform, Kardec's bust was adorned with laurels. The flags of different countries and flower garlands hung from the ceiling (see Figure 7.1). Spiritist mottos could be read on shields and wall hangings distributed all about the room, such as: 'In immobile eternity, Spirits subsist, while material beings pass away'.[16] Public sessions started at 9 pm and lasted until almost midnight. This schedule allowed the attendance of the working class; the audience the spiritists were most eager to reach.

As the historian John Warne Monroe has shown, after Kardec's death and especially from the 1880s on, spiritism split into two currents: one that emphasized the moral doctrine of spiritism, while the other highlighted its scientific character.[17] In France, the philosopher Léon Denis (1846–1927) represented the tendency that prioritized the moral side of spiritism, while the engineer Gabriel Delanne (1857–1926) was interested in promoting a positivist, scientific approach to spiritist phenomena.[18] Both Denis and Delanne attended the event in Barcelona, where the confrontation between the moral and the scientific aspects of spiritism was present. As a follower of Delanne's positivist approach, Torres-Solanot handed out Delanne's book *Le spiritisme devant la science* [Spiritism in the face of science] (1885) to the journalists covering the event. Yet the moral side of Kardec's doctrine was the more prominent at the congress.

16 *Primer Congreso Internacional Espiritista*, p. 60.
17 John Warne Monroe, *Laboratories of Faith. Mesmerism, Spiritism, and Occultism in Modern France* (Ithaca/London: Cornell University Press, 2008), pp. 226–7.
18 Sophie Lachapelle, *Investigating the Supernatural: From Spiritism and Occultism to Psychical Research and Metapsychics in France, 1853–1931* (Baltimore: Johns Hopkins University Press, 2011), pp. 30–52.

General non-expert audiences, the speakers and their communications – mostly centred on the moral lessons offered by spiritism – and the room decor – brimming with consoling spiritist mottos – set the tone of the meetings. In a society where social injustice was a tangible reality, highlighting the moral and humanitarian aspects of spiritism worked better, as propaganda, than emphasizing its 'scientific' features. It is important to note that during the meetings there were no displays of spectacular phenomena or demonstrations of mediumship similar to the shows performed by stage mediums. It was not the aim of the congress to stun the audience with séances. Speakers tried to convince the public of the benefits of the doctrine only through rhetoric and praise.

In terms of attendance, the meetings were a success. The theatre was packed every day, even exceeding the 2,000-person limit for the closing session.[19] The organizers were extremely pleased with the range of social classes represented by those who attended: from labourers to university professors and physicians. They interpreted this diversity as complementing the decoration of the room by emphasizing the social equality sought by spiritism.[20] While they described their achievement in terms of class egalitarianism, local newspapers highlighted the high attendance by women.

Historians generally agree that most followers of the spiritist and spiritualist movements were women, although almost all of the leaders were men.[21] Critics of spiritism at the time took advantage of its predominance among women to question the rational character that Kardec claimed his doctrine had. From this point of view, women's involvement in spiritism was attributed to their penchant for the marvellous and for mysticism. Kardec himself regarded this tendency as an obstacle for his doctrine.[22] As we argue in the next section, not only did many women share the spiritist project to change society, but in Barcelona spiritism became a platform for women to express themselves. Accordingly, the predominance of women at the congress should not be interpreted as a result of a feminine penchant for the marvellous. On the contrary, it should be explained by the alliance between spiritism and emerging feminism.

Apart from the praise and reviews of the event in spiritist journals, two general daily newspapers covered the event: *La Vanguardia* – liberal and bourgeois – and *La Publicidad* – republican, anticlerical and representative of the

19 'Noticias varias', *El Día*, 14 February 1888, 2.

20 *Primer Congreso Internacional Espiritista*, pp. 59–60.

21 A gender analysis of spiritualism and spiritism in Alex Owen, *The Darkened Room: Women, Power, and Spiritualism in Late Victorian England* (London: Virago, 1989); Janet Oppenheim, *The Other World: Spiritualism and Psychical Research in England, 1850–1914* (Cambridge: Cambridge University Press, 1985); Nicole Edelman, *Voyantes, guérisseuses et visionnaires en France: 1785–1914* (Paris: Albin Michel, 1995); Ann Braude, *Radical Spirits. Spiritualism and Woman's Rights in Nineteenth-century America* (Bloomington: Indiana University Press, 2001); Lynn L. Sharp, *Secular Spirituality: Reincarnation and Spiritism in Nineteenth-century France* (Lanham: Lexington Books, 2006) and Heather Wolffram, *The Stepchildren of Science. Psychical Research and Parapsychology in Germany, c. 1870–1939* (Amsterdam/New York: Rodopi, 2009).

22 Monroe, *Laboratories of Faith*, p. 120.

moderate Catalan left-wing.[23] In *La Vanguardia* the journalist kept his distance but reported the event with respect and treated it as an intellectual one.[24] Meanwhile *La Publicidad* looked down on the spiritists with contempt. It denounced the congress as 'an act of propaganda of a comforting doctrine', assuming that spiritism was a religious sect.[25] Yet *La Publicidad* did publish some spiritist diatribes against the Roman Catholic Church.

In conclusion, the congress helped to put Barcelona on the map of spiritism. At the national level, Barcelona was consolidated as the spiritist capital of Spain, confirming the Centro Barcelonés de Estudios Psicológicos as the movement's leading Spanish institution. Not surprisingly, most Spanish representatives at future international spiritist congresses would be linked to that centre. At the international level, the congress contributed to strengthening the connections between Barcelona and Paris, and to encouraging an intercontinental collaborative network of spiritist propaganda with Latin America.[26] French spiritists held a meeting the following year, at which they commemorated the congress in Barcelona.

Public sites for women

The first civil associations of women which formed around republican and left-wing parties are not well known in Spain. Until recently our historical understanding of early Spanish feminism was inaccurate because scholars applied Anglo-American interpretative frameworks to the Spanish context. As Mary Nash argues, Spanish historiography has tended to portray historical feminism from the perspective of liberal struggles based on the notions of individual political rights, equality and individual freedom; a framework that cannot be applied directly to Mediterranean Europe.[27] At first, feminism in Spain did not focus on demands for women's suffrage or other political rights because the political context was not conducive to such an orientation.

Our research led us to identify, as other scholars do, the first women's civic associations in Catalonia as those that fought the domination of the male clergy.[28]

23 *La Vanguardia*, 9 September 1888, 2; 10 September 1888, 1–2; 11 September 1888, 2–3; 21 September 1888, 1 and *La Publicidad*, 10 September 1888, 1.

24 It was a policy of the newspaper to report scientific and philosophical news with an impartial tone. See Matiana González-Silva and Néstor Herran, 'Ideology, elitism and social commitment: Alternative images of science in two *fin de siècle* Barcelona newspapers', *Centaurus*, 51 (2) (2009): 97–115.

25 *La Publicidad*, 10 September 1888, 1.

26 By the beginning of the twentieth century, strong connections with Latin America can easily be seen by looking at some Barcelona-based spiritist journals. For instance, *Luz y Unión* managed a network of correspondents who reported from Argentina, the Dominican Republic, Brazil, Cuba, El Salvador, Costa Rica and Nicaragua. The congress enhanced this existent intercontinental spiritist network.

27 Mary Nash, 'Political culture, Catalan nationalism, and the Women's movement in early Twentieth Century Spain', *Women's Studies International Forum*, 19 (1–2) (1996): 45–54, pp. 45–7.

28 This connection has also been pointed out by Horta and Ramos (see footnote 5) and especially by Pere Sánchez Ferré, 'Els orígens del feminisme a Catalunya: 1870–1926', *L'Avenç*, 222 and 223 (1998): 8–13; 6–11.

That was the context that could bring together an important sector of women in Barcelona at the end of the nineteenth century. When Teresa Claramunt,[29] an anarchist, tried to form a women's workers association, she found no support among female workers.[30] It seems they were not prepared to organize themselves autonomously before 1900. Claramunt had to collaborate with an organization that was neither of women workers nor anarchist: the Sociedad Autónoma de Mujeres. That society, which espoused feminism and anticlerical thought, was born in the bosom of the spiritist movement. It offered a social setting in which women of different political tendencies could come together; not workers but republicans and thinkers on the left. It attracted lower-middle-class women who wanted to fight for women's educational rights as a way to combat the domination of the church over women.

In this section, we detail the sites of action of this women's league and its struggle for a more emancipated female citizenship in relation to the spiritist movement. From an urban perspective, we account for their very public presence in the city and the male boundaries they broke.

The spiritists were no more radical in defending egalitarianism and women's rights than the freemasons or the freethinkers in their discourses; but it is a matter of fact that spiritist societies were more equal in terms of participation.[31] Séances and working groups were open to both men and women from the beginning of the movement, while it was a controversial issue in freemasonry and not at all a common practice among trade unions.[32] Spiritism offered a space where men and women worked together for common goals. Yet in this space women could also experiment with positions of authority and power that were denied them in society at large.[33] The empowerment that spiritism afforded Amalia Domingo, tangible in the history of her life, is a good example, as we will see below. Within the movement, women could also acquire a

29 Teresa Claramunt (1862–1932) was a textile worker. The role that Spanish women played in the initial labour rights demands has scarcely been researched. After her death, she was buried in Amalia Domingo's niche in the civil graveyard at Montjuïc. See Maria Amalia Pradas Baena, *Teresa Claramunt, la virgen roja barcelonesa* (Barcelona: Virus, 2006), p. 93.

30 Laura Vicenta Villanueva, 'Teresa Claramunt (1862–1931) propagandista i activista llibertària', *Arraona. Revista d'història*, 28 (2004): 144–63, p. 151.

31 According to the spiritist doctrine, everybody could be reincarnated as a man or woman indistinctly. Perhaps that explains why male spiritists acted in more egalitarian ways than other freethinkers. See Nicole Edelman, 'Spiritisme et politique', *Revue d'histoire du XIXe siècle*, 28 (2004): 149–61, p. 152.

32 Several papers discuss the participation of women in freemason societies. On Hispanic contexts see: Pere Sánchez Ferré, 'Mujer, feminismo y masonería en la Cataluña urbana de la Restauración', in José Antonio Ferrer Benimeli (ed.), *Masonería, política y sociedad*. Vol. 2 (Zaragoza: Centro de Estudios Históricos de la Masonería Española, 1989), pp. 929–45. María José Lacalzada de Mateo, 'La mitad femenina "para" la masonería y "en" masonería (1868–1936): balance y perspectivas', *Investigaciones históricas: Época moderna y contemporánea*, 23 (2003): 117–39.

33 Marco Pasi, 'The Modernity of Occultism: Reflections on Some Crucial Aspects' in Wouter J. Hanegraaff and Joyce Pijnenburg (eds), *Hermes in the Academy: Ten Years' Study of Western Esotericism at the University of Amsterdam* (Amsterdam: Amsterdam University Press, 2009), 59–74.

collective sense of themselves as citizens, so they could become participants in the public sphere of the city.[34]

However, spiritist groups could not completely escape from the reproduction of gendered roles. The Centro Barcelonés, for instance, had a female philanthropic group that supervised weekly donations of food and clothes to poor people. In the nineteenth century, this type of presiding over charitable acts was part of the role of upper-class women. So the spiritists were reproducing that traditional image of the woman as a 'charitable' being. Female spiritists maintained this gendered class identity.

The urban life of a female medium

The case of Amalia Domingo (1835–1909) is highly illustrative of the engagement of some spiritist women with the doctrine. In fact, she was one of the local leaders of spiritism in Barcelona. She is also an example of one type of 'mediumship' bound to the political and propaganda enterprises of the movement and not to the production of spiritist phenomena.[35] She considered herself a medium as she could anticipate things, saw the *périsprit*,[36] wrote under the inspiration of spirits and had private communications with them, but she never participated as a medium in a séance. She was not a medium of séances or one who could be used for spiritist demonstrations. Yet, she used the products of private communication with the other world to publicize the moral principles of spiritism. Moreover, like other women in close contact with the spiritist milieu, she also fought for women's rights. She supported the new feminism in Barcelona city, although she is hardly known for that.

Amalia Domingo was born in Seville, to an unknown father and her mother died when she was young.[37] Due to her bad eyesight and lack of family, neighbours suggested she marry or join a convent. Instead, she went to Madrid and tried to make a living writing poems and sewing. After two years of searching for spiritual solace in the Roman Catholic and Protestant Churches – living off their alms partly – she came into contact with spiritism. One of the doctors who treated her partial blindness altruistically gave her a copy of *El Criterio*

34 See also this assumption in Pere Sánchez Ferré, *La maçoneria a Catalunya (1868–1947)*. Vol. 1 (Premià de Mar: Clavell Cultura, 2008), p. 172.

35 Amalia Domingo's female leadership can be compared to that of Emma Hardinge Britten (1823–1899), the English-born medium and famous female leader of modern spiritualism in the United States. In the 1860s, Britten gave up séances to focus on giving public lectures on spiritual matters. She was politically active and supported Abraham Lincoln's re-election in 1864. She was also one of the founders of the Theosophical Society. See: Jeffrey D. Lavoie, *The Theosophical Society: The History of a Spiritualist Movement* (Florida: Brown Walker Press, 2012), pp. 294–315.

36 In the spiritist doctrine, the human being is understood as a triple union between spirit, *périsprit* and body. Kardec defined the *périsprit* as a fluidic element that united the two other substances, stored one's personality and survived after the death of the body.

37 On her life, see her memoirs: Amalia Domingo Soler, *Memorias de una mujer* (Barcelona: Amelia Boudet, 1990).

Espiritista, a spiritist journal which finally provided her with spiritual comfort and a platform from which to fight for her socio-political ideals. She began to read Kardec, to attend the meetings of the Spanish Spiritist Society and to write for different spiritist journals. The success of her writing in defence of spiritism allowed her to travel and offered her an opportunity to settle in Gràcia (Barcelona). Lluís Llach, president of the spiritist centre La Buena Nueva, offered her board and lodging in exchange for writing in favour of spiritism. She became a famous propagandist and the spiritist community provided her with a stipend for the rest of her life in return.[38]

Once Amalia Domingo had settled in Gràcia, the spiritist editor Juan Torrents suggested she publish a spiritist journal for women. So she founded the spiritualist newspaper *La Luz del Porvenir* (1879–1900), a weekly journal exclusively written by women and especially addressed to women. In her own words, the aim of the journal was to 'awaken [moral] feelings'[39] by means of tales from spirits with love as a conquering hero, probably thinking of her female audience. Nevertheless, the voice of Amalia Domingo denouncing women's oppression – more than defending alternative roles – comes through in the tales that spirits told her. She especially condemned female subordination to 'oppressive' husbands and clergymen. As Christine Arkinstall points out, Domingo's writings were 'strategically designed to arouse emotional identification and sympathy for transformative sociopolitical purposes'.[40] Other women who worked for the journal wrote on women's educational rights and claimed a place for women in the social regeneration of Spain. The periodical became a platform for these voices.

Amalia Domingo's journal also built up a supporting economic network among the huge Spanish spiritist family. From time to time, the journal recommended a woman for a job or raised campaign funds to help a widow. In general, spiritist journals were useful platforms where people could ask for, and receive, aid. *La Luz del Porvenir* also received donations that were distributed among the poor.[41] Letters from women in Latin America show that the journal was read there and their subscriptions were also crucial. When the colonial war in Cuba and Puerto Rico erupted in 1898, *La Luz del Porvenir* ended due to a shortage of funds.[42] The journal joined up with another local spiritist publication, *La Unión*, and Amalia Domingo continued to run it.

38 She published countless essays praising spiritism, clarifying some parts of the doctrine and criticizing the Catholic Church; one of her favourite topics. Amalia Domingo's arguments with clergymen who attacked spiritism became famous and were broadly published in republican and freethinking newspapers. See for instance: Domingo, 'Impresiones en la catedral de Barcelona al oír al Padre Sallarés refutar el espiritismo', *La Luz del Porvenir*, 6 (2) (1884), 9–13; 6 (10), 73–7.

39 Amalia Domingo, *La Luz del Porvenir*, 6 (2) (1884), 10.

40 Arkinstall, *Spanish female writers*, p. 36.

41 For instance, between 1881 and 1891, 9,272 'pesetas' were given as alms thanks to the journal. Domingo, *Memorias de una mujer*, p. 242.

42 Pere Sánchez Ferré, 'Los neoespiritualismos ante la crisis española de entresiglos. Espiritismo y teosofía' in Ferrer Benimeli (ed.), *La masonería española y la crisis colonial del 98*. Vol. I (Zaragoza: Centro de Estudios Históricos de la Masonería Española, 1999), 3–20.

Figure 7.2 Amalia Domingo's funeral parade leaving her home in Gràcia neighbourhood (1909).

The day of Amalia Domingo's funeral in 1909 a parade was organized (see Figure 7.2). Apart from spiritists, representatives of secular female associations marched alongside theosophists, freethinkers and other advocates of secular education such as members of the Institución Libre de Enseñanza.[43] This sizeable

43 'Funeral', *Luz y Union*, 10 (5) (1909): 136.

and varied participation shows the multiple audiences she connected with, since she formed part of the hub of initiatives that fought for the secularization of the city.

Spiritists and the secular women's league

In 1890, Amalia Domingo participated in the founding of a society that was representative of early secular feminism in Barcelona: the Sociedad Autónoma de Mujeres (Autonomous Women's Society), which in 1892 became the Sociedad Progresiva Femenina (Progressive Feminine Society). The two other founders of this feminist society were the anarchist Teresa Claramunt and a freemason and freethinker, Ángeles López de Ayala.

The society was the first left-wing women's association in Barcelona. Initially located in Gràcia, where Domingo and López de Ayala lived, the headquarters was later established on Ferlandina Street, where the Centro Barcelonés de Estudios Psicológicos was also located. The spiritists gave them physical and symbolic space within their ranks. For more than a decade, the Sociedad Progresiva Femenina and the Centro Barcelonés de Estudios Psicológicos worked hand in hand to promote the social values they shared. The propaganda committees of these institutions organized weekly meetings to plan civil and charity-oriented activities to recruit followers, both for spiritism and for feminism.[44] The struggle against illiteracy and for secular education for both genders justified their alliance.

Those women were not a homogeneous group, however. Some of them, such as López de Ayala and Claramunt, embraced anarchist ideas and were more radical in their approaches; meanwhile Domingo and other female spiritists identified themselves as apolitical.[45] Not all of them believed in spirits, but they all wanted a secular society and shared basic claims for women in the fields of education and labour rights.[46] The feminism they promoted did not advocate female suffrage or women's economic emancipation, although they believed in the equal intellectual capacities of women. Those specific aims were not taken up in Spain until the 1930s.

Regarding educational goals, they supported two types of schools, as did male spiritists. On the one hand, they offered free instruction to children and workers in their centres by both male and female spiritist teachers. On the other hand, they contributed to the funding of private secular schools, as did freethinkers and others who supported a 'rationalist' education.[47] One of the

44 'Agrupaciones', *Luz y Union*, 2 (9) (1900): 94.

45 In the freethinker press *Las Dominicales del Libre Pensamiento* appears Amalia Domingo's regret of the combative positions of Ángeles López de Ayala, despite they were friends. *Las Dominicales*, 24 February 1898, 4.

46 See the works of María Dolores Ramos cited in footnote 5.

47 Secular schools were based on libertarian educational principles such as coeducation. The best known was Ferrer Guardia's 'Modern School'.

most ambitious of the women's projects was the creation of the Fomento de la Instrucción Libre (Promoting Free Instruction), in Sant Pau Street, 31 (Raval) where working women could attend evening classes for free.[48] Another initiative was organized in the circle La Buena Nueva (Gràcia) where Belén de Sagarra, a rationalist, freemason and spiritist teacher, gave lessons to both boys and girls. The school was financed by a spiritist, but had to close when the benefactor died.[49] The lack of continuity of schools that depended on private patrons was a common problem.[50] It was also a common tactic of the spiritists not to call these centres 'spiritist schools' in order to get funding as well as to attract pupils.

Another initiative that shows the commitment of spiritists to female emancipation was the awarding of a prize for 'work and virtue' in 1901. To dignify the value of female workers, the Centro Barcelonés sought to award a young 'honest' woman worker who could prove to be sustaining her family only by means of her labour.[51] In the mid-nineteenth century, a female worker in the Catalan textile industry was often considered as morally degenerate and promiscuous.[52] So the award illustrates the wish to change that image.

The women in the Sociedad Progresiva de Mujeres actually occupied some public male positions by giving public speeches. They spoke not only at the spiritist centres, but also at mass freethinker events in the middle of Barcelona, joining male anarchists and libertarian thinkers on stage.[53] Around 1900, Domingo, Sagarra and López de Ayala gave regular talks at such meetings. They tried to encourage women to join the libertarian cause.[54] Their speeches applauded women's social role in the family as secular educators, and in society as 'social mothers' of secularism within the regeneration framework of Krausism.[55]

A circus on Plaça Catalunya was one of the freethinkers' favourite places for speeches at the end of the nineteenth century. That emblematic square, nowadays the heart of the city, was at the time just an unpaved space, with few buildings around it. Yet it was ideally situated to connect with the intellectual

48 Pradas Baena, *Teresa Claramunt, la virgen roja barcelonesa*, p. 111; Ajuntament de Barcelona, *The Legacy of Working Women in the Historic Old Quarter of Barcelona* (Barcelona: Ajuntament de Barcelona, 2009), p. 44.

49 Domingo, *Memorias de una mujer*, pp. 272–3.

50 *Lumen* (1905), p. 282

51 'De colaboración', *Luz y Unión*, 2 (56) (1901): 566–7.

52 See Mary Nash, 'Identidad cultural de género, discurso de la domesticidad y la definición del trabajo de las mujeres en la España del siglo XIX', in Georges Duby and Michelle Perrot (eds), *Historia de las mujeres en Occidente. Tomo IV* (Madrid: Taurus, 1993), pp. 591–3.

53 Some of their speeches were published in *Las Dominicales* and in Amalia Domingo's journal, which facilitated the circulation of different claims and demands within close but different circles.

54 See, for instance, their participation in *Las Dominicales*, 26 March 1892, 2 and 25 August 1895, 3.

55 The Krausist movement was a philosophical and political trend with liberal and progressive characteristics that spread across Spain in the mid-nineteenth century. Krausism saw the woman as part of the social regeneration of Spanish society through her role as mother. For instance, Krausism advocated the instruction of women in their maternity duties. Elías Díaz García, *La filosofía social del krausismo español* (Madrid: Debate, 1989), pp. 9–53.

bourgeoisie from the upper parts of the city as well as the workers and artisans from the centre of the old town. Male and female freethinkers helped turn the space into a site of social participation and intellectual exchange where people from different social strata and of both sexes intermingled.

Female spiritists also moved into another gendered space. In the Barcelona of 1900 it was not common for women to enter a graveyard during a burial; women could not accompany a coffin to the tomb. When Amalia Domingo expressed her wish to go to the funeral of José Fernández Colavida, known as the 'Spanish Kardec', spiritists discouraged her and only one male friend encouraged her. At the risk of being frowned upon, she entered the cemetery anyway with some other spiritist women and she read a poem to honour the spiritist leader.[56] Her desire was to mourn a friend and erect a monumental tomb to praise spiritism. The result was that those spiritist women broke a gendered Spanish tradition and entered a space that, under those circumstances, had been reserved for men until then.

A year after Amalia Domingo's death, in 1910, the Sociedad Progresiva Femenina led by López de Ayala and in collaboration with two left-wing women's leagues was able to organize a mass demonstration of women. The protest was in favour of a law that prevented the establishing of new religious congregations for two years. Since there had been a march against this law in Madrid, some women in Barcelona wanted to demonstrate that Catholic women were not the only voice in Spain.[57] According to the newspaper *Nuevo Mundo*, 5,000 women marched and handed over 22,000 signatures in support of the law to the civil governor.[58] The demonstration was scheduled for a Sunday, at 4 pm and to be held in Urquinaona square under the slogan 'freedom of conscience'.[59] As Figure 7.3 shows, a significant part of the female citizenship wished not to be identified as Roman Catholic. The photography would have been a good depiction to counteract the image of Spanish women as 'beatas' – 'lay sisters', said in a mockery – which was also very present among male anticlerical groups.[60] However, the picture was not carried by the mainstream daily press. Since historiography has dedicated little space to women taking the leading role in pictures, and even less when they adopt dissent attitudes and lead a public cause, its reproduction here aims to fill that gap.

56 Domingo, *Memorias de una mujer*, pp. 231–2.

57 Luz San Feliu, 'Del laicismo al sufragismo. Marcos conceptuales y estrategias de actuación del feminismo republicano entre los siglos XIX y XX', *Pasado y Memoria. Revista de Historia Contemporánea*, 7 (2008): 59–78.

58 'Manifestación femenina en Barcelona', *Nuevo Mundo*, 27 (862) (1910); Sánchez Ferré, *Els orígens del feminisme*, 223, pp. 8–9.

59 *La Vanguardia*, 10 July 1910, 3; 'Por la libertad de conciencia. La manifestación femenina', *La Publicidad*, 11 July 1910, 2.

60 See María Pilar Salomón, 'Beatas sojuzgadas por el clero: la imagen de las mujeres en el discurso anticlerical en la España del primer tercio del siglo XX', *Feminismo/s*, 2 (2003): 41–58.

Figure 7.3 Anticlerical demonstration of women (1910).

The urban practice of secular charity

With few exceptions, the spiritists considered themselves apolitical. Still, their behaviour in everyday life was highly political. It was not only their public speeches explicitly on secularization that clashed with the church. They questioned mandatory church rituals at a time when the boundary between the religious and the secular was not well defined in Spanish society. Civil burials and civil marriages were encouraged between them, thus challenging the social order. Their civil ceremonies acted as part-ritual, part-propaganda against Roman Catholicism.[61]

61 Ángel Aguarod, 'Consecuencia', *La Revelación*, 32 (4) (1903): 60–1.

The spiritists also engaged in public polemics with some priests, using the media of freethinkers to counteract the discourse certain clergymen gave in popular churches of Barcelona. Spiritism did not intend to be a secret or discreet society as freemasonry was. They tried to forcefully propagate their doctrine and encouraged public discussion of their principles in athenaeums, journals and newspapers.

For a while, groups in Barcelona led publishing on spiritism in Spain. They edited all of Kardec's books in affordable editions and distributed spiritist pamphlets around the city.[62] However, around 1915, they lost the support of the two publishing houses that had mostly printed their publications: Juan Torrents and Carbonell y Esteva. Popularizing ideas through printed works was expensive and publishing houses could not afford the expenses without sufficient sales.[63]

Despite the decline, between the 1880s and World War I, the spiritists were an important group in Barcelona in the struggle to secularize the city. They staged rallies in libertarian athenaeums and in theatres. At least twice a month, spiritist societies organized dances, music recitals, theatre performances, poetry evenings, talks and children's entertainment open to everyone. Every year they celebrated and honoured Kardec together with distinguished local spiritists and mediums, taking flowers to the graves of spiritists in the civil cemetery of Montjuïc.

Apart from the means of propaganda already mentioned, some of their members also visited hospitals and prisons to give spiritual relief to the patients and inmates. They also sponsored philanthropic events. They collected money, distributed clothes and organized banquets for poor people. On one occasion, they held a banquet for around 400 poor people. The meal was served by 50 spiritists, who also gave morally uplifting speeches and distributed pamphlets (see Figure 7.4). According to one spiritist source, 2,000 people attended the event, including two choruses invited to enliven the crowd and some members of republican and working-class parties such as Fraternidad Republicana (Republican Fraternity) and Federación Obrera (Workers Federation).[64]

The spiritists tried to attract people through this kind of acts where their moral doctrine was espoused. These charitable acts were an excellent propaganda tool; they reflected the spiritist moral view of life and served as a way for members to improve themselves in accordance with their doctrine. They served the same purpose as charity did, as practised and understood in Roman Catholicism. We call this type of practice 'secular charity' as it was emulating Roman Catholic practices but with anticlerical aims. Such practices reshaped Barcelona's urban space, transforming the city into a place where spiritual and social needs found alternatives to the orthodox ways of addressing them, such as those fostered by the church through its education and care institutions.

62 For instance, in 1895, 8,000 handouts titled 'Interesting for all' were distributed in the streets. 'Boletín del Centro barcelonés de estudios psicológicos', *Revista de Estudios Psicológicos*, 17 (1985): 115.

63 B. Coris, 'Pesimismo ultramarino', *Luz, Unión y Verdad*, 27 (10) (1916): 297.

64 'En honor de Allan Kardec y Miguel Vives', *Luz y Unión*, 11 (6) (1910): 169–78.

Aspecto del salón comedor en el acto de comer los que habían servido la comida á lós pobres.

Figure 7.4 Spiritists after a banquet organized for the poor (1908).

It seems fair to speculate that they did actually succeed in reaching some of those who suffered from spiritual anxieties but rejected the Catholic Church. They transmitted the hope of eternal life in spiritual evolution through reincarnation. The doctrine resonated with the public due to its messages of justice and retribution. The spiritist belief system held that one's current material conditions were a consequence of one's moral behaviour in previous lives. Spiritism became an attractive alternative to Roman Catholicism to the extent that it was able to show that a morally correct life would be rewarded without the guidance of the Roman Catholic Church. As we have already mentioned, spiritists used the term 'science' to lend credibility to their doctrine and give it an appearance of being progressive, modern and positivist.

Catalan spiritists staged all these propaganda events by means of collective funding. They lacked in their circles bankers or rich industrialists to support their associations. Neither did they have a prominent scientist or personality to promote them, although the famous architect Antoni Gaudí (1852–1926) was sympathetic to their ideas in his youth.[65]

In the next section we will deal with a remarkable enterprise patronized by spiritists: a clinic. This project bound the movement to the funding of heterodox therapeutic regimes: the use of magnetism and hypnosis.

65 The young architect Antoni Gaudí, moving in the republican circles of the city, was a friend of spiritists, freemasons and freethinkers. The Masonic and heterodox symbols he used in his architecture are well known before his supposed conversion to Catholicism.

A charity clinic sponsored by spiritists

One of the most innovative ventures that the spiritists sponsored in Barcelona was a clinic that offered hypnotic and magnetic treatments. We analyse this venture to show the spiritists' commitment to providing free health care to all. At the same time, the success of the clinic shows the popularity of heterodox medical practices in the heart of the city.

In 1895, the Barcelona spiritists sponsored a charitable clinic run by two physicians and spiritist sympathizers from Barcelona: Víctor Melcior Farré (1860–1929), who became its director, and José Cembrano Ferrer. The clinic was known as Clínica Hidro-Magnética de Caridad (Hydro-Magnetic Charity Clinic). The only treatments used were hypnotic suggestion, practised by Melcior, and magnetic healing, offered by Cembrano.[66] It was located in a building that belonged to a Barcelona spiritist journal, the *Revista de Estudios Psicológicos* (Journal of Psychological Studies) in the Raval.[67] It was a strategic location from which to offer health care as the neighbourhood was already a 'care' centre of the city with charity and non-charity clinics, hospices, orphanages and a hospital (see Chapter 4). The property also housed a spiritist library and a room for séances. Melcior and Cembrano attended patients at the clinic without charge, six days a week for three hours a day. The clinic closed in 1905, it seems due to Melcior's exhaustion.

Towards the end of the nineteenth century, hypnotism was becoming popular among the Spanish population thanks to hypnosis stage shows. By then, Britain and France had successfully marginalized the practice of hypnosis by quacks and lay healers, demarcating a medical use of hypnotic suggestion. Yet, that was not the case in Germany or Spain.[68] During the First International Spiritist Congress, Barcelona also held an International Medical Congress at which some Spanish physicians discussed how to legitimize hypnosis within the medical field. They complained that the use of hypnotism in spiritist séances discredited its medical uses.[69] Those physicians would not have imagined that a few years later a charitable clinic sponsored by Barcelona spiritists would at the very least contribute to popularizing the medical use of hypnotic suggestion, and may be said to have demarcated the field entirely. Thanks to the success of the clinic, spiritists were able to popularize magnetic and hypnotherapeutic treatments.

66 The distinction between hypnotism and animal magnetism was not recognized by spiritists in general, who considered that the former was just a modern name to refer to magnetism, understood under the mesmeric paradigm (see, for instance, Torres-Solanot in *Primer Congreso Internacional*, p. 9).

67 At number 10, Doctor Dou street; very near to Barcelona's famous central boulevard 'Las Ramblas'.

68 Andrea Graus, 'Hypnosis in Spain (1895–1905): From spectacle to medical treatment of mediumship', *Studies in History and Philosophy of Science, Part C: Studies in History and Philosophy of Biological and Biomedical Sciences*, 48 (2014): 85–93; Heather Wolffram '"An object of vulgar curiosity": Legitimizing medical hypnosis in imperial Germany', *Journal of the History of Medicine and Allied Sciences*, 67 (1) (2012): 149–76.

69 'Congreso médico', *La Exposición*, 67 (1889): 6.

Hypnotic suggestion was not seen by ordinary citizens as a medical cure when the clinic started. Melcior himself mentioned the difficulties they had to face in gaining the trust of the public at the beginning. But it seems that both magnetic and hypnotic treatments worked well enough to make the clinic a success. During the ten years of its existence, the clinic treated an average of 20 patients per day.[70] Considering that it opened six days a week that means that they received and treated nearly 500 patients a month. According to Melcior, the patients were initially 'hysteric, superstitious and ignorant people', but soon also 'physicians, soldiers, employees and some lawyers' became interested in the 'new cure'.[71] Most of them were from Barcelona and its surroundings, but thanks to the propaganda in spiritist journals, letters from Latin America also reached the physicians asking for long-distance cures and written suggestions.[72] The number of patients grew so fast that long queues formed in the street. Once, the civil governor of the province even had to send an officer to prevent public disorder. Local authorities also came to inspect the clinic but could not close it. As the clinic was run by two formal graduate physicians, they could not accuse them of irregular practice or charlatanism. The spiritists did not miss the opportunity to celebrate this recognition by the political authorities.[73]

Around 1890 hypnotherapy was only accessible to a wealthy segment of Spanish society. As it lacked scientific legitimization in Spain, it was only practised in a few private clinics and asylums. This fact made the Hydro Magnetic Charity Clinic more valuable in the eyes of the spiritists. By financing such a charitable health institution, where no drugs or expensive medical instruments had to be paid for, the spiritists saw a chance to vindicate themselves as public benefactors. The governing board of the *Revista de Estudios Psicológicos* was very proud of their decision to finance the project. Even though it did not offer spiritist therapies, they considered it a superb way to promote spiritism.[74] According to them, the efficiency of the hypnotic and magnetic treatments testified to the truth of their own doctrine. From their point of view, the success of such a charitable institution proved their commitment to equality and scientific progress. Moreover, by financing the Hydro-Magnetic Charity Clinic, the spiritists attributed to themselves the role of introducers of magnetic and hypnotic medical procedures into Barcelona.[75] They considered that these therapeutic techniques were at the forefront of medical knowledge and would soon be adopted by all physicians.

70 In 1895, the journal was asking other physicians to join the project. A spiritist manufacturer was also asked to construct a mesmeric bucket for collective treatment. 'Clínica hidro-magnética', *Revista de Estudios Psicológicos*, 17 (1895): 145.

71 Víctor Melcior, *¿Puede considerarse la voluntad como una fuerza medicatriz?* (Madrid: Establecimiento Tipográfico de Enrique Teodoro, 1908), p. 88.

72 'Notas clínicas del consultorio hidro-magnético', *Revista de Estudios Psicológicos*, 17 (1875), 125. Agustín María Ramos, 'Clínica. Cartas', *Revista de Estudios Psicológicos*, 17 (1897): 125.

73 'Clínica hidro-magnética', *Revista de Estudios Psicológicos*, 17 (1895): 210.

74 'La Clínica Hidro-Magnética', *Revista de Estudios Psicológicos*, 17 (1895): 114.

75 'Clínica hidro-magnética', *Revista de Estudios Psicológicos*, 17 (1895): 182.

Pursuing that end, Melcior offered public lectures to show how hypnotherapy worked. He presented fully recovered patients to large audiences, mostly formed of spiritists but also with some sceptic physicians and pharmacists present.[76] He explained the cases in clinical terms, naming the hypnotic suggestions given to each patient. Hence, the charitable clinic became one of the few sites in Barcelona were physicians could learn about medical hypnosis and see the results in former patients. For Melcior himself, the clinic was also a place for learning through practice, as hypnotherapy was not taught at Spanish universities.[77]

In conclusion, by sponsoring the Charity Clinic, Barcelona spiritists helped to legitimize the medical use of hypnosis to the population and some physicians. The long queues outside the clinic and the attendance of some white-collar professionals looking for help, as reported by Melcior, prove the success of this project among people from different social strata, many of them non-spiritists. By offering free health care, the spiritists were able to present themselves as independent public benefactors.

Conclusions

Urban spaces contribute to the construction of social organizations and the shaping of civil society, which in turn transform cities. In part, the formation of social classes is a spatial process,[78] and so is the formation of civil associations such as the spiritists. If we consider the city of Barcelona as the locus of production of the local spiritist movement, we can analyse the specific practices that promoted the secularization of the city. As organized actors in the city, they contributed to the urban 'marketplace' of ideas as well as to transform the urban space.

In this chapter we have seen different initiatives related to the spiritists' urban areas of influence, mainly in the Raval and Gràcia. These two quarters, not coincidentally, were traditional working-class neighbourhoods. Rather than acting as a sectarian movement, the spiritists aimed to offer citizens of Barcelona alternatives to the dominant Roman Catholic social and health care institutions, regardless of their creed or ideological orientation. Their philanthropy through secular charity was especially directed at a new type of urban poverty resulting from industrialization. This new urban sociological context provided an excellent setting for the development of the movement.

76 Víctor Melcior, 'Notas clínicas del consultorio médico hidro-magnético', *Revista de Estudios Psicológicos*, 17 (1895): 126–7.

77 Antonio Diéguez Gómez, 'Hipnotismo y medicina mental en la España del siglo XIX', in Luis Montiel and Ángel González de Pablo (eds), *En ningún lugar, en parte alguna: Estudios sobre la historia del magnetismo animal y del hipnotismo* (Madrid: Frenia, 2003), 197–228.

78 José Luis Oyón, *La quiebra de la ciudad popular. Espacio urbano, inmigración y anarquismo en la Barcelona de entreguerras, 1914–1936* (Barcelona: El Serbal, 2008).

Our research shows that, in Barcelona, the process of the secularization of society was not led by academics or politicians. Instead, it was in the hands of civil society. The spiritists were generating and reproducing anticlerical thought and, in less explicit ways, republican politics. Under the flag of being a 'modern' religion grounded on leading scientific theories that had nothing to do with Roman Catholicism, they contributed to anticlerical discourse.

In accordance with their aim of secularizing society, the spiritists of Barcelona promoted secular schools and supported secular laws in parliament. They also sponsored an alternative health care project: the Hydro-Magnetic Charity Clinic. By sponsoring charitable projects that were directed at the general population, spiritists were able to present themselves as public benefactors in the city. Although most of the initiatives did not involve spiritist practices, such as séances, the spiritists perceived them as a practical means of promoting their own movement, since the activities promoted the secular and progressive values defended by them.

In material terms, they literally transformed an urban space that remains with us nowadays: the graveyard. Spiritists fought for a secular space in cemeteries and financed mausoleums for their local leaders that still embellish the civil cemetery of Montjuïc. Today, Amalia Domingo's grave is still a site of devotion.

Finally, as we have argued, the first women's civil association was born in the bosom of the spiritist movement. The spiritists enabled the empowerment of women as there was more equality in their societies. Thanks to our urban approach, we have also shown how some women spiritists and freethinkers entered men's public realms in Barcelona – for instance, Amalia Domingo's attendance at a burial or the speeches by women in Plaça Catalunya. Even if those women were not taking over those sites for feminist reasons, they did open the spaces up to women.

8 Anatomy of an urban underworld

A medical geography of the Barrio Chino

Alfons Zarzoso and José Pardo-Tomás[1]

In 1932 a report was published under the title 'El Barrio Chino de Barcelona (Distrito V)' amongst the pages of the magazine *A.C. Documentos de Actividad Contemporánea*.[2] The condemnation was radical indeed, on both an architectural and urban level, denouncing the deplorable state of the streets and dwellings of this infamous part of the city. Symptomatically, the unsigned text was written shot through with medico-sanitary terms and metaphors: 'prophylaxis', 'the cancer of Barcelona', 'a focus of infection', 'degenerated populace' and so on. The article climaxed in a less-than-subtle hint to the authorities of the city that they would do well to 'find interesting the suggestions that, for this preventative purpose, resulted from the Architect's Congress which had taken place in Barcelona'.[3] Indeed, as the same magazine had announced in its previous issue, Barcelona had hosted the preparatory meeting of the congress on urbanism that was to be held in Moscow.[4]

One of the pictures published on the occasion of this conference (see Figure 8.1, left) shows the President of Catalonia giving an official reception, in one of the Gothic halls of the Palau de la Generalitat, to the members of the Comité international pour la réalisation de l'architecture contemporaine (CIRPAC) and its Spanish offspring, the aforementioned GATEPAC (see Figure 8.1, left). Francesc Macià (1859–1933), the elderly gentleman in the centre of the photograph, was then the venerable and respected figure who embodied the republican Catalan government, after years of frequently clandestine

1 We would like to thank Diego Armus and Joan Ramon Resina for their valuable comments on an earlier draft of this chapter.

2 Antonio Pizza and Josep Maria Rovira, *GATCPAC. Una nova arquitectura per a una nova Ciutat 1929–1939* (Barcelona: COAC Publicacions, 2007). *A.C. Documentos de Actividad Contemporánea* was the magazine in which the Grupo de Artistas y Técnicos Españoles para el Progreso de la Arquitectura Contemporánea (GATEPAC) articulated their ideas.

3 'El Barrio Chino de Barcelona (Distrito V)', *A.C. Documentos de Actividad Contemporánea*, 6 (1932): 31–3, quote p. 33.

4 'Congresos Internacionales de Arquitectura Moderna. Internationale Kongrese für Neues Bauen. Preparatory meeting for the Moscow Congress on Urban Planning. Barcelona, 29, 30 and 31 March 1932', *A.C. Documentos de Actividad Contemporánea*, 5 (1932): 38–40.

Figure 8.1 On the left, the reception of the CIRPAC commission (Comité international pour la réalisation de l'architecture contemporaine) at the Palau de la Generalitat (Barcelona, April 1932). Sert is the first from the left and behind him other young architects appear. In the front row, next to him, Le Corbusier is accompanied by the counsellors of the Government of the republican Generalitat, with President Macià (fourth from left) and Mayor Aiguader (sixth from left). On the right, picture of DCA under construction planned by Sert, Torres-Amat and Subirana, *c* 1934.

political struggle. Macià had many years before been a military engineer and active in the building sector.[5]

Sixth from the left is the mayor of Barcelona: Dr Jaume Aiguader (1882–1943). Since the beginning of the century – the doctor had shifted from the libertarian sympathies of his student years towards Catalanist republicanism – Aiguader had emerged as an advocate of social medicine, the post-hygienist medical paradigm which would be adopted by many Catalan doctors.[6] The successful support for social medicine was due, among other things, to the media effort his supporters made to spread their doctrine, beginning with the doctors

5 In 1893, Macià had founded the company *Sociedad Batlle, Macià y Cía.*, which was to construct, amongst others, the first reinforced concrete buildings in Catalonia employing the Monier patent. Two years after its constitution the company had more than 40 projects under way all over Catalonia: Laura Valls, 'El mamut de la Ciutadella (1907): biografia d'un objecte urbà', *Afers*, 30 (80) (2015): 243-68, p. 254. From 1899 on, Macià dissociated himself from the company.

6 To cite two of Aiguader's conference titles, separated by a period of 20 years, both held at the Ateneo Barcelonés and published the same year they were given: *Aspecte social de les infeccions sexuals en el matrimoni* (Barcelona: Tipografia 'La Academia', 1912) and *El problema de l'habitació obrera a Barcelona* (Barcelona: Publicacions de l'Institut Municipal d'Higiene, 1932).

themselves.[7] In fact, Aiguader had already been a leading advocate of social medicine for decades, through articles in the daily press, conferences in cultural associations and societies and especially through the direction and writing of the series *Monografies Mèdiques*.[8] This social medicine advocated, among other things, direct political commitment by physicians themselves for the solution of hygienic, health and medical issues of social groups and of the most deprived or needy urban areas.[9]

Shifting our attention to the left corner of the photograph we can see several young Barcelona architects, with Josep Lluís Sert (1902–1983) in the front row alongside Charles Édouard Jeanneret-Gris (1887–1965), better known as Le Corbusier. These young architects, mostly belonging to educated bourgeois Catalan families, considered him their mentor and enthusiastically adhered to the tenets of his rationalist beliefs on architecture and urban planning. Together, they would soon convince the authorities in the photograph to sponsor an urban reform plan for Barcelona, inspired by the Swiss master. This would be a radical plan which would raise fierce social and professional opposition and that would be only poorly applied, but which for some of them would provide an excellent springboard for their brilliant careers.

Indeed, a few months after this meeting, the so-called 'Macià Plan', an ambitious urban planning project for the city of Barcelona and its metropolitan area, was presented. The plan was a declaration of the result of planning principles, fruit of the strange – or not – alliance between the elderly President Macià, the doctor Mayor Aiguader and young architects from well-to-do families, dazzled by Le Corbusier.[10] The idea that drove the project was to transform Barcelona, which for half a century had been an industrial city, into a modern capital, equipped with a port, an airport and other communication channels which would rationalize entrepreneurial activity and open up the city for residents by

7 An example of contemporary end-of-century hygienism in Barcelona is to be found in the works of Pedro García Faria (1858–1927), chief exponent of Spanish sanitary engineering: *Saneamiento de Barcelona. Condiciones higiénicas de la urbe: su mejoramiento, disminución de la mortalidad de sus habitantes y aumento de la vida media de los mismos* (Barcelona: E.T. de N. Ramírez, 1884); *Insalubridad de las viviendas de Barcelona* (Barcelona: Impr. de J. Balmas Planas, 1890); *Medios de aminorar las enfermedades y mortalidad en Barcelona* (Barcelona: Adm. de Industria e Invenciones, 1894). See: M.A. Miranda, 'Pedro García Faria, Civil Engineer (and architect)', *Scripta Nova. Revista electrónica de geografía y ciencias sociales*, 10 (2006), http://www.ub.es/geocrit/sn/sn-221.htm (last accessed 17 October 2014).

8 Enrique Perdiguero, José Pardo-Tomás and Àlvar Martínez-Vidal, 'Physicians as a Public for the Popularization of Medicine in Interwar Catalonia: The *Monografies Mèdiques* Series', in F. Papanelopoulou, A. Nieto-Galan and E. Perdiguero (eds), *Popularizing Science and Technology in the European Periphery, 1800–2000* (Aldershot: Ashgate, 2009), 195–215.

9 Iris Borowy, 'International social medicine between the wars. Positioning a volatile concept', *Hygeia Internationalis*, 6 (2) (2007): 13–35; Iris Borowy and Anne Hardy (eds), *Of Medicine and Men: Biographies and Ideas in European Social Medicine between the World Wars* (Frankfurt/M.: Peter Lang, 2008).

10 Le Corbusier had recently returned from Soviet Russia, disappointed at having failed in his attempt to sell Stalin his projects, convinced that the utopian world he heralded would not come about in the USSR but could nevertheless be pursued in other areas such as southern Europe. Alexander Tzonis, *Le Corbusier: The Poetics of Machine and Metaphor* (London/New York: Thames & Hudson, 2002).

enhancing facility of movement, modern dwellings and communications in the city in line with the rationalistic and utopian vision of 'modern life' so characteristic of the interwar period in Europe and North America.[11]

This conception of urbanism had a special plan for what was then, from an administrative point of view, the Fifth District of the city (see Figure 8.2). It was a densely populated sector, with some neighbourhoods that constituted an archetype of houses, streets and 'unhealthy' people from portuary zones.[12] But the Fifth District also contained, close to the port, a neighbourhood that was then known by the name of Barrio Chino: a maze of narrow shadowy alleys, with buildings too high, too narrow and too shabby, where a diverse array of night-life and a high ratio of brothels were concentrated, defining it as the quintessence of the underworld of the city. The special plan for the Barrio Chino was none other than the demolition of almost all buildings, destroying the whole network of streets, alleys and courtyards, creating wide avenues flanked by new housing, geometrized and stripped of all ornament that were not pure lines of concrete, glass and iron materials that the new architects were so enthused about. The shared dream amongst authorities and planners was the eradication of the physical, human and geographical configuration of the entire area through urban sanitation and materialized in a plan which contemplated the most dramatic destruction and presented explicitly, inspired by a surgical simile: 'We must remove the infected core as would a surgeon from a sick body'.[13]

The radical redesign of this area of the Fifth District leaves no doubt about the medical and architectural idea behind the pursued objective – to remove the social and medical problems of the slums of Barcelona – and urban debauchery was seen as the optimal solution. There was no room for relics of the past such as the winding, narrow and dirty streets of the Barrio Chino; even less so for their unhealthy homes and yards. And for its inhabitants and the activities carried out there? Their future remained unclear, but the sanatorium, other confinement or forcible transfer were the pathways with which until then the authorities had followed when attempting to address that particular problem.

11 Francesc Roca, 'El Plan Macià i el regional planning (1932–1937)', *Materials ACCAT*, 9 (2008); Hermínia Pujol, 'Dues visions de Barcelona als anys 30. El Plan de distribució en zones del territorio català i el Plan Macià (1932–1934)', *Treballs de la Societat Catalana de Geografia*, 43 (12) (1997): 81–116; Salvador Tarragó, 'El Plan Macià o la nova Barcelona 1931–1938', *Cuadernos de Arquitectura y Urbanismo*, 90 (1972): 24–36.

12 It was not, however, the first sanitary reform plan that had been set forth for the Fifth District. See: Vicente Casals et al., 'Manuel Mujica Millán y el urbanismo novecentista en Cataluña, 1917–1927', *Biblio 3W. Revista Bibliográfica de Geografía y Ciencias Sociales*, 16 (925) (2011), http://www.ub.es/geocrit/b3w-925.htm (last accessed 1 October 2014).

13 GATEPAC, display panel 'Sanejament Districte V' exhibited during the presentation of the 'Plan Macià' for Barcelona at the IV meeting of the CIAM (*Congrès international d'architecture moderne*), held in Athens, 1933. Reproduced in: Antonio Pizza, 'Representaciones de la Ciudad Funcional', in *A.C. La revista del G.A.T.E.P.A.C. 1931–1937* (Madrid: Museo Nacional Centro de Arte Reina Sofía, 2009), 191–220, p. 220.

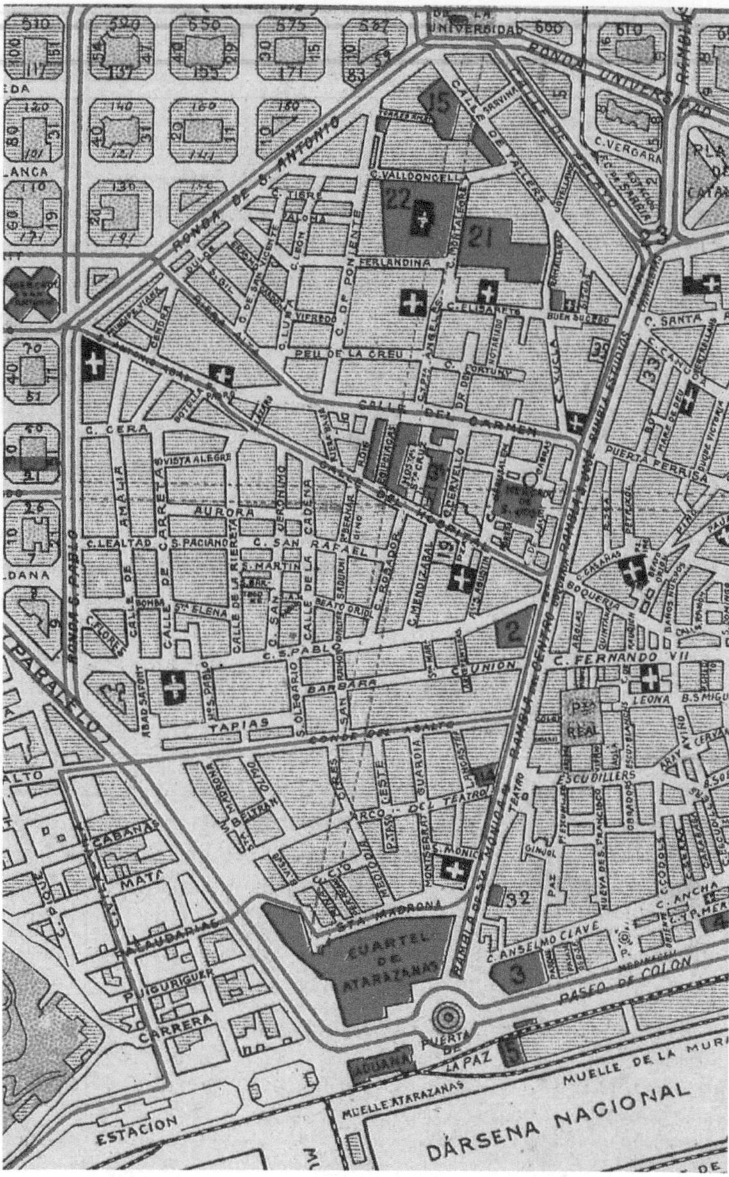

Figure 8.2 Barcelona, map of the streets of the Fifth District, 1916.

Unsurprisingly, the plan was fiercely attacked by property owners and did not receive clear political support until the arrival of the new republican government of the Popular Front in February 1936, on the eve of the Civil War. This appalling conflict finally demolished the dreams of planners and politicians: the vision proclaiming the design of a modern, ambitious and healthy city that involved the destruction of the ailing urban fabric barely materialized, or at least which would not come about until much later, with a whole new set of managers and players and well mixed results.[14]

The only action carried out in the Fifth District under the Macià Plan before the Civil War was the construction and operation of the anti-tuberculosis Central Dispensary (DCA) (see Figure 8.1, right). The design and building of the DCA, located at the northern limit of the Fifth District, was commissioned precisely to Josep Lluís Sert, together with his even younger colleagues Josep Torres Clavé (1906–1939) and Joan Baptista Subirana (1904–1978).[15] The three architects raised the project from reading the work of Lluís Sayé, *Les noves orientacions de la lluita antituberculosa i la seva aplicació a Catalunya* (1933).[16] The tisiologist Sayé noted, apart from highlighting the role of poor nutrition, the importance of the structural housing conditions in the spread of the disease.[17] The tuberculosis dispensary allowed a structuring of tisiology as a medical speciality.[18]

What we would like to propose in this chapter is, in fact, to look back at the two decades prior to the early 1930s. So far we have presented the point of arrival, which is already right on the final frontier of the period covered by this book. In the pages that follow we aim to explain what the Barrio Chino was like at the time, what was happening there and why those in the photograph

14 In the late 1940s and early 1950s, paradoxically – or maybe not so much – under the Franco regime authorities were to carry out drastic urban remodelling in much the same fashion as envisaged by the Catalanist republicans in the 1930s: the widespread destruction of streets and entire blocks, the opening of avenues, radical sanitation measures of the most deteriorated zones and so on. Authorities in the pre-Olympic Barcelona (1992) would continue with the 'surgery', now sheltered beneath the less drastic metaphorical umbrella of 'esponjament' (opening up/ventilation/porosity/permeability). On the historical continuities of the actions developed by authorities, in the name of the interests of international real estate, beyond the dictatorship or the democratic periods, see: Manuel Delgado, *La ciudad mentirosa: fraude y miseria del 'modelo Barcelona'* (Madrid: Los Libros de la Catarata, 2007), p. 33.

15 Antonio Pizza, *Dispensario antituberculoso de Barcelona, 1933–1937: J.Ll. Sert, J.B. Subirana y J. Torres Clavé* (Almería: Colegio de Arquitectos de Almería, 2003). Boris Berenguer, *Estudi de dos edificis: Dispensari Central Antituberculós de Barcelona i Casa Bloc* (Master thesis, UPC, 2011).

16 Sayé's text appeared in numbers 68 and 69 of 'Monografies Mèdiques', the collection created in 1926 by Jaume Aiguader. About Sayé: Josep Cornudella, *Estudi biogràfic del profesor Lluís Sayé Sempere* (Barcelona: Reial Acadèmia de Medicina, 1979).

17 Celia Miralles Buil, *La tuberculose et l'habitat populaire à Barcelone dans les années 1930* (Master thesis, Université Lyon 2, 2008); *La tuberculose dans l'espace social Barcelonais, 1929–1936* (PhD thesis, Université Lyon 2 and UPC, 2014).

18 Alfons Zarzoso et al., 'Especialización y divulgación de la medicina en la ciudad de Barcelona (1868–1938)', in *Actas del XV Congreso de la SEHM. Transmisión del conocimiento médico e internacionalización de las prácticas sanitarias: una reflexión histórica* (Ciudad Real: SEHM-UCLM, 2011), 279–305. We understand this action in the framework of the medical police deployed by the Generalitat de Catalunya, as studied by Carles Hervàs, *Sanitat a Catalunya durant la República i la Guerra Civil: política i organització sanitàries. L'impacte del conflicte bèlic* (PhD thesis, Universitat Pompeu Fabra, 2004). As regards the material produced by the GATEPAC and the reflection of those proposals in the medical press, see: Celia Miralles Buil, 'Controlar la ciudad para eliminar la endemia: el caso de la prevención antituberculosa en Barcelona en el primer tercio del siglo XX', in P. Fraile and Q. Bonastra (eds), *El control del espacio y los espacios de control* (Barcelona: Ediciones del Serbal, 2015), 131–143.

would wish to destroy it. This wish was nurtured by a desire to sanitize, in line with the most recent trends in urbanism and social medicine. It was promoted by a mesocracy of professional engineers, doctors and architects who were in direct contact with the republican authorities while they themselves did not have any political or institutional responsibilities.[19]

What was the neighbourhood like? How did residents and visitors face health problems? How had they tried to control – regardless of any radical urban proposals – this area inhabited by generally poor people of such varied nature, which was at the same time frequented by members of almost all social classes who did not actually live there but sought fun, sex, music, entertainment and night-life until dawn?

In the following section, we will proceed to present a vision of the urban geography of the Barrio Chino. In the third section, we descend to the micro-geography of a single street (Conde de Asalto) and its immediate neighbourhood to observe not only the diverse world based on show business and sex, but also the consequences thereof, both hygienic and sanitary. In the fourth section, we have chosen one of the premises of that street, the anatomical museum run by the show business entrepreneur Francesc Roca, to reflect on the ambiguities and limitations that the regime of control that social medicine attempted to impose from above was facing, from its role as prescribers and expert advisors to municipal, regional or national authorities, as well as some of the appropriations the end users of these policies made of the content, materials or speeches that were destined for them.

The geography of a district: Literary itineraries and sanitary problems

Urban geography, urban planning and architectural design are paramount to understanding medical cultures centred in such a particular area of the city as the 'underworld'. Therefore, our approach will resort to the kind of sources in which the dimension of public representation is notable: media reports, literary excerpts, images or advertisements promoting medical and health services. Naturally, the assertions of medical experts and planners regarding any proposed control, sanitation or the combatting of disease are also present in the analysis, but these sources have historically been given privilege on these issues, so our wish here is to relativize their weight in understanding the complex world of medical cultures in coexistence or dispute in an area and a specific historical moment.[20] An approach to the geography of the urban district under study

19 Jaume Valentines, *Tecnocràcia i catalanisme tècnic a Catalunya als anys 1930: els enginyers industrials, de l'organització del taller a la racionalització de l'estat* (PhD thesis, Universitat Autònoma de Barcelona, 2012).

20 We have attempted to apply the category of analysis of 'medical cultures' from a Gramscian viewpoint, for another period in the history of Barcelona: Alfons Zarzoso, 'El pluralismo médico a través de la correspondencia privada en la Cataluña del siglo XVIII', *Dynamis*, 21 (2001): 409–33.

should avoid giving priority to certain aspects over others when considering the confusing world lived daily in the pre-civil war Barrio Chino. The chronology, the physical space and the social actors go beyond the ideological forms of intervention of public authorities defined in terms of 'social medicine' that we are used to seeing.[21] On the other hand, taking into account the special characteristics of Barcelona, we wish to point out the usefulness of comparing this case with those of other European industrial and 'non-capitals' such as Lyon, Marseille, Hamburg or Turin.[22]

In the first third of the twentieth century, the Barrio Chino was a scenario characterized by the complex coexistence of diverse urban subcultures, from libertarian anarchism to the trade in and consumption of drugs, the emergence of musical styles and genres such as jazz, or the constant renewal in the settings of prostitution, be they heterosexual or homosexual. The narrow streets of the Barrio Chino spread between buildings crowded with subdivided homes that housed families of disparate composition and condition, but almost always large; dozens of workshops and small factories where largely immigrant workers toiled, cafés, cabarets, taverns and other spaces in which alternative political and social topics were discussed, and where throbbed the city night-life, with the permanent presence of the worlds of prostitution and drugs, music, dance and gambling.[23]

At the same time, those streets and surrounding areas were still the place of the old institutions devoted to the care of the poor and sick, not to mention the Queen Amalia prison, reserved exclusively for women from 1904 when male prisoners were transferred to the new Modelo Prison, built outside the walls of the old city.[24] So, if on the one hand the Fifth District was the geography of the traditional institutions of social control, on the other, it was also the geography of social tumult: there were both male and female prostitution, begging, drug use (especially alcohol and cocaine), subversive activities and so on. From the perspective of the ruling bourgeoisie, this was an infected area, a place that should be cleansed completely. Moreover, serious health consequences marked

21 Dorothy Porter (ed.), *The History of Public Health and the Modern State* (Amsterdam: Rodopi, 1994); Dorothy Porter, *Health, Civilization and the State* (London: Routledge, 1999).

22 Patrice de Moncan and Claude Heurteux (eds), *Villes haussmanniennes: Bordeaux, Lille, Lyon, Marseille* (Paris: Mécène, 2003); Guido Montanari, 'Memoria e genius loci. Il caso di Torino da città fabbrica a "città degli eventi"', *Nuvole*, 47 (2003): 1–12; Elena Manzo (ed.), *La città che si rinnova. Architettura e scienze umane tra storia e attualità: prospettive di analisi a confronto* (Milan: Franco Angeli, 2012), Matthew Jefferies, *Hamburg. A Cultural and Literary History. Cities of the Imagination* (Oxford: Signal Books, 2010); '"A City in Distress"? Paul Broecker and the New Architecture of Hamburg', in M. Gee, T. Kirk and J. Steward (eds), *The City in Central Europe: Culture and Society from 1800 to the Present* (Aldershot: Ashgate, 1999), 9–26.

23 José Luis Oyón, *Barcelona 1930: un atlas social* (Barcelona: Edicions UPC, 2001); *La quiebra de la ciudad popular: espacio urbano, inmigración y anarquismo en la Barcelona de entreguerras, 1914–1936* (Barcelona: El Serbal, 2008).

24 Òscar Montero, *Normativització a la presó Model de Barcelona abans de 1936* (PhD thesis, Universitat Autònoma de Barcelona, 2014); Elsa Plaza, *Desmontando el caso de La Vampira del raval. Misoginia y clasismo en la Barcelona modernista* (Barcelona: Icaria, 2014), pp. 147–65.

the sanitary degradation of the whole area: venereal disease, tuberculosis and sundry ailments associated with poverty and overcrowding spread throughout the district during this period. The bourgeoisie in command of institutions resorted to strategies based on their idea of civic regeneration and the recovery of a morally healthy urban landscape. Social medicine and urban planning were the basis for action and the formulation of projects to give an answer to a reality that stubbornly persisted despite all the hype of regulations, institutions and norms.

Beyond the regulatory documents, policy statements and speeches of medical statisticians and sanitary engineers, the use of literary and journalistic sources allows us to capture the cityscape of the time. The novels, stories and news reports devoted to different aspects of life in the Barrio Chino constituted a genuine literary genre in those early decades of the century. Not only in Spanish and Catalan literature, but also in some literary works published abroad.[25] Short novels in Spanish such as *Sangre en las Drassanes* (1926) by Francisco Madrid (1900–1952) or literary monuments in Catalan literature such as *Vida privada* (1932) by Josep Maria de Sagarra (1894–1961); newspaper reports such as those by the aforementioned Francisco Madrid, or those by Domènec de Bellmunt (1903–1993) compiled in a volume together with his short novel *L'àngel bohemi* (1935) or those gathered by Josep M. Planes (1907–1936), in *Nits de Barcelona* (1931).[26] These and other reports were published throughout the interwar period in *L'Esquella de la Torratxa*, *El Escándalo*, *D'Ací i d'Allà*, *El Diluvio*, *La Rambla* and *La Publicitat*, publications which reflected various ideological positions and should be analysed as an expression of the Barrio Chino's visibility to the public and in the troubled political arena of the moment, from the bourgeois and conservative Catholic positions to Catalan anarchists, through various modes of republicanism, all in an unstoppable upsurge since the beginning of the century, before the collapse of the traditional dynastic parties (conservative and liberal).[27] For reasons of space, we will present three examples taken from these sources that portray three different scenes, which represented the Barrio Chino for some authors and their readers in different social spheres.

The first of our examples dates from 1929. That year, the Association of Hoteliers in Catalonia sponsored, on the occasion of the International Exhibition, the publishing of what were known as the *Guías P.I.C.S.*, tourist guides with practical and artistic information about the city, with maps and detailed routes. The edition of this year incorporated a section with a variety of itineraries for the city. The eighth of these routes ran around the neighbourhoods

25 Joan Ramon Resina, *La vocació de modernitat de Barcelona. Auge i declivi d'una imatge urbana* (Barcelona: Cercle de Lectors-Galàxia Gutenberg, 2008), pp. 115–45.

26 Jordi Castellanos, 'La descoberta literaria del Districte Cinquè', in Margarida Casacuberta and Marina Gustà (eds), *Narratives urbanes. La construcció literaria de Barcelona* (Barcelona: Fundació Antoni Tàpies, 2008), 83–108.

27 An example of historical analysis based on this literary production can be found in Robert A. Davidson, *Jazz Age Barcelona* (Toronto: University of Toronto Press, 2009).

of Hospital and Sant Pau, in the Fifth District. By taking as a starting point Carme Street from the corner of Egipcíaques, the guide talks about the 'severe overcrowding at the Hospital de Santa Cruz'. The tour ran through the streets of Roig, Malnom and Picalquers, 'very interesting for their picturesque squalor, the incarnation of a portrait by Balzac'. Turning back into Carme, the guide limits itself to highlighting 'a stately home of the early nineteenth century' commissioned by the manufacturer Erasme of Gonima (1746–1821). The route continued through the Plaza de Pedró up to San Antonio Abad, where a plaque on the wall of the church reminded the sightseer that it had once been the Hospital de Ardientes. The route continued down the streets of Salvador and Reina Amalia, where, as mentioned previously, the women's prison was to be found, though the only thing apparently worth mentioning in the guide was a house built by another nineteenth-century manufacturer, and added: 'adjacent streets are those of labourers losing their economic relevance; rather like the working neighbourhoods as would have been the Barcelona factory districts at the time mechanization began to develop'. Down the Calle de la Cera tourists wandered to the hospital, from which one could access the space which, at the time, had relinquished almost all hospital functions. Returning to the street, towards the Rambles, the guide reminds us of the importance of the Teatro Romea and the artistic allures of the church of St Augustine. Skirting the church, one came to the street of San Pablo, which the guide recommended strolling in its entirety, highlighting a farm shop from 1778 and the Romanesque Sant Pau del Camp.

This seventh route dedicated to Las Rambles included constant entries and exits to and from its tributary streets and peeked out, in its final stretch, onto Conde de Asalto and, on leaving the Plaza del Teatro, the tour left behind the Portal de la Paz to enter, however briefly, the Parallel and skirt the Drassanes (shipyards) building. The character and content of the tour differs little from the previous itinerary. It described routes to follow during the day, amongst the lively hubbub of the inhabitants but, of course, without stopping to discuss with them their miserable living conditions or their squalid dwellings.

Cosmetic operations for the sake of tourism have historical origins and are not merely the product of present times. In this seemingly neutral rhetoric – celebrating background noise – the historical reasons for a port and segregated neighbourhood were deliberately forestalled as a grand receiver of permanent immigration (at that time mainly from Catalonia and adjacent Spanish regions), a schizophrenia of daylight and darkness. A living space for workers, artisans, prostitutes and criminals; night-time entertainment, crime and shootouts between labour activists and the employers' gunmen, and misery, a great deal of misery.[28]

28 *Guia P.I.C.S.*, Barcelona, Catalònia, 1929; Paco Villar, *Historia y leyenda del Barrio Chino (1900–1992): Crónica y documentos de los bajos fondos de Barcelona* (Barcelona: La Campana, 1996); Chris Ealham, 'An Imagined Geography: Ideology, Urban Space, and Protest in the Creation of Barcelona's 'Chinatown', c. 1835–1936', *IRISH*, 50 (3) (2005): 373–97. On the Parallel, see the articles dealing with the cultural context of the shows, theatre circus and cinema architecture and the world of prostitution in the Fifth District in *El Paral·lel, 1894–1939* (Barcelona: CCCB, 2012).

The second example we would like to present stands, in a way, in a line similar to that of Barcelona hoteliers, but does so through literary fiction from one of the most gifted pens of the Catalan language of the time, the aristocrat Josep Maria de Sagarra, as described in his novel *Vida privada* (1932): 'An excursion to the sleazier neighbourhood known as the Barrio Chino' located chronologically, as we shall see in the quote, on the eve of the 1929 Exhibition. Unlike the guide, however, Sagarra shows us another reality, a true 'geography of evil'.[29] His is a nocturnal fiction starring eight representatives of the wealthier classes of the city, men and women alike, in which the Barrio Chino is presented as a museum, access to which is gained on foot once the cars have been left at the Lion d'Or, on the Rambles, near the Arc del Teatre. A museum or, as it would now be called, a theme park, with ammoniacal odours arousing the visitors' fascinated interest and the eagerness 'to see who knows what'. Sagarra reviewed the aim of the visit with irony:

> We are on that stretch of Calle del Cid carefully arranged by the Sociedad de Atracción de Forasteros [Society for the Attraction of Foreigners]. Then, those neighbourhoods were preparing for the great exhibition and operators were scouring the Barrio Chino for Chinese, blacks, the truculently inverted and women taken from the dissection room of the Hospital, half butchered, a green skirt and shawl turning them into gypsies, two pieces of desalted cod feigning breasts; so attired they would then be nailed to the door of the most strategic houses. [...] As for the men, they were of all kinds, from sailors, mechanics to ordinary workers, paedophiles with painted lips, cheeks caked with plaster and eyes heavy with mascara. Among the quirky wag their tails a kind of poor, the ragged and pickpockets, only to be found in those neighbourhoods [...] In this part of the city were the humble also to be seen, not picturesque in the slightest, as one might see in other parts; but the others [...].[30]

Sagarra also provides a literary visit to certain spaces of sociability:

> The taverns were smeared with the extract of orujo; this alcoholic excretion of stones is perhaps the best thing in the world to depict the life sentence of the heart, an irremediable brutishness. Besides the taverns, shops selling condoms, with their blue and red rubber pears, amongst other rubberware, glass and fabric, along with packets of cotton wool, all alluding in such an unlyrical way to the catastrophes of sex [...]

29 Gary McDonogh, 'The geography of evil: Barcelona's Barrio Chino', *Anthropological Quarterly*, 60 (4) (1987): 174–85.

30 Josep Maria de Sagarra, *Vida privada* (Barcelona: Catalonia, 1932). This and the following paragraphs are our translation into English.

Finally, Sagarra takes a shot at explaining an alleged acceptance of that 'natural order':

> Despite the accumulation of pathological material, humanity present there seemed to squeeze the juice from life with great naturalness and an oblivion that could be perfectly taken for optimism [...] ladies and gentlemen stopping at the door of 'La Criolla' would command very little attention. The neighbourhood was accustomed to this kind of visit and enough literature on the festering appearance of the Fifth District had already seen the light of day, and both the foreign and the native curious were received at 'La Criolla' in a perfectly normal and correct manner.

Visitors to this unique museum left it behind on their stroll back to the Rambles, where they would recover their cars to return to the geographical normality of their Eixample.

The *Noucentisme* movement argued with literary force that, from the beginning of the century, was hatching in the fields of medicalization and transformative proposals to be imposed upon the landscape of the Fifth District.[31] Xavier Pla recalls that one of the glosses written in 1906 by Eugeni d'Ors (1881–1954), considered the most representative intellectual of the *Noucentisme*, sought, with envy, an earthquake for Barcelona as San Francisco had experienced, to finish off the old city at a stroke.[32] This idea, linked to other types of intervention in the area, was to re-emerge throughout the first third of the century from different perspectives, including the literary. For example, Carles Soldevila (1892–1967), in a passage from his sexually ambiguous novel *Eva* (1931), defends the concept of urban space as an area of transit. The project for wide streets in a regular pattern would facilitate movement – commercial, recreational, athletic or sanitary – radically contrasting with the winding streets of the Fifth District 'full of onlookers, shouts, voices, smells and bands of children; as if they were the corridors of a formidable dwelling' angering ladies trying to get to the orphanage on Montealegre Street making them express aloud the dominant bourgeois thought, 'I can't imagine why they are waiting so long to knock all of this down [...] how disgusting!'.[33]

31 *Noucentisme* was a cultural and ideological movement starring the Catalan bourgeoisie in the first third of the century, characterized by a 'modernizing' programme with Catalanist roots reflected in literature, art, science and political economy by means of governing structures such as the *Mancomunitat de Catalunya*. Rosa Cabré, Montserrat Jufresa and Jordi Malé (eds), *Polis i nació: política i literatura (1900–1939)* (Barcelona: Aula Carles Riba-Societat Catalana d'Estudis Clàssics, 2003).

32 Eugeni d'Ors, *Glosari 1906–1907* (Barcelona: Quaderns Crema, 1986); Carlos X. Ardavín, Eloy E. Merino and Xavier Pla (eds), *Oceanografía de Xènius* (Kassel: Edition Reichenberger, 2005).

33 Núria Santamaria, 'Les Barcelones imaginades de Carles Soldevila', in Casacuberta and Gustà (eds), *Narratives urbanes*, 127–144. McDonogh reminds us of the theme of the international tourist attraction of the Barrio Chino and recollects the critical words of Avel·lí Artís in the article 'Legend of Barcelona: "If we clear up the rougher parts of the city, where will tourists go?"', *L'Opinió*, 22 October 1933, 1–4.

The last example we would like to expose in this wandering comes from a publication that appeared two years after the CIRPAC meeting in Barcelona with which this chapter began. The journalist Domènec Bellmunt published one of the stories that, from the pages of *Mirador, La Rambla* or other Catalan magazines of the period,[34] served their readers a seductive yet gruesome image of 'the catacombs of Barcelona'.[35] The article was titled *La 'Creu Roja' de l'amor* (The 'Red Cross' of Love) and was dedicated to the dispensary battling against venereal diseases to be found on the Parallel, the western frontier of the Barrio Chino, led by Dr Antoni Peyrí (1889–1973). Bellmunt presented the work of the authorities of the Generalitat whose venereal treatment service Peyrí himself ran – as an 'efficient and perfect' organization whilst also recalling the existence of another clinic in the portuary zone – the southern boundary of the Barrio Chino – and a third in Sant Andreu, a working-class district on the north-eastern outskirts of the city.

Syphilis was present in many ways in the popular imagination. Carles Soldevila gives a sample in *Fanny* (1929), the story of the survival of a young bourgeois girl who has fallen into social disgrace: yet another trip from the bourgeois Eixample to the reproachable Fifth District. The self-sufficient Fanny tries to rebuild her life as a dancer in a club; fearful from the rumours spread on infection by syphilis, she goes to the Library of Catalonia to look up the corresponding entry in the *Enciclopedia Espasa*. In turn, Sagarra, in *Vida privada*, speaking of the children of wealthy families in Barcelona, highlights that 'the conversation amongst most young men revolved around car chassis, bodywork and gonorrhoea'.[36]

By the time *'Creu Roja' de l'amor* was published, for the 12 previous years Domènec Bellmunt had already been writing – in *La Publicitat*, organ of the Acció Catalana party, a splinter group of the Lliga – the news reports that would win him the title of 'father of modern Catalan journalism'.[37] After years of Parisian exile due to the dictatorship, his return to Barcelona in 1928 is notable for his activity as a reporter for the Catalan press, but also for his official entry into the Generalitat Health Service, having passed the corresponding public examinations. The article singing the praises of the personification of Peyrí in the programme of social medicine fighting venereal disease would hence not appear to be a mere coincidence.

So between contemplative museum visits and dreams of radical action, the Fifth District survived the conditions imposed by social and political reformers.

34 For a different perspective see the radical republican press on Barcelona's underworld: Joan Lluís Marfany, 'Catalanistes i lerrouxistes', *Recerques*, 29 (1994): 41–60; Enric Ucelay Da Cal and Dorsey Boatwright, 'La dona del "Barrio Chino". The image of rough areas and the magazine "El Escándalo"', *L'Avenç*, 79 (1984): 26–34; McDonogh, 'The geography of evil'; Davidson, pp. 27–50 and 104–40.

35 The title of the book that he would publish in 1930 was a compilation of some of his most successful reports from the previous decade.

36 Josep M. de Sagarra, *Vida privada* (Barcelona, 1932), p. 165.

37 Francesc Canosa, 'Apunt biogràfic. Domènec de Bellmunt: un espia doble i una missió', in Domènec de Bellmunt, *La Barcelona pecadora* (Barcelona: A contracorrent, 2009), 163–94, p. 163.

The interventions that had been suggested during the first third of the twentieth century had a variety of effects.[38] Republican and Catalan journalists exerted a strong influence on the institutions of city government and on the medical profession. Social medicine was not directed to a mere healthfulness of the district, but rather to its medicalization from above, with its clinics and health propaganda – with special attention as to the treatment of syphilis and tuberculosis – as well as its urban planning, in collusion with architects and engineers and supported by those forces installed in political power. But within this equation, in which everything might appear to fit, something fundamental was lacking: the limits set by the effort to medicalize and the moralizing discourse. The diverse population that inhabited or frequented the Barrio Chino certainly met with various manifestations of controlling pressure from social medicine, but also with so much else, abiding side by side in the same urban space.[39]

A microgeography: Conde de Asalto from the perspective of a *Flâneur*[40]

In the geography of the Barrio Chino, Conde de Asalto was the main human river that connected the Rambles with the Parallel along which strollers would swim or sail on board the riverboat running through it, Tram N°48, nourishing numerous tributaries and alleys with its passengers, especially on the southern side. The scene that lay before them should be described house by house, premise by premise; you would find a panorama of offices and clinics, rubberware and bathing establishments, brothels, photographic studios, anatomical museums, music and picture houses where films where shown.

Beginning with the clinics we would come across: the Clínica Gallego, at n° 18; the Instituto Médico Femenino at n° 23; the Clínica San Antonio, at n° 47; and the clinic La Corona, at n° 95. Apart from the consultancies – normally advertised as 'urinary' – of Doctor Paulino Alcántara (Unió, n° 19), the Clínica Balart (Unió, n° 7) and 'La Japonesa', founded by Doctor Holgado in 1924 (Arc del Teatre, n° 1). Octavi Carulla had a urology practice at Unió, n° 10; Francesc Imbert, at Principal at Gravina, n° 8; Joan Morros at San Pablo, n° 63. Doctor Josep Mascaró had a 'women's ailments' practice at n° 47 in Calle Hospital, and Josep M. Segalà at n° 15, Calle Barberà. Brothels also existed with their own consultants, such as that described by Francisco Madrid in 'The commercial organisation of a brothel', published in 1925 in *El Escándalo*

38 Silvia Sánchez Estrada, *Clases populares y anarcosindicalismo. Barcelona, 1917–1923* (Entremons: UPF, 2013), pp. 20–53.

39 Celia Marin Vega, *Antropografía del Barrio Chino. Arquitectura o Revolución (1925–1938)* (Master thesis, Universitat Politècnica de Catalunya, 2011).

40 On the cognitive status of the notion of the *flâneur*, developed by Walter Benjamin in the light of 1920s and 1930s modernism, and how productive *flânerie* can be, see Karl Schlögel, *Moscow 1937* (Cambridge: Polity Press, 2008).

and in the following year in the compilation of 23 of his articles titled *Sangre en Atarazanas*.[41] Madrid presents his readers with the description of a brothel in Calle Arco del Teatro accentuating its 'all modern advances', among which there was listed 'a medical clinic' established by 'Mr. Ugarte', the proprietor of the business itself.[42]

The establishments known as 'rubberware and bathing' (that is, offering condoms and a washing of the private parts) were sited preferably in the side-streets, close to the main flow: La Mundial (Espalter, n° 6) 'Open until 3 a.m', with 'Immediate bathing and irrigation service'; La Clínica Oriental and El Cupido (both at San Pablo, n° 110); three 'rubberware' establishments in Calle San Ramón: La Manota at n° 1, La Previsión at n° 6 and La Favorita at n° 10; La Normanda was to be found in Calle San Oleguer; La Especial in Barberá; La Cosmopolita in Calle Robadors, practically beside the La Bola de Oro clinic.

Abortion 'clinics', sited in various parts of the district (and town), were obviously clandestine, but made an appearance in literary and journalistic sources. Bellmunt, for example, published in 1930 a report-story called 'human meat for sale' in which the main character, pregnant at 17, aborted following what the author calls a 'medicinal treatment'. Bellmunt claims 'this business of abortion was promoted thanks to the clandestinity of quacks [...] ' and concludes with a selection of terms, entirely symptomatic of his mentality, that 'young people should be aware of all the perils of sexual anarchism and the advantages of prevention of venereal diseases', proposing the introduction 'in girls' schools, to the teaching of real life'.[43]

Photographic studios located here, such as the 'Estudio Venus' at n° 42 Conde de Asalto, could not help but become disreputable and show the ambivalent relationship between artistic, medical and pornographic photography. The theatrical, musical, and from a certain time onwards, cinematic shows were definitely the main attraction of the street. For example, the Eden Concert Hall which, from 1914, became the paradigm of theatrical premises providing quite luxurious *variétés* to select clients who wished to 'walk on the wild side' in the Barrio Chino, with the help of entrepreneurs such as Francesc Buxó.[44] Some aspects of life in these places appear in the lyrics of the songs that were sung in variety shows there. Perhaps one of the most significant is the issue of drug use and, especially, cocaine, which had made a swift leap from the operating theatres of modern clinics in the late nineteenth century to the dance halls on Conde de Asalto or the Parallel, into which the aforementioned street flowed. Cocaine use seems to have peaked, coming to a head in 1926 when Joan Viladomat

41 Barcelona: Ediciones La Flecha, 1926.

42 A recent edition exists, translated into Catalan: *Sang a les Drassanes* (Barcelona: A contracorrent, 2010); quote is from pp. 42–3. Paco Madrid is considered to have penned the denomination Barrio Chino in one of his reports from the same years.

43 Domènec de Bellmunt, 'Carn humana per vendre', in *La Barcelona pecadora* (Barcelona: A contracorrent, 2009), pp. 33–4. Sagarra reflects upon abortion in *Vida privada*, p. 302.

44 The short novel by Francesc de Bellmunt *L'Àngel bohemi*, published in 1935, is a magnificent example. Re-edited in: *Barcelona pecadora* (Barcelona: A contracorrent, 2009), 125–62.

wrote the lyrics to 'Cocaine Tango', which became a commercial hit.[45] So it is not surprising that the issue of addiction eventually became part of the discourse of social control drawn from medicine. A discourse of control not always stated specifically by the medical press or formulated by the action of dispensaries, as discussed in the following section.

Social control: The ambiguous discourse in anatomical museums

As it has been said, anatomical museums were a paramount component of the geography of spectacle in the Barrio Chino. They played an important role as a means of popularizing a particular medical culture. The publicity inserted in the general press and the records of the Arxiu Municipal Contemporari of Barcelona allow us to set the exact location and physical features of those *museos anatómicos*, which were distinctive not only by their situation in the area but also because of the role they played as a means of popularizing a particular medical culture. Actually, those spaces were the setting for a cultural event that went beyond the *flâneur* who visited the neighbourhood, a complex expression which mixed media, genres and styles, a business initiative and, simultaneously, a form of campaigning against disease. But we have direct and unique sources for only one of those collections: the Roca Museum. Significantly, we have a record of the different items of the collection, which is something really unusual in the European context of the time, rich in such establishments, but that has experienced a great loss of that kind of material exhibits. The *ephemera* produced by the Roca family, which has also survived, allows us to know the variety of resources deployed by the owners in that space: an anatomical collection, some movies, a book store with sex education books and drugs for sexually transmitted diseases, many posters, flyers, ads, and so on (see Figure 8.3).

According to the *Diario del Museo*, published by the owners of the Roca Museum also located in the Conde de Asalto, the museum had been open to the public in Barcelona since 1925. The documented remains of the wreckage of the museum after the Spanish Civil War showed little more than scattered signs of an institution that had emerged from the world of illusionists and music, fairs and travelling shows from towards the end of the previous century. Francesc Roca, musician, magician, ventriloquist, over the years was to become yet another mainstay of showbiz; a business that would eventually include one of the anatomical museums and freak shows that were all the rage in those decades.[46] Billboards surviving from the period suggest that the museum was part of a spectacle in which the owner combined a series of entertainments based on

45 Jaume Collell, *El músic de l'americana vermella. Joan Viladomat i la Barcelona descordada dels anys vint* (Barcelona: RBA, 2013), p. 113.

46 On the intended audience and general contents of the Roca Museum, see: Alfons Zarzoso and José Pardo-Tomás, 'Fall and Rise of the Roca Museum: Owners, Meanings and Audiences of an Anatomical Collection from Barcelona to Antwerp, 1922–2012', in Rina Knoeff and Robert Zwijnenberg (eds), *The Fate of Anatomical Collections* (Farnham: Ashgate, 2015), 161–76.

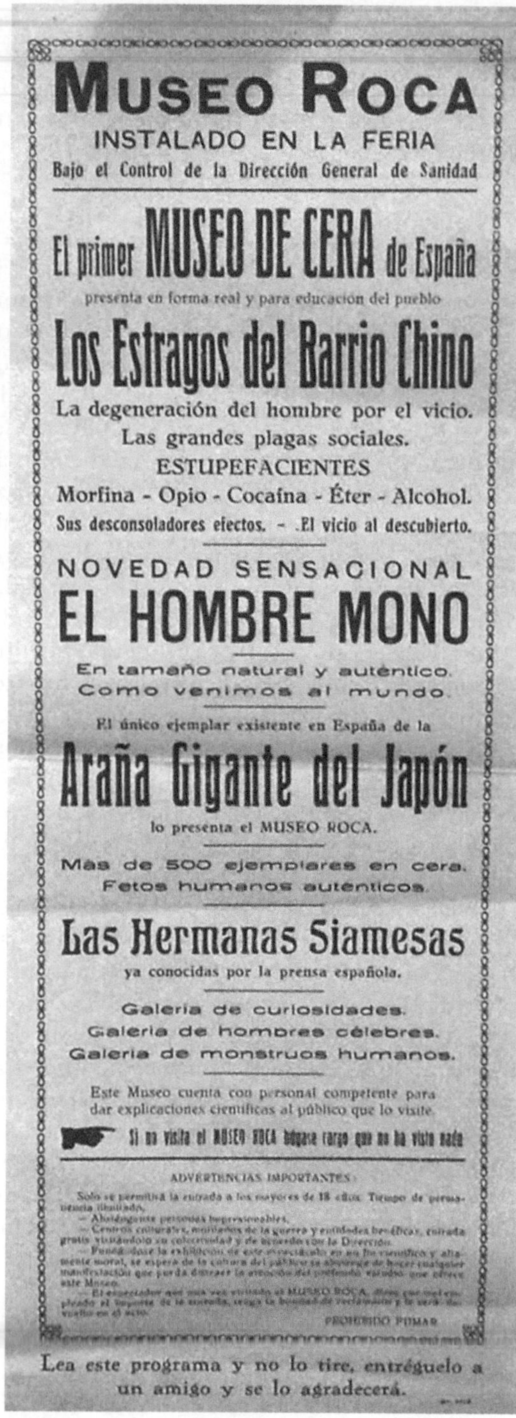

Figure 8.3 Ephemera from the Roca Museum, Barcelona, *c* 1920s.

the exhibition of objects and artefacts, musical sketches and circus acts, rarities and monstrosities both human and animal, scientific film screenings and the sale of informative books on themes related to sex.[47]

The collection of anatomical models had a fundamental weight in the order, division and development of artistic elements of the museum. The exhibits focused on vice and the terrible consequences they could have on the wrong-doer's body and on their family. Drugs and alcohol, as well as sexually transmitted diseases, were a central theme of the collection, alongside a systematic unveiling of the naked female body. These anatomical models explicitly detailed the process of fertilization, gestation and human reproduction. The projection of films such as 'How we come into the world' would reinforce the argument for information expressed and consistently claimed by Roca the entrepreneur. The capacity to attract custom was increased by producing numerous models and exhibits with vivid reproductions of human embryos and foetuses, and especially those with congenital defects.

The nature and content of the Roca show looks set to order the conduct of visitors with learning by means of imagining the terrible consequences of certain types of behaviour. Thus the idea of giving a pedagogical and disciplinarian significance to the exhibits would seem plausible, given the primarily male audience as well as the location chosen by various museums and anatomical collections in the Fifth District from the late nineteenth century onwards and especially during the first third of the twentieth century.

Sagarra's novel illustrates the existence and the *raison d'être* for these premises in a literary context characteristic of the deflowering of the young son of a wealthy family in a brothel in Calle Barberá, in the lower part of the Barrio Chino. A commonplace circumstance in many periods: urban youth discovering sex in a seedy world. Sagarra explores the cultural motives that lead from youthful sexual palpitations to a loss of focus. The irresistible impulse must face misunderstanding, taboos, repression and prejudices created by men, the adults, represent an ideological order which prohibits and does not explain, promoting denial and obscurity. Given this confusion, the options available to uninformed youngsters are limited:

> They wish the teenager to believe in the existence of sin; the case is presented specifically, with the associated horrifying material consequences. Certain pedagogies use compelling images, catastrophes of secret diseases vividly represented, the most repugnant secretions, deformations and intolerable pain. But it is useless; there comes a time when shame or cowardice

47 Lucenay was the author of many of such books that could be purchased at the Museo Roca. For bestsellers from the library of sexual promogulation in the 1930s see: Raquel Álvarez Peláez, 'Publicaciones sobre sexualidad en la España del primer tercio del siglo XX: entre la medicina y la pornografía', *Hispania*, 64 (218) (2004): 947–60; Richard Cleminson, 'El libro "Homosexualidad" del Dr. Martin de Lucenay: entre el conocimiento científico y la recepción pública de la ciencia sexológica en España a principios del siglo XX', *Hispania*, 64 (218) (2004): 961–86.

> pass; the temptation is too cruel and the bare skin of Siegfried will leap over any flame.[48]

It is difficult to clearly envisage the public attending these shows. The *Diario del Museo* shows a photograph in which a sizeable crowd fills Conde de Asalto, at the doors of what is presumably the museum (see Figure 8.4). The mass, many with their backs to the observer, is dominated by soldier's caps and the berets of members of the popular classes. Nevertheless it cannot be assured that the photograph corresponds exactly to the given description. In *Verdades*, a publication displaying an identical layout, font and publicity as the *Diario del Museo*, in the same picture, enlarged, you can see that people are about to enter premises where the film 'Los Averiados' is being shown. It seems, therefore, that the business project led by Roca would include, besides the wax exhibits at the anatomical museum, film screenings of daring sexual content, presented as part of the social medicine campaign against venereal diseases. This kind of cinema and specifically 'Los Averiados',[49] as press cuttings from 1933 show, were projected in major cinemas in the city for several weeks and then went on to smaller theatres located in less central positions.[50]

Significantly, 'Los Averiados' and, logically, the Museo Roca, being one of the places where the film was projected, received fierce criticism from Catalan conservatives, as evidenced by the review published by the doctor and priest Pere Tarrés. Tarrés questioned the educational purpose defended by those promoting the film, as well as the creators of other types of products marketed by people like Roca:

> The 'pseudo-scientific', the unscrupulous, unsuccessful writers in noble literature, have taken advantage of a moment in which much of the world of youth revolves around issues related to sexuality, introducing into the big market – with evocative and catchy titles to explain his way, under a roughly cut scientific flag – all sorts of immoralities which in a cruder form they would not have been allowed to publish. What is the purpose? Just to earn a few miserable pesetas.

Tarrés thus condemns the artifice of entrepreneurs – who were qualified by the doctor priest as 'true film Pharisees', 'hypocrites of cinematic art', 'new masters (?)

48 Sagarra, *Vida privada*, pp. 337–8.

49 Palmira González, *El cine en Barcelona. Una generación histórica: 1906–1923* (PhD thesis, Universitat de Barcelona, 1984); Luisa Suárez Carmona, *El cinema i la constitució d'un públic popular a Barcelona. El cas del Paral·lel* (PhD thesis, Universitat de Girona, 2011).

50 For further information on the social context of this and similar films, see: Francisco J. Navarro, *A la revolución por la cultura: prácticas culturales y sociabilidad libertarias en el País Valenciano (1931–1939)* (Valencia: Universitat de València, 2004); M. Antonia Paz Rebollo and José Cabeza San Deogracias, 'La realidad que vieron los españoles. El cine de no-ficción durante la Segunda República Española (1931–1936)', *Hispania*, 70 (236) (2010): 737–64; M. Antonia Paz Rebollo and Julio Montero Díaz, 'Las películas censuradas durante la Segunda República. Valores y temores de la sociedad republicana española (1931–1936)', *Estudios sobre el Mensaje Periodístico*, 16 (2010): 369–93.

Diario del Museo

AÑO II - NUM. 315 ADMON.: MUSEO ROCA PRECIO: 15 céntimos

¡¡Los peligros de la ignorancia!!

La ignorancia es uno de los males que conduce, por lo general, a la degeneración de la raza.

¡Cuánta y cuánta gente se encuentra hoy en la miseria, en el hospital, en un manicomio, precisamente por haber ignorado una serie de cosas cuya necesidad de ser conocidas de todo mortal comienza ya desde que el muchacho sale de la escuela para enfrentarse con el árido y difícil camino de la vida!

¿Quiere usted

Un grito

La mejor lección

Una visión de la realidad

Algo importante

No le hará reir

Figure 8.4 Front page and picture of the audience attending the museum published by Roca in his *Diario del Museo*, *c* 1935.

of the spirit of lust' – who 'claim that watching graphic exhibitions of such calamities, submitted supposedly as if they were a scientific discipline, our youth "should be good" and never go to visit the houses of prostitution'. The Catalan patriot holds his hands up to the heavens, ' [...] and we are tired of so much "scientific immorality"! Science must never serve as a pretext for such barely contained passions! [...] Friends, do not be fooled! Let us strive with all our might against this "pseudo-science", a dishonour to all lands of the world'.[51]

Conclusion

Throughout this chapter we have proposed a complex vision from multiple historical sources and narrative genres in an attempt to understand the Barrio Chino, the neighbourhood inhabited by working people and immigrants, fraught with misery and poverty, lit by the music and electric lights of shows and frequented by members of most of Barcelona's social strata. Our view has attempted to highlight three key ideas. First, an eclectic use of historical and literary sources has allowed us to show the complexity of the medical cultures concentrated in the neighbourhood streets before the risks of the lifestyles that prevailed there. Second, we have noted these tensions by closing in on the geography of a major urban artery, the Calle Conde de Asalto, where a variety of experts were available to the public, amongst which the even more diversely profane displayed their wares. Finally, we attempt to sketch the limits of the social medicine intervention from the ambiguity shown by the materials exhibited at the Museo Roca. Nevertheless, this chapter is written from an office lodged in the heart of what was the Fifth District, which today remains a vivid neighbourhood inhabited by tens of thousands. It is therefore unavoidable, as historians committed to the present, that we try to understand, to brief

51 Pere Tarrés Claret, 'Glossa. Los Averiados', *Flama*, 87, 20 August 1933, 6.

the reader on what has happened and what is still happening in this slice of Barcelona.

As already noted, under the municipal councils of the Franco regime the plan of radical surgery proposed by the Catalan republicans began to be partially carried out: the demolition of the headquarters of the Drassanes and from 1953 on, in the toughest part of the Barrio Chino, entire streets would disappear, such as Arco de Civés, the opening up of García Morato avenue, rupturing the continuity of Conde de Asalto and the associated side-streets.[52] The dictatorship ended; in 1979 the surgical operation was to continue at municipal level until the closing of the task, thrown into the fire of the pre-Olympic reforms and executed in the post-Olympic era, by a City Council run by a left-wing local government. The renaming of the district as 'El Raval' dates from this period, as well as the demolition of entire blocks to open up the Rambla del Raval, a continuation of the breach opened from the south by the Franco regime, the euphemistic term 'esponjament' describing the destroying of homes and opening of 'hard squares' which included aggressive architecture for tourism businesses (for example a cylindrical ten-storey four star hotel) or respectable cultural institutions (MACBA, the Film Library) and so on.

All this happened in a new social context and transpired with little obvious foresight over the influx of extra-European immigration during the last 20 years. Again, then, experts (planners, architects, doctors, journalists and officials) have acted 'on behalf of' the common inhabitants of the Raval, with little contact with the authentic, changing and dynamic reality of those who live or work there, who carry on in their daily struggle to inhabit the place and to combat health issues and other difficulties.[53] Problems have worsened since 2008: not only due to the so-called crisis, which has served to implement severe cuts in cultural, social security and health system, but also due to a change in the municipal government. Since 2011 the Catalan right is in power and it has proven very susceptible to the tourist lobby's successive assaults on the old Fifth District.[54]

52 Olga Sendra Ferrer, *Barcelona (1950–1976): Ciudad de márgenes. La construcción de la ciudad condal a través de la narrativa y la fotografía de posguerra* (PhD thesis, Princeton University, 2012); McDonogh, 'The geography of evil'.

53 Gary McDonogh, 'Discourses of the City: Urban Planning and Urban Problems in Post-Transitional Barcelona', in Setha Low (ed.), *Theorizing the City* (New Brunswick: Rutgers, 1999), 340–76. Francesc X. Hernández, Mercè Tatjer and Mercè Vidal, *Passat i present de Barcelona III. Materials per l'estudi del medi urbà* (Barcelona: Universitat de Barcelona-ICE, 1991).

54 Miquel Fernández González, *Matar al 'Chino'. Entre la revolución urbanística y el asedio urbano en el barrio del Raval de Barcelona* (PhD thesis, Universitat de Barcelona, 2012); 'La invención del espacio público como territorio para la excepción. El caso del Barrio Chino de Barcelona', *Revista Crítica Penal y Poder*, 3 (2012): 21–35.

Part III

Networks – experts and amateurs

9 The sky above the city

Observatories, amateurs and urban astronomy

Antoni Roca-Rosell and Pedro Ruiz-Castell[1]

Josep Comas i Solà (1868–1937), the highly renowned Catalan astronomer, died on 2 December 1937 in Barcelona, while the city was suffering the horrors of the Spanish Civil War; in particular, the bombardments by the Italian Fascist Aviation on the civilian population. Despite this desperate situation, the whole city came out to honour its local hero. His public funeral became an impressive tribute to a man who had 'placed' the name of Barcelona in the heavens, after his discovery of a minor planet – orbiting the Sun in the Asteroid belt – that was named after the city.[2] The ceremony was presided over by local political authorities who accompanied the funeral procession, together with the municipal police force and the city brass band. There were also representatives of the military and of different media – newspapers and periodicals for the general public, scientific journals and radio stations – and leaders of different cultural and educational institutions, union members, students and the general public.[3]

Comas i Solà had pursued his scientific career by combining highly technical astronomical observations with a successful popularization programme. Between 1915 and 1930 he discovered 11 asteroids, which led to a general recognition of his scientific excellence.[4] Much of his astronomical research

1 The Spanish Ministry of Science and Innovation has partially contributed to this research through the projects *Matemáticas, ingeniería, patrimonio y sostenibilidad en el mundo moderno y contemporáneo* (HAR2013–44643-R) and *Física, cultura y política en España* (HAR2011–27308). We are very grateful to Oliver Hochadel, Agustí Nieto-Galan, Charlotte Bigg and Maria Rentetzi for their helpful comments and suggestions on earlier drafts of this chapter.

2 José Comas y Solá, 'Nuevo planeta', *La Vanguardia*, 16 February 1921, 8.

3 See: 'El entierro de D. José Comas y Solá', *La Vanguardia*, 4 December 1937, 3; 'La muerte de Comas y Solá', *La Vanguardia*, 5 December 1937, 3; and Alberto Carsí Lacasa et al., *José Comas Solá* (Barcelona: Oficinas de Propaganda C.N.T.–F.A.I.–JJ.LL. Comité Regional de Cataluña, [1937]).

4 In fact, some other discoveries were perhaps more remarkable. Ralph Lorenz analysed the observation of Comas Solà and concluded that, in effect, he was the first astronomer to describe the atmosphere of Titan. Nevertheless, his observation entailed considerable difficulty and, thus, no other astronomer was able to repeat it. In that sense, the attribution of the discovery to Gerard Kuiper in 1944 is reasonable, given that his work could be more easily reproduced. See: Ralph Lorenz, 'Did Comas Solà discover Titan's atmosphere?', *Astronomy & Geophysics*, 38, 3 (1997): 16–18.

was conducted at the Fabra Observatory, one of the most relevant sites for astronomical observation that were created in Barcelona during these years. Since Comas i Solà was the director of the observatory, it soon became an urban site for conveying credibility, authority and expertise. Comas i Solà was convinced that astronomy – and science in general – was a useful and versatile cultural resource to transmit moral values and improve society. His popularization programme led him to collaborate on a regular basis with the newspaper *La Vanguardia*, in which he published weekly popular science articles from 1893 to his death. From bourgeois to working-class audiences, he was also an active public lecturer in cultural societies and institutions.

The case of Comas i Solà is particularly useful for depicting the important role of science – and more specifically astronomy – in Catalan and Spanish culture during the late nineteenth and early twentieth centuries. Furthermore, it may be used to analyse how local scientists shaped the social, political, economic and cultural values of the city. At the same time, the city influenced the development of astronomy, which was based on a set of regimes of observation and circulation of knowledge that included new discourses and forms of social organization. In urban spaces like the Fabra Observatory, metropolitan science was not only a form of knowledge and a matter of technical sophistication, but also a practical activity that involved the management of educational and civic spaces in the city.[5]

The history of Barcelona during the late nineteenth and early twentieth century cannot be written without taking into account the role of astronomy, and neither can the emergence and development of astronomy in the city be understood without considering the local context and the urban public. To this end, several facets of astronomical practice will be addressed regarding the creation of several observational sites and amateur societies, as well as the development of a programme for the popularization of astronomy that was particularly useful for the discipline of citizens in a growing industrial city.

Astronomy in nineteenth-century Barcelona

The professional status of astronomy in Barcelona during the second half of the nineteenth century was controversial. As the forestry engineer Rafael Puig i Valls bitterly complained in an article published in *La Vanguardia*, the university had no chair devoted to the subject and only one cabinet of topography and geodesy with a small telescope and some theodolites and tachymeters, while the Reial Acadèmia de Ciències i Arts de Barcelona (Royal Academy of Sciences and Arts of Barcelona, RACAB) possessed only some damaged astronomical instruments.[6]

5 See: Samuel Alberti, 'The Status of Museums: Authority, Identity, and Material Culture', in David N. Livingstone and Charles W.J. Withers (eds), *Geographies of Nineteenth-Century Science* (Chicago/ London: The University of Chicago Press, 2011), 51–72.

6 Rafael Puig y Valls, 'La astronomía en Barcelona', *La Vanguardia*, 22 December 1889, 1–2.

However, the RACAB managed to play a crucial role in the development of astronomy and meteorology in the city.[7] The exceptional events of the mid-nineteenth century – the solar eclipses of 1842, 1850 and 1860 – attracted the interest of some fellows of the Academy such as Onofre Jaume Novellas and Llorenç Presas.[8] Later, in 1883, under the presidency of the military officer and mathematician Angel del Romero, the RACAB strove to renew its activities. This led to a refurbishment of its building, which included a natural history cabinet and the transformation of the roof into an astronomical and meteorological observatory.

The creation of these new observational sites was perceived by the local elites as particularly necessary. On the one hand, astronomy had proved useful for the unification of time all over the world, which had a great impact on the development of railway networks during the second half of the nineteenth century.[9] Meteorology, on the other hand, was particularly relevant for public health issues, mainly in the cities. Therefore, Catalan cultural elites were keen to build such observatories. As they saw it, all important cities and intellectual centres abroad had such institutions, which offered a very useful service to citizens.[10] This agenda may well be identified with the spread of a heterogeneous *regenerationist* movement throughout Spain. This movement, which was a legacy of nineteenth-century Spanish liberal ideas, sprang up during the last quarter of the century and figured largely in the press, literature and art.[11] Also known

7 Founded in 1764 with the name of the Experimental Physico-Mathematical Conference, and with the aim of developing a course of Experimental Physics, it was soon recognized by the King with a Royal title. In 1770, it adopted a new name: the Royal Academy of *Natural* Sciences and Arts. In this early period, Astronomy was not its main activity. However, the members of the Academy began to be involved in astronomical observations and theoretical astronomy during the following decades. See: Agustí Nieto-Galan and Antoni Roca-Rosell (eds), *La Reial Acadèmia de Ciències i Arts de Barcelona. Història, ciència i societat* (Barcelona: Reial Acadèmia de Ciències i Arts de Barcelona – Institut d'Estudis Catalans, 2000); and Antoni Roca-Rosell, *Reial Acadèmia de Ciències i Arts de Barcelona. 250 anys d'història* (Barcelona: Reial Acadèmia de Ciències i Arts de Barcelona, 2014).

8 Carles Puig Pla, 'L'observació científica i l'enregistrament fotogràfic de l'eclipsi solar del 28 de juliol del 1851: un cas de col·laboració d'institucions barcelonines', in Carles Puig Pla et al. (eds), *III Trobada d'Història de la Ciència i de la Tècnica* (Barcelona: SCHCT, 1995), 287–302; Carles Puig Pla, 'Llorenç Presas i Puig 1811–1875. La Matemàtica Aplicada', in Josep M. Camarasa and Antoni Roca-Rosell (eds), *Ciència i Tècnica als Països Catalans* (Barcelona: Fundació Catalana per a la Recerca, 1995), 147–80; Francesc Barca Salom, *Onofre Jaume Novellas i Alavau (Torelló, 1787–Barcelona 1849). Matemàtiques i astronomia durant la revolució liberal* (Barcelona: SCHCT, 2005).

9 On public timekeeping and the development of time zones to coordinate schedules of railroad companies, see: Ian R. Bartky, *Selling the True Time: Nineteenth-century Timekeeping in America* (Stanford: Stanford University Press, 2000).

10 See, for example: 'Crónica', *La Vanguardia*, 17 November 1883, 1; 'Real Academia de Ciencias y Artes', *La Vanguardia*, 31 March 1901, 4–5; José Comas Solá, 'Notas Científicas – Un nuevo Observatorio Astronómico', *La Vanguardia*, 17 May 1902, 4.

11 See, for instance: Pedro Cerezo Galán, *El mal del siglo. El conflicto entre Ilustración y Romanticismo en la crisis finisecular del siglo XIX* (Madrid: Biblioteca Nueva, 2003). In general terms, *regenerationists* desired a global public works programme to create the infrastructure needed for a modern economy. See: Sebastian Balfour, *El fin del imperio español (1898–1923)* (Barcelona: Crítica, 1997); Juan Pablo Fusi, *Un siglo de España: La cultura* (Madrid: Marcial Pons, 1999). The basis for this programme was

in Spain as *Europeanization*, its aim was to emulate successful initiatives from abroad for the modernization of the country.

In 1883, the RACAB appointed José Joaquín Lánderer (1841–1922) to serve as supervisor of the new observatories. The astronomical observatory had to include an official time service, while the meteorological observatory had to coordinate a Catalan network of climatological stations in the region. An amateur scientist in several fields, mainly geology and astronomy, Lánderer was born in Valencia and lived in Tortosa, in the South of Catalonia, where he was a landowner.[12] Being a member of several societies, he was fully integrated into the European community of amateur astronomers. Lánderer's project was ambitious and covered not only astronomy and meteorology, but also some fields of geophysics, telluric currents and earth magnetism. In accordance with his plans, some instruments were purchased in France. Nevertheless, the building work on both observatories suffered delay and Lánderer resigned from his position.

The project had been partly funded by the Diputació de Barcelona (Barcelona Provincial Council) and the City Council.[13] Both institutions urged the RACAB to find a way of organizing the scientific services they intended to provide. The Academy asked one of its members, Josep Ricart Giralt, for the use of his chronometric office for the time service. Thus, in 1890, the new official time service in Barcelona was established to deal with the determination and unification of civil time.[14] A few years later, after the building was eventually opened, observations and calculations were conducted at the RACAB. In this new context, the Academy appointed Eduard Fontserè (1870–1970) to be in charge of the observatory, just after he had submitted his doctoral thesis in physics.[15]

education and science. Consequently, several reforms were promoted in the country in the name of progress, based on reformist ideals. See, for example: Alfredo Baratas Díaz, 'La cultura científica en la Restauración', in Manuel Suárez Cortina (ed.), *La cultura española en la Restauración* (Santander: Sociedad Menéndez Pelayo, 1999), 279–95; and José Luis Peset and Elena Hernández-Sandoica, 'La recepción de la cultura científica en la España del siglo XX: la Universidad', in Antonio Morales Moya (ed.), *Las Claves de la España del siglo XX. La cultura* (Madrid: España Nuevo Milenio, 2001), 127–51.

12 Rodolfo Gozalo and Víctor Navarro, 'Josep Joaquim Lànderer i Climent. València, 1841–Tortosa, 1922', in Camarasa and Roca Rosell (eds), *Ciència i Tècnica als Països Catalans*, 459–92.

13 On the building of the RACAB, see: Joan Puig-Pey Saurí et al., *L'edifici de la Reial Acadèmia de Ciències i Arts de Barcelona: Un testimoni viu de 250 anys d'història urbana* (Barcelona: Reial Acadèmia de Ciències i Arts de Barcelona, 2014).

14 In Spain, a unified official time was declared in 1900, due to begin in 1901. The different time references were established in the following years. See: Pere Planesas, 'La hora oficial en España y sus cambios', in *Anuario de Observatorio Astronómico de Madrid para el año 2013* (Madrid: Instituto Geográfico Nacional, 2013), 373–402.

15 Fontserè wished to become a teacher at the university, but no positions were available. Working for the RACAB, at least, ensured him an income of which to survive. On Fontserè, see: Antoni Roca Rosell, Josep Batlló Ortiz and Joan Arús Dumenjó, *Biografia del doctor Eduard Fontserè i Riba (1870–1970)* (Barcelona: Associació Catalana de Meteorologia, 2004).

Figure 9.1 Photograph of the RACAB building with its two towers, 1914.

The astronomical and meteorological services were finally set up in 1894 on the roof terrace of the RACAB on the Rambla, the famous boulevard in the heart of Barcelona – considered a centre of urban life.[16] The new RACAB building had two towers to be used as an observatory: one for the meridian circle – to determine the path of stars at the local meridian in order to calculate local time – and another for the equatorial telescope – for astronomical observations in general. Nevertheless, during the decade that elapsed between the initial project and the completion of the observatory, circumstances had changed dramatically. The extension of telegraph lines, electric lighting and pollution had all grown considerably in the city, creating adverse conditions for astronomical observations and even worse conditions for geophysical measurements.

Projecting a new astronomical observatory

As David Aubin pointed out, the modernization of Paris in the second half of the nineteenth century clashed with the increasing precision required for astronomical studies in the observatory of the city.[17] This led to an intense public debate regarding the possibility of moving the observatory outside the city to preserve isolation from noise, dirt, temperature gradients, vibration and lighting. Thus, the growth of modern cities seemed to affect the ideal of precision required for astronomical observations, in which the high sensitivity of astronomical instruments played a crucial role. The astronomical culture of high precision was closely related to the determination and unification of time, regarded as a crucial issue for the correct social functioning of cities, and Barcelona was no exception. In fact, some authors complained about how, despite the fact that the RACAB provided the general public with a standard mean time, the lack of a unified time in the city reflected 'the anarchy that ruled in Barcelona'.[18]

The city needed a properly equipped astronomical observatory. But where should it be placed? Barcelona lies on the Mediterranean coast and is surrounded by the mountains of the Serra de Collserola. Such a peculiar topography clearly influenced the decision. The highest point was the Tibidabo hill, an exceptional location for a new observatory, as the air was much cleaner than in the city centre. Tibidabo had acquired a very popular image during the 1888 Universal Exhibition, when a provisional construction was erected there for the visit of the Queen, and several official excursions were organized.[19]

16 This came together with the creation in 1893 of Free Chairs in several arts and sciences, including astronomy, subsidized by the Spanish Society for the Protection of Science (Sociedad Española Protectora de la Ciencia).

17 David Aubin, ' The fading star of the Paris Observatory in the nineteenth century: Astronomers urban culture of circulation and observation', *Osiris*, 18 (2003): 79–100.

18 Rafael Puig y Valls, 'La determinación de la hora media en Barcelona', *La Vanguardia*, 21 November 1891, 4 5. Many factors contributed to this sense of crisis in Spain around 1900 such as epidemics, floods, wars (the loss of the American and Asian colonies), mutinies, civil disorders, and the insecurity of the middle classes due to the emergence of an organized working class. On the significant role that anarchism played in Barcelona, see Chapter 6 in this volume.

19 See: 'Crónica local y regional', *La Dinastía*, 18 May 1888, 2; and 'Las bodas de plata del funicular del Tibidabo', *La Energía Eléctrica*, 28 (21), 10 November 1926, 289.

The idea to set up a new observatory was discussed during the last decades of the nineteenth century. There is evidence that it was debated in informal meetings of several RACAB members at the workshop of the watchmaker Adolf Juillard. Fontserè, who used to attend these *tertulias* or casual get-togethers, was officially requested by the RACAB to devise a project for the new observatory. He designed an astronomical, meteorological and seismological observatory to promote scientific services, including the official local time service, a meteorological service to study regional meteorology and to establish weather forecasting, and a seismological service to study the seismicity of Catalonia.

The project for the new observatory on top of the Tibidabo hill became a topic of general interest that was reported and discussed in the local press. Blueprints of the project appeared in newspapers.[20] Yet not everybody agreed; some local experts claimed that meteorological observatories in the heart of the old city were especially useful. They would provide data of the atmosphere that could be related to hygienic conditions – and therefore had to be obtained from the environment in which people lived. In that sense, the meteorological observatories located at the University of Barcelona and the Experimental Agricultural School were to prove more effective.[21] Therefore, while some supported the Tibidabo project, others argued against such an initiative, which was not considered to be of priority, given the urgent need for a modern sewage system and new schools for children.[22]

Finally, the project of the new so-called Observatory of Barcelona was submitted to the Diputació in 1895. Although promoters of the observatory were convinced that private support would also be necessary, its institutional backing was crucial.[23] The Diputació, however, did not approve any subsidy for the new observatory, as there was no guarantee of either the local council or any private initiative being financially involved.[24] Moreover, the Provincial Council regarded Tibidabo as the ideal place for the construction of an expiatory church, a tribute paid by the owners of the mountain to the Italian priest John Bosco for his visit to Barcelona in 1886. Nevertheless, the RACAB pursued the project while seeking further support.

From a scientific point of view, the Tibidabo hill seemed to be the best location for the new observatory. It had been redefined during the last decade of the nineteenth century, and had become a symbol of the ambitions of the

20 Dionisio Puig, 'El Observatorio del Tibidabo', *La Vanguardia*, 11 May 1895, 4; 'Un observatorio más', *La Vanguardia*, 10 May 1895, 4.

21 Note that these authors even claimed that meteorological observatories should be run by physicians. See: Puig, 'El Observatorio del Tibidabo'.

22 'Algo más acerca del Observatorio del Tibidabo', *La Vanguardia*, 19 May 1895, 1–2. Urban reformers were particularly concerned with improving the sanitary conditions. One of the most relevant figures was Pere García Faria, whose interest in health matters in areas affected by epidemics led him to develop the sanitation project for Barcelona. See: Miguel Ángel Miranda González, 'Pedro García Faria, ingeniero de Caminos (y arquitecto)', *Scripta Nova. Revista electrónica de geografía y ciencias sociales*, 10 (221), 15 September 2006, http://www.ub.es/geocrit/sn/sn-221.htm.

23 In fact, the RACAB project was influenced by the news that arrived from America regarding the creation of new observatories, such as the Lick Observatory, built from a bequest by James Lick.

24 'En la Diputación', *La Dinastía*, 15 January 1896, 2.

Catalan bourgeoisie, particularly in terms of how modernity should be incorporated into the city. This perception also influenced the final decision to set up a new observatory there. Similarly to what George Wise has noted for Albany's Dudley Observatory, the new observatory had to become 'an attempt at civic astronomy', a place to understand the heavens and to proclaim the city's proud place in the world – a site to see from and a spot to be seen.[25]

In other words, the Tibidabo hill was not only included as part of the city, but also became the centre of a new urban and metropolitan identity promoted by the upper classes and intellectual elites of Barcelona. It became a symbol of modernity and the place where several cutting-edge innovations were introduced, including the first cable railway, in an attempt to promote a new image of the city. As Garriga Bosch pointed out:

> Tibidabo was a reason of pride and prestige and an opportunity for political, religious, scientific, cultural and social promotion. The identity shaped was . . . a deliberate task of construction and assimilation of a new urban and social attitude . . . that channelled the ideals of a society devoted to the introduction of modernity into the city . . . The conquest of Tibidabo was in the social imagination a symbol of the modern expression of the city and of the ambition of the bourgeois classes.[26]

During the last decades of the nineteenth century, the upper classes of Barcelona had already found a 'natural' space on the lower mountainside of the Tibidabo hill to be used not only for second residences, but also for the organization of private events.[27] This new urban model constituted an intelligent and rational urban development, in which the management of transport, leisure and health issues played a central role.[28]

The Sociedad Anónima El Tibidabo company was in charge of urbanizing the hill and setting up a tram and a cable railway, the latter being the first one in Spain, which was officially opened in October 1901. This initiative was part of a larger project and was presented as a crucial element for the growth and progress of the city:

> It seems pointless to exhort the advantages that, in every respect, the rail and the urbanization of the hill must provide to the people of Barcelona

25 See, for instance: George Wise, *Civic Astronomy: Albany's Dudley Observatory, 1852–2002* (Dordrecht: Kluwer, 2004), p. 4.

26 Sergi Garriga Bosch, *El Tibidabo i la concepció metropolitana de Barcelona* (Master Thesis, Universitat Politècnica de Catalunya, 2012).

27 Note that the popular classes frequented Paral·lel Avenue and therefore found it more accessible to go to Montjuïc hill.

28 See, for instance: Antonio Torrents y Monner, 'El Tibidabo y la higiene ó el barrio modelo', *La Vanguardia*, 1 September 1903, 1–2; and Manuel Albi y Morera 'El Tibidabo y la higiene', *La Vanguardia*, 2 September 1903, 1. On the introduction of innovations for entertainment purposes and how such a project was related to land speculation, the promotion of tourism and hygienist and civilizing ideals through the development of transport networks, see Chapter 5 in this volume.

> [...] The work that has been done here is great, immense, not only from a hygienic, but also from a moral point of view; it is, indeed, mainly an educational and moral act.[29]

Thus, the construction of an astronomical observatory on top of the Tibidabo hill formed part of a broader and more ambitious enterprise. Urbanization had rendered astronomical observations increasingly difficult, forcing observatories to move to distant and isolated locations. However, in Barcelona, it opened up new possibilities for astronomical studies not far from the city centre.

The Fabra Observatory

In 1902, Camil Fabra (1833–1902), a textile entrepreneur and former mayor of Barcelona, made a private donation to the RACAB to secure the construction of the observatory.[30] Therefore, a new name was given to the projected observatory to honour the donor, who died the same year. After a period of complaints about the slowness of the works,[31] the Fabra Observatory was finally inaugurated in 1904. Following Eduard Fontserè's original plan, a design was created by the architect Josep Domènech Estapà (1858–1917), who was also member of the RACAB and had been in charge of the restoration of the building of the Academy.[32] The historicist architecture he developed was similar to that of other Spanish observatories erected during these years – such as the Cartuja Observatory, founded in 1902.[33] In fact, the neoclassical elements shared by these observational sites may be understood as an attempt to gain credibility and authority from the Royal Observatory of Madrid, the main astronomical institution in Spain – established in 1790 and one of the masterpieces of Spanish neoclassical architecture.[34]

Despite his rejection of 'modernism',[35] Domènech Estapà was one of the most influential architects in Barcelona. He was known for his eclecticism. Amongst his constructions one can find the Palace of Justice, built for the 1888 Universal Exhibition; the Cárcel Modelo (Model Prison); the Clinic Hospital; and the Faculty of Medicine. Therefore no one questioned his ability to carry

29 'Ferrocarril Funicular del Tibidabo', *La ilustración artística: periódico semanal de literatura, artes y ciencias*, 20 (1033), 14 October 1901, 7.

30 For a complete account of the origins of the Fabra Observatory, see: Antoni Roca-Rosell, *La Física en la Cataluña finisecular. El joven Fontserè y su época* (PhD thesis, Universidad Autónoma de Madrid, 1990).

31 'Notas bibliográficas – Anuario de la asociación de Arquitectos de Cataluña para 1903', *Revista de la Asociación Artístico-Arqueológica-Barcelonesa*, 4 (1903–1905): 122.

32 'El Observatorio Fabra', *La ilustración artística: periódico semanal de literatura, artes y ciencias*, 21 (1061), 28 April 1902, 7.

33 On the Cartuja Observatory, see: Manuel Espinar, José Antonio Esquivel and José Antonio Peña (eds), *Historia del observatorio de Cartuja, 1902–2002. Nuevas investigaciones* [CD-ROM] (Granada: Ayuntamiento de Granada, 2002).

34 On the Royal Observatory of Madrid, see: Manuel López Arroyo, *El Real Observatorio Astronómico de Madrid (1785–1975)* (Madrid: Dirección General del Instituto Geográfico Nacional, 2004).

35 In Spain, 'modernism' referred to the art movement known as 'Art Nouveau' in France and 'Jugendstil' in Germany.

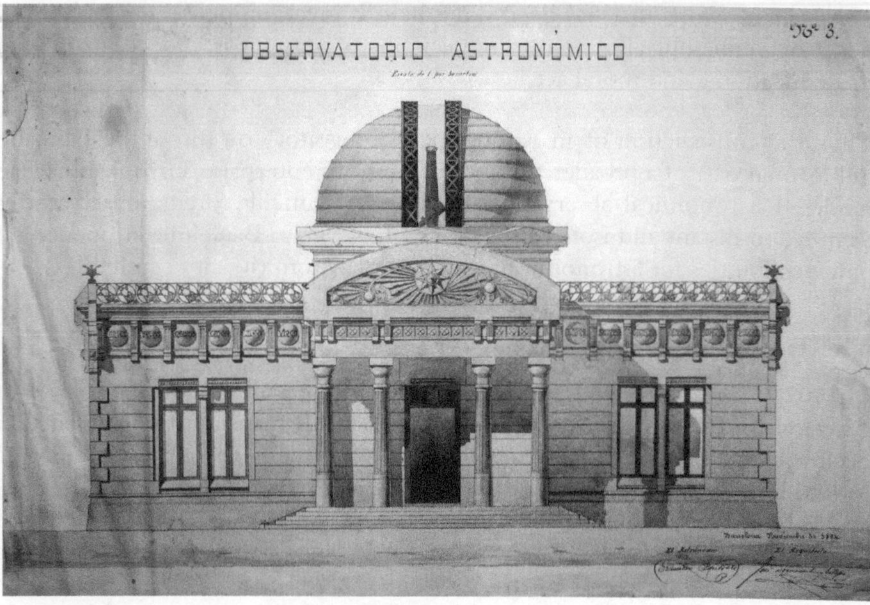

Figure 9.2 Josep Domènech's 1895 project.

out the project he had originally helped to conceive.[36] Fontserè, however, was marginalized. In fact, the directorship of the new observatory was left in the hands of Comas i Solà.

Comas i Solà had been elected to the RACAB in 1900. He was a very active astronomer and had published his observations in several specialized international journals. In the previous decade, Comas i Solà and Fontserè had been close friends while undergraduates at the Science Faculty of the University of Barcelona, but they followed different paths in life. Fontserè held a modest position in the RACAB while trying to become a professor at the university, a position he finally achieved in 1900. He was only elected as a member of the RACAB in 1909. Comas i Solà, on the other hand, had accepted a position as an astronomer at a private observatory in Sant Feliu de Guíxols, on the north coast of Catalonia. Comas i Solà's family was wealthy and he finally decided to leave his position in Sant Feliu and return to Barcelona, where he continued his astronomical activity in a private observatory set up at his home, named *Vil·la Urània*, and where he made important observations of Mars and Jupiter, as well as of asteroids, comets and other celestial bodies.

The directorship of the new astronomical observatory was an exceptional opportunity for Comas i Solà. The Fabra Observatory was to host the most

36 The contributions of Domènech Estapà are analysed in Ramon Graus, *Modernització tècnica i arquitectura a Catalunya 1903–1929* (PhD Thesis, Universitat Politècnica de Catalunya, 2012).

powerful equatorial telescope in Spain. Moreover, Comas i Solà was also to form a research group, probably the earliest in the exact sciences in Barcelona. Many of the discoveries he made – including the minor planets – required an accurate analysis of photographic plates and calculations, and in the 1920s the Fabra Observatory contracted the physicists Joaquim Febrer (1893–1970), Isidre Pòlit (1880–1958) and the mathematician Assumpció Serret (1894–1934) to deal with this work.[37]

The Fabra Observatory had to be built on top of a hill that was now fully integrated into the city and could easily be reached by the cable railway. Comas i Solà therefore had the opportunity to develop a programme here for the popularization of astronomy, which would put the new observatory at the centre of scientific sociability, making it a central site for the cultural and intellectual life of the city – not unlike what had occurred at the Paris Observatory in the nineteenth century.[38] Indeed, Comas i Solà's scheme to popularize astronomy at the Fabra Observatory became a success. According to him it even transcended class barriers:

> The poorest social classes have so far taken a very active part in this wonderful scientific movement [...] On the days of public lectures, under the great equatorial dome of the Observatory, one may see there many people of both sexes from all social classes showing deep respect for the event, the feeling of fraternity among the audience, the good manners and behaviour of all without exception, and the huge interest they all show for what they hear and see.[39]

Astronomy had captured the public imagination and was seen by authors such as Comas i Solà as an activity that could help to elevate the soul. In fact, he saw his popular astronomy programme as part of a peaceful crusade of public good and progress, a particularly useful tool for preaching morality and a social remedy in times of constant political strife. In Comas i Solà's words, astronomical knowledge helped human beings not only to find their place in the universe and experience intellectual delight, but also to be noble of mind and heart, to annihilate hatred and moral turpitude and to leave the fierce struggles of humanity behind.[40]

Captivated by the effects that science and technology had on their lives, the citizens of Barcelona visited the Fabra Observatory to listen to Comas i Solà, who intended to help them to better understand such changes. In fact, the Fabra Observatory was to play a crucial role, not only in the political economy of both professional and public science, but also in the shaping of the ideal of a modern city.[41]

37 Febrer and Pòlit became professors at the Faculty of Sciences of Barcelona. Serret, who died prematurely in 1934, was probably the first woman to become a professional astronomer in Barcelona.

38 See: Aubin, 'The Fading Star'.

39 José Comas Solá, 'Admiración y agradecimiento', *La Vanguardia*, 17 November 1904, 4.

40 Ibid.

41 See, for instance: 'El Observatorio en Tibidabo', *La Correspondencia de España: diario universal de noticias*, 10 March 1902, 1; and Orión, 'El Observatorio Astronómico', *La Vanguardia*, 4 December 1903, 4.

Figure 9.3 Comas i Solà during the visit of a group of women from the Instituto de Barcelona, 1911.

The University of Barcelona Observatory

Comas i Solà catered for the general public not only at the Fabra Observatory, but also at his private observatory Vil·la Urània. Indeed, the turn of the century was characterized by the flourishing of both private and institutional observatories. One of the reasons was the establishment of international scientific collaborative networks, leading to decentralization. In the particular case of astronomy, this meant a significant proliferation of observing sites.[42]

One of these new sites in Barcelona was to be found in the university.[43] In spite of the efforts to provide full astronomical training to science students,

42 The foundation of the International Union for Cooperation in Solar Research in 1904 illustrates the institutionalization of international collaboration in astronomy and astrophysics in a period characterized by the promotion of cooperative scientific research. See, for example: George E. Hale and Charles D. Perrine, 'International co-operation in solar research', *Astrophysical Journal*, 20 (5) (1904): 301–5.

43 During the first half of the nineteenth century, Spanish universities had taught Physical Astronomy as part of general physics courses. It was only in 1900 that a new subject, named Astronomy of the Planetary System, was added to the syllabuses. See: Antonio Moreno González, 'De la física como medio a la física como fin. Un episodio entre la Ilustración y la crisis del 98', in José M. Sánchez Ron (ed.), *Ciencia y sociedad en España* (Madrid: El Arquero – CSIC, 1988), 27–70, p. 69.

nineteenth-century Spanish university courses on astronomy were poor.[44] Barcelona was the first city in Spain to build an astronomical observatory attached to a university. Earlier attempts in previous centuries had been unsuccessful.[45] Some universities had meteorological stations, but astronomical observations were unusual and no specific buildings were set up to permanently house astronomical equipment and make systematic observations. The University of Barcelona Observatory was promoted by Ignacio Tarazona Blanch (1854–1924) and was conceived almost exclusively for the students' practical instruction.[46] The observatory was erected in the gardens of the new building of the University of Barcelona. The building was designed by the architect Elies Rogent in the early 1880s and was one of the first constructions outside the old medieval walls in the new district of Eixample.[47] This was an area of transition towards the new part of the city, in which the bourgeoisie gradually built their homes.

The Faculty of Sciences of the University of Barcelona owned not only an outstanding Repsold theodolite, but also a topographic theodolite and a sidereal chronometer, both of them acquired in 1899.[48] When Tarazona took up a position at the University of Barcelona, he sought to assemble these instruments. A permanent installation of such instruments was expected to help students to solve problems concerning cosmography and spherical astronomy, particularly those related to the apparent and real movements of stars. Bureaucracy delayed progress and it was not until February 1904 that the construction of the observatory buildings began.[49] The observatory's main instrument was to be an equatorial telescope from the Irish firm owned by Howard Grubb that was delivered to Barcelona in January 1905, soon after the University of Barcelona had completed the observatory.

44 Academic scholars repeatedly complained about the lack of facilities for practical training and education in the subject. See: *Memoria del Claustro de Profesores de la Facultad de Ciencias de la Universidad de Madrid, 25 de Febrero de 1898* [Archivo General de la Administración, Sección Educación. Caja 8102. Legajo 8874–1]. Cited in: Baratas Díaz, 'La cultura científica en la Restauración'.

45 A good example is the failed attempt in 1790 by the Chancellor of the University of Valencia to equip the institution with an astronomical observatory. See: Víctor Navarro Brotons, 'El conreu de les ciències', in *Cinc Segles i un dia* (Valencia: Universitat de València, 2000), 119–26.

46 Ignacio Tarazona received his scientific education at the universities of Valencia and Madrid during the late 1870s. He obtained his doctorate at the Central University of Madrid in 1880. He lectured at the University of Valencia from 1887 to 1898, and was in charge of the Meteorological Station of the university. Five years later, Tarazona moved to the University of Barcelona, where he was appointed Professor of Cosmography and Geophysics. He later added to this Chair that of Spherical Astronomy and Geodesy, before moving back to the University of Valencia.

47 See: Pere Hereu, *L'arquitectura d'Elies Rogent, Barcelona* (Barcelona: Col·legi d'Arquitectes de Catalunya, 1986); Carmen Isasa (ed.), *Elies Rogent i la Universitat de Barcelona* (Barcelona: Generalitat de Catalunya, 1988); and Mireia Freixa, 'Elies Rogent i la construcció del Paranimf de la Universitat de Barcelona', *Barcelona quaderns d'història*, 14 (2008): 63–80.

48 Ignacio Tarazona Blanch, *La fotografía solar* (Valencia: Tipografía Moderna, 1909), p. 60.

49 On Tarazona and the processes of building the astronomical observatories of the universities of Barcelona and Valencia, see: Pedro Ruiz-Castell, *Astronomy and Astrophysics (1850–1914)* (Newcastle: Cambridge Scholars Publishing, 2008), pp. 89–108.

Astronomy was to find its place in the academic world in Barcelona, in a fluid dialogue with the city. A new set of observational techniques and astronomical instruments as well as disciplining practices and forms of social organization were introduced to the university and taught to students. Nevertheless, access to the observatory was restricted to university and secondary school students.

Astronomical societies and amateur astronomy

The establishment of new observational sites in the city was also a result of the rapid increase in the number of amateur astronomers during the second half of the nineteenth century, and therefore the creation of a large and heterogeneous community of amateur scientists. Indeed, amateurs played a crucial role in the development of astronomy during this period.[50] Most of them were self-taught, and they conducted some important original research, which was duly recognized by the international scientific community. This was particularly the case for planetary, lunar and solar observations, at a time when government support for astronomical research was still very poor.[51] A good example is that of the new generation of young Catalan scientists who pursued their careers during these years on their own initiative, with a very weak institutional support. This had been the case of individuals such as Comas i Solà and Fontserè, the importance of their work having been acknowledged by the 1890s in various French astronomical journals – particularly those under the influence of Flammarion.[52]

By the end of the nineteenth century, the number of amateur astronomers in Barcelona had grown to such an extent that several authors demanded the foundation of an astronomical society in the city, under the institutional auspices of the RACAB and/or the assistance of individuals such as Comas i Solà.[53] However, it was not until 1910 that the Astronomical Society of Barcelona (SAB) was founded.[54] This initiative resembled those developed in many other

50 For an international perspective, see for instance: Allan Chapman, *The Victorian Amateur Astronomer: Independent Astronomical Research in Britain 1820–1920* (Chichester: Praxis, 1998); and John Lankford, *American Astronomy: Community, Careers, and Power, 1859–1940* (Chicago: University of Chicago Press, 1977).

51 See: Antoni Roca Rosell and Víctor Navarro Brotons, 'Reflexions sobre la ciència i la tècnica a la Restauració', in Georgina Blanes, Lluís Garrigós, Antoni Roca Rosell and Jon Arrizabalaga (eds), *Actes de les IV Trobada d'Història de la Ciència i de la Tècnica* (Alcoi/Barcelona: Societat Catalana d'Història de la Ciència i de la Tècnica, 1997), 181–90.

52 See, for instance, references to Lànderer, Comas i Solà, and Fontserè in: Bouquet de la Grye, 'Les Progrès de l'Astronomie en 1891', *Bulletin de la Société Astronomique de France*, 6 (1892): 151–3, 155; 'Annuaire Astronomique pour 1892', *L'Astronomie*, 11 (1892): 2; and Camille Flammarion, 'Les Progrès de la Société Astronomique de France', *Bulletin de la Société Astronomique de France*, 6 (1892): 159–63, here p. 161. See also references to Lànderer and Comas i Solà in: Camille Flammarion, 'Nouvelles observations sur la planète Mars', *L'Astronomie*, 11 (1892): 36–382, here pp. 371 and 377. On Flammarion, see for instance: Bernadette Bensaude-Vincent, 'Camille Flammarion: prestige de la science populaire', *Romantisme*, 19 (65) (1989): 93–104.

53 J. Canudas y Bordas, 'El cielo en la actualidad', *La Vanguardia*, 27 March 1897, 1.

54 'Sociedad Astronómica de Barcelona. Actas. Juntas Generales', Fons personal de Salvador Raurich, Biblioteca de Catalunya, Barcelona, Ms 4688.

cities, following the foundation in 1887 of the Société Astronomique de France and the creation in 1890 of the British Astronomical Association – organizations that not only promoted cooperation between professional and amateur astronomers, but also encouraged original contributions from amateurs to astronomical research and played a crucial role in the popularization of astronomy.[55] After its first year of existence, the SAB had about 400 members and was soon expected to be equipped with an observatory for the use of its members. The new observatory was to house instruments donated by Rafael Patxot, whose observatory at Sant Feliu de Guíxols was to be re-erected in Barcelona.[56]

Under the auspices of Fontserè and the musician and amateur astronomer Salvador Raurich (1869–1945), the SAB was intended to be a meeting point and the institution in charge of coordinating Spanish, and particularly Catalan, astronomical and meteorological amateurs. However, the foundation of this society exacerbated existing tensions and led to serious conflicts. As pointed out above, the marginalization of Fontserè from the new Fabra Observatory had been controversial, and had an impact in the organization of astronomy in the city. The consequence was a community of local astronomers divided into two groups centred on the figures of Fontserè and Comas i Solà, who had unsuccessfully attempted to control the new astronomical society.[57]

A few months after the foundation of the SAB, Comas i Solà accused Raurich of campaigning against him and his observations of Mars in the society's journal. Raurich was a prominent figure in the public sphere through his articles published in such periodicals as *Por esos mundos* and *Nuevo mundo*. What was at stake here was not only Comas i Solà's scientific status and prestige, but also his public reputation as a major scientific authority. As the references were not explicit, however, the Governing Body of the SAB – to which Raurich belonged – asked Comas i Solà to overlook Raurich's comments.

However, Comas i Solà left the society, which in his opinion had been set up specifically to antagonize him.[58] One year later, the Astronomical Society of

55 See: Camille Flammarion, 'La Société Astronomique de France', *Bulletin de la Société Astronomique de France*, 15 (1901): 205–217; and British Astronomical Association, 'The history of the British Astronomical Association. The first fifty years', *British Astronomical Association Memoirs*, 42 (1) (1989): 7–10.

56 The instruments were a double equatorial (visual and photographic) of 230 mm (9 inches) aperture and several accessories for spectroscopic, photographic, and micrometric work. See: 'Notes: The Astronomical Society of Barcelona', *Journal of the British Astronomical Association*, 22 (2) (1911–1912): 111–12. See also: Manuel Castellet (ed.), *Rigor científic, catalanitat indefallent: Rafael Patxot i Jubert (1872–1964)* (Barcelona: Institut d'Estudis Catalans, 2014).

57 This tension among the Catalan astronomical community is briefly addressed in Antoni Roca-Rosell and José M. Sánchez Ron, *Esteban Terradas (1883–1950). Ciencia y técnica en la España contemporánea* (Madrid: Instituto Nacional de Técnica Aeroespacial, 1990), 54–60.

58 Antoni Roca-Rosell, 'Una vida consagrada a la ciencia', in Antoni Roca-Rosell (ed.), *Josep Comas i Solà, astrònom i divulgador* (Barcelona: Ajuntament de Barcelona, 2004), 13–42. See also: Pedro Ruiz-Castell, 'Priority claims and public disputes in astronomy: E.M. Antoniadi, J. Comas i Solà and the search for authority and social prestige in the early twentieth century', *The British Journal for the History of Science*, 44 (4) (2011): 509–31.

Spain was founded, promoted and presided by Comas i Solà. An offshoot of the SAB, this new society aimed to encourage cooperation between amateurs and the few local professional astronomers.[59] The governing body of this society, later known as the Astronomical Society of Spain and America (SADEYA), was composed of 12 people, 10 of them living in Barcelona.

This situation eventually proved to be dysfunctional. There was not enough activity in astronomy for both societies to survive. The SAB soon made meteorology its main activity, partly encouraged by Fontserè's own interests and personal relationships with other important Spanish meteorologists. The SADEYA, on the other hand, had a higher level of participation and proved more committed to astronomy. While the SAB was dissolved in the 1920s, after the instauration of the Catalan Meteorological Service, the SADEYA has survived until today.

During the first decades of the twentieth century, both astronomical societies competed and played a vital role in the development of amateur astronomy and the popularization of the discipline in Barcelona. Several initiatives were launched in the city during these years as part of their popularization programmes, providing common ground for the interaction of a small group of experts, a few amateur astronomers and the general public. A good example is the Exhibition of Lunar Studies held at the University of Barcelona in 1912, under the auspices of the SAB.[60] The exhibition presented drawings, maps and photographs of the Moon, as well as astronomical instruments and some information about several astronomical societies.[61] The exhibition was a considerable success in Barcelona, with an attendance of tens of thousands, including visitors from almost all educational centres and cultural institutions in the city.[62]

Similarly, in 1915 the SADEYA began the organization of a Fiesta of the Sun, held every June – coinciding with the summer solstice. This festival, originally promoted in Paris by Camille Flammarion in 1904, has been organized every year by the Société Astronomique de France since then.[63] Open to the general public, several events were arranged, including astronomical lectures, projection of celestial images, social gatherings and banquets, as well as a guided visit to the Fabra Observatory.[64] Both societies organized public lectures in different places in

59 See: 'Notiz', *Astronomische Nachrichten*, 188 (1911): 15–6.

60 See: Agustí Nieto-Galan, '" . . . Not fundamental in a state of full civilization": the Sociedad Astronómica de Barcelona (1910–1921) and its Popularization Programme', *Annals of Science*, 66 (4) (2009): 497–528.

61 Salvador Raurich, 'Exposición General de Estudios Lunares, mayo–junio 1912', *Boletín de la Sociedad Astronómica de Barcelona*, 19 (1912): 532–4.

62 See: Salvador Raurich, 'Correspondence: The Astronomical Society of Barcelona', *Journal of the British Astronomical Association*, 22 (5) (1911–1912): 239–40; and Salvador Raurich, 'Memoria presentada por la "Sociedad Astronómica de Barcelona" al "Congreso de Sociedades Astronómicas" celebrado en el Observatorio Nacional de París en 22–24 de junio de 1914', *Boletín de la Sociedad Astronómica de Barcelona*, 43 (1914): 505–9.

63 See: 'General Notes – La Fête du Soleil', *Popular Astronomy*, 12 (1904), 500.

64 See, for instance: 'Notas locales', *La Vanguardia*, 26 June 1917, 2; Emigdio Fernández, 'IV Fiesta del Sol en Barcelona', *Revista de la Sociedad Astronómica de España y América*, 9 (67) (1919): 53–7; and 'Informaciones de Barcelona', *La Vanguardia*, 9 June 1921, 4.

the city on a regular basis, which were addressed to a wide range of audiences – including women.[65] One of the high points of these popularization initiatives was the organization of an International Astronomical Exhibition in 1921. It was the SADEYA that organized this event in the Palace of Industry – built for the 1888 Universal Exhibition in the Parc de la Ciutadella. Fully supported by the City Council, thousands of people visited the astronomical exhibition, in what could be interpreted as a rehearsal for the 1929 International Exhibition.[66]

The development of astronomy in Barcelona also benefitted from the overall development of the city in the second half of the nineteenth century. The press may be identified as one of these factors.[67] The last decades of the century brought about the definitive consolidation of Catalan newspapers, particularly in Barcelona.[68] Some members of the SADEYA were convinced of the guiding role that the press played in the public sphere, and accordingly they called on it for some support and publicity – particularly from local newspapers – of their popular activities.[69] Such agenda soon included new mass media technologies, particularly after the consolidation of commercial radio broadcasting in the 1920s.[70]

Furthermore, several scientific industries started to flourish in Barcelona in the late nineteenth and early twentieth century. This led to a whole set of alliances that benefitted the development of astronomy and its public cultivation. Some of these alliances were made explicit in 1921, when tribute was paid to Comas i Solà during the International Astronomical Exhibition.[71] Good examples of this are the public endorsements by scientific instrument makers such as Miguel Hidalgo Vega and José Vilaplana Enrich, and by watchmakers such as Emilio T. Besses and José Novelles Rizo.[72]

65 See, for instance, the review of the conference held in May 1921 by Comas i Solà at the premises of Acción Femenina: 'Informaciones de Barcelona', *La Vanguardia*, 27 May 1921, 3.

66 Emigdio Fernández, 'Discurso de Inauguración', *Revista de la Sociedad Astronómica de España y América*, 11 (1921): 78–9; 'La Sociedad Astronómica. Homenaje al Sr. Comas y Solá', *La Vanguardia*, 23 November 1921, 22; and 'Discurso de gracias de Don José Comas Solá', *Revista de la Sociedad Astronómica de España y América*, 11 (80) (1921): 109–11.

67 On the role played by nineteenth-century periodicals as mediators in the circulation of scientific knowledge, see: Geoffrey Cantor et al., *Science in the Nineteenth Century Periodical: Reading the Magazine of Nature* (Cambridge: Cambridge University Press, 2004).

68 Josep Francesc Valls, *Prensa y burguesía en el XIX español* (Barcelona: Anthropos, 1988), p. 197.

69 See, for instance: Fernández, 'IV Fiesta del Sol en Barcelona', p. 55.

70 On the lecture programming of Radio Barcelona and how astronomy found its place here, see Chapter 10 in this volume.

71 Antoni Roca-Rosell, 'L'exposició organitzada per la Societat Astronòmica d'Espanya i Amèrica el 1921. Josep Comas i Solà i el Regeneracionisme Científic', in Pasqual Bernat (ed.), *Astres i meteors. Estudis sobre història de l'astronomia i de la meteorologia* (Vic: Patronat d'Estudis Osonencs; Calvià: Edicions Talaiot, 2014), 229–44.

72 See: 'Lista de las adhesiones recibidas por la Sociedad Astronómica de España y América, con motivo del homenaje dedicado a don José Comas Solá, reunidas en el álbum que se entregó a dicho señor', *Revista de la Sociedad Astronómica de España y América*, 11 (80) (1921): 112–3; and 'Lista de las adhesiones recibidas por la Sociedad Astronómica de España y América, con motivo del homenaje dedicado a don José Comas Solá, reunidas en el álbum que se entregó a dicho señor', *Revista de la Sociedad Astronómica de España y América*, 12 (81) (1922): 15–16.

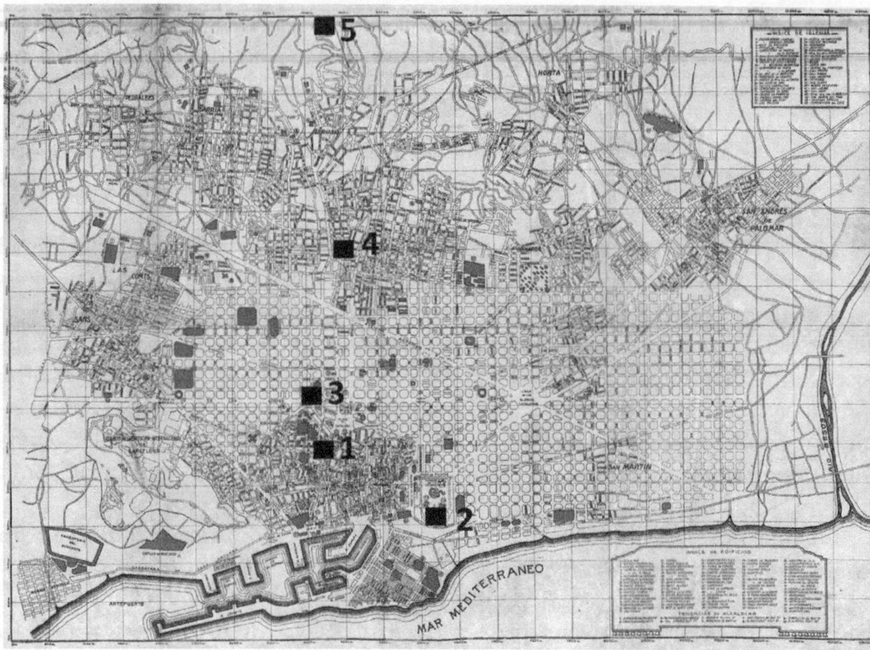

Figure 9.4 Map of Barcelona, 1929. 1. RACAB. 2. Palace of Industry (Parc de la Ciutadella). 3. University of Barcelona. 4. Vil·la Urània. 5. Fabra Observatory.

The optician Federico Font de la Vall, one of the pioneers in the development of telephony in Barcelona in the late 1870s, was also amongst those who publicly endorsed the tribute to the Catalan astronomer.[73] His name represents an example of the synergies between astronomy and the optical and electrical industries developed in Barcelona during the last decades of the nineteenth century. It suffices to mention the businesses of Juan Roca, Manuel Olió Pigem and Laureano Ychasmendi Errasti. The former owned a shop in Barcelona that in 1910 became the first one in the city making astronomical telescopes.[74]

Before that, Olió Pigem had set up a shop in the city named the Instituto de Óptica Olió, where glasses, microscopes, electrical apparatuses, meteorological instruments and telescopes were sold. When his brother in law, Laureano Ychasmendi, joined the business, it was renamed the Instituto de Óptica Olió e Ychasmendi.[75] By the late 1890s, Olió Pigem was one of the most active amateur astronomers of the city. The back room of his shop soon became the place where regular astronomical gatherings were held. In fact, it was there that the idea for the Astronomical Society of Spain was born and put forward.

73 See: Horacio Capel Sáez, 'Estado, administración municipal y empresa privada en la organización de las redes telefónicas en las ciudades españolas (1877–1923)', *GeoCrítica*, 100 (1994): 5–61, p. 6; and Angel Calvo, *Historia de Telefónica, 1924–1975. Primeras décadas: tecnología, economía y política* (Madrid: Fundación Telefónica, 2010), p. 31.

74 Josep M. Oliver, *Historia de la astronomía amateur* (Madrid: Sirius, 1997), p. 26.

75 Ibid., pp. 152–5.

Conclusion

The 1921 tribute to Comas i Solà may serve as a better understanding and an example of the status that astronomy achieved in Barcelona during the late nineteenth and early twentieth centuries. Astronomy had become a public good, as a result of the programme of public action led by the two astronomical societies founded in the city and by individuals such as Comas i Solà. Such a programme was an appropriation of similar programmes of popularization of astronomy that were successfully developed in countries like France and Britain, adapted to and shaped by the specificities of a growing city such as Barcelona.[76] In fact, it was presented as a useful tool for instructing citizens and guiding them in morality, particularly in troubling times. Science, and astronomy in particular, was publicized in terms of human and social regeneration, conveying tranquillity and providing the basis for the progress and welfare of modern society.[77]

It was thus that Barcelona witnessed the blossoming of new urban spaces in which astronomical knowledge and objects circulated widely. The access to and impact of astronomy went beyond social classes and gender. Public lectures brought astronomy to cultural associations and educational institutions, for the delight of those interested in astronomy. Furthermore, new observational sites were set up, both under institutional and private auspices. All this meant the consolidation of an important network of institutional and amateur astronomers in the city, most of whom were involved in popularization tasks.

As pointed out above, Comas i Solà stood out in this context, publishing regular articles in the daily press and lecturing at many associations and institutions. Furthermore, he opened up the Fabra Observatory to the citizens of Barcelona, integrating this site into the network of new civic spaces that not only brought together research, instruction, spectacle, popularization and leisure, but also contributed to the distribution of credibility, authority and expertise. Indeed, men and women from all social classes were welcomed in this building, located on top of a hill that had been integrated into the city and transformed into a symbol of urban modernity.

In summary, astronomy developed in Barcelona during the late nineteenth and early twentieth centuries as a system that was formed not only by astronomers, academic institutions and astronomical associations, but also by new urban spaces and observational sites that were engaged both in research and popularization, the press and new urban industries. This was a scheme that was shaped by the needs of a growing modern and industrial city, and not only proved to be very appealing for the general public, but also led to a very successful development of astronomy, understood in the broad sense.

76 The interrelationship between astronomy and the city in the context of a cultural history of science has been analysed in few occasions. See, for example, Mari E.W. Williams, 'Astronomy in London: 1860–1900', *Quarterly Journal of the Royal Astronomical Society*, 28 (1) (1987): 10–26.

77 See, for example: Antonio Vela, 'Un recuerdo de la exposición astronómica de Barcelona', *Revista de la Sociedad Astronómica de España y América*, 11 (80) (1921): 79–80.

10 The city in waves

Radio Barcelona and urban everyday life

Meritxell Guzmán and Carlos Tabernero[1]

On 14 November 1924, Darío Rumeu, the mayor of Barcelona, along with numerous political, military and religious authorities, officially inaugurated Radio Barcelona. The momentous event took place at the station's original studio, on the sixth floor of the Grand Hotel Colón on Plaça Catalunya, in the heart of the city. Two 29-metre antennas crowned the roof of the building, heralding the beginning of commercial radio broadcasting in Spain. Indeed, Radio Barcelona operated under the EAJ-1 official license, where the number 1 denoted the first official permit for a radio station. This permit had been issued exactly four months before, on 14 July, by the National Communications Office, according to the earliest Regulations of Radio Broadcasting in Spain, which had been approved on 24 June. The station finally received permission to start regular broadcasting on 12 November, although test transmissions had already been carried out for some time. And soon enough, its waves were piercing the air of the city, day by day, for those who had a radio set, from 6 to 11 pm.[2]

The establishment of commercial radio stations in the midst of the roaring 1920s constituted a defining episode in the long chain of transformations of the physical and epistemic structures of contemporary societies.[3] In this chapter, we will explore the entry of commercial radio broadcasting in the city of Barcelona, from the founding of Radio Barcelona in 1924 to the celebration of the International Exhibition in 1929.[4] We aim to contribute to the historical

1 We are particularly grateful to Arne Schirrmacher and John Postill for their invaluable help with their critical comments, as well as to Jim Cason for his extremely kind and quick copy-editing.

2 *La Vanguardia*, 14 October and 14 November 1924, 23 and 4 respectively; Juan de Gastambide, 'La emisora del Hotel Colón', *Radiosola*, October 1924, 9–10; 'Inauguración de las "Emisiones Radio-Barcelona"', *RBM*, 16 November 1924; the daily and weekly schedules published respectively in *La Vanguardia* and *RBM* in 1924 and 1925. See also references in note 4.

3 For an overview of the early development of commercial radio broadcasting, see: Asa Briggs, *The History of Broadcasting in the United Kingdom*, vol. 1: *The Birth of Broadcasting* (Oxford: Oxford University Press, 1961); Robert L. Hilliard, *Radio Broadcasting – An Introduction of the Sound Medium* (New York: Longman, 1985), Susan J. Douglas, *Inventing American Broadcasting* (Baltimore: Johns Hopkins University Press, 1987).

4 For the history of early radio broadcasting in Spain, and particularly in Catalonia, see: José Manuel Salillas and Agustín Rafel Llena, *50 años de EAJ 15: Radio Associació de Catalunya, Radio España de Barcelona* (Barcelona: Picazo, 1980); Rosa Franquet i Calvet, 'Els primers anys de les emissores

understanding of how the early spread of radio technology and the associated transformation of communication practices both reflected and helped shape contemporary Barcelona in the first third of the twentieth century.[5]

Radio was an early technology that allowed mass broadcast media to enter private domestic consumption spaces. This had a bearing on processes of production, circulation and management of knowledge, and therefore on the interests and uses of this technology on the part of corporate and institutional actors. We will not only focus on the industrialists and policy-makers who played central roles in the founding and early development of Radio Barcelona; we will also look at the contribution to these occurrences of amateur radio enthusiasts, aficionados and early listeners.[6] To do so, we will draw primarily on the content of *Radio Barcelona* magazine (1924–1938; denoted as *RBM* throughout the chapter); its forerunner publication, *Radiosola* (1923–1924); another radio magazine strongly linked to Radio Barcelona, *Radio-Lot* (1925–1929);[7] and what was published in relation to radio, and particularly to Radio Barcelona, in the leading Catalan newspaper, *La Vanguardia*, from 1923 to 1929.

The publications associated with radio stations in the early days of radio broadcasting constitute a fundamental research source despite the tensions generated between these textual sources and the aural nature of the medium analysed.[8] The engineer and radio buff José M. Guillén García founded *Radiosola*

barcelonines', *Annals del periodisme català* 23 (1993): 19–31; Rosa Franquet i Calvet, *Ràdio Barcelona: Setanta anys d'història (1924–1994)* (Barcelona: Diputació de Barcelona/Col·legi de Periodistes, 1994); Rosa Franquet i Calvet, *Història de la ràdio a Catalunya al segle XX: de la ràdio de galena a la ràdio digital* (Barcelona: Generalitat de Catalunya, 2001); Jaume Guillamet, *Història de la premsa, la ràdio i la televisió a Catalunya* (Barcelona: La Campana, 1994); Lorenzo Díaz, *La radio en España, 1923–1997* (Madrid: Alianza, 1997); Armand Balsebre, *Historia de la radio en España* (Madrid: Cátedra, 2001); Ángel Faus Belaus, *La radio en España (1896–1977)* (Madrid: Taurus Historia, 2007); and Meritxell Guzmán, *La ciència a la ràdio en la revista de Ràdio Barcelona (1923–1930)* (Master thesis: Universitat Autònoma de Barcelona, 2013).

5 The beginning and early spread of radio broadcasting can be located in urban contexts owing to the concentration of (economic, political) power, the proximity of factories and services, and the related easier access to infrastructure development. However, practices related to radio broadcasting would rapidly grow as well beyond metropolitan areas, as incidentally intended – and clearly stated, for instance, in the early days of Radio Barcelona: Joaquín Arrarás, 'La T.S.H. y el problema del campo', *RBM*, 16 November 1924, 6; José M. Guillén García, 'Proyectos y realidades', *RBM*, 19 September 1925, 6–7; Rafael del Caño, 'Apostillas a unos rumores', *RBM*, 19 September 1925, 13.

6 For an extensive discussion on media and their socio-cultural traits in contemporary societies, see John B. Thompson, *The Media and Modernity. A Social Theory of the Media* (Cambridge: Polity Press, 1995). Regarding the consideration of media as a wide range of everyday practices, see also Nick Couldry, 'Theorising media as practice', *Social Semiotics*, 14 (2) (2004): 115–32; and Birgit Bräuchler and John Postill (eds), *Theorising Media as Practice* (Oxford: Berghahn, 2010).

7 *Radio-Lot* was founded and directed by Eduard Rifà i Anglada, owner of an electrical household appliances distribution and manufacturing company, incidentally one of the co-founders and member of the board of directors of Radio Barcelona.

8 '[O]ur object of study no longer exists. The early pioneers of radio [...] lamented the 'ghastly impermanence' of their medium. Radio, and later television, developed first as live systems of transmission and recording technologies came later. [T]he vast bulk of output perished in the moment of transmission. The fleeting, unrecorded character of early radio seems obstinately to resist the possibility

on September 1923. It was edited and printed at Imperio Printer's, owned by Eduard Solà, also a radio enthusiast. The magazine focused on radio technology and was intended for enthusiasts. Not much later, on July 1924, *Radiosola* gave way to a new publication, *RBM*. This new magazine was devised as the official organ of the Asociación Nacional de Radiodifusión (National Radio Broadcasting Association – ANR hereafter), which had been established in February of that same year. Guillén also directed *RBM*, which was funded by advertising and subscribers' fees. Most content referred to radio technology and broadcasting concerning the industries involved (electricity and communications), the technological feats in the field achieved in the context of the city of Barcelona, and practical and technical information for enthusiasts, users and listeners. As a publication linked to the station, it also included programme scheduling and transcripts. Additionally, it included pieces referring to or taken directly from other radio stations and magazines (from Spain and other countries). Yet, in all, *RBM* worked basically as a one-way popularizing and advertising platform for radio technology and broadcasting, and so, for the institutions and corporations behind the venture of Radio Barcelona.

With this background established, we will look into the processes leading to the foundation and early development of Radio Barcelona and the visible transformations brought by the spread of radio production and consumption. As commercial audio content started crowding streets and households, we will examine, in the next two sections, the development of distinctive communication practices related to radio broadcasting and listening. We will discuss the construction of contexts of production, circulation and appropriation of knowledge, and the related power struggles involved in the inception and diffusion of this technology as a commercial medium. These dynamics were permeated with powerfully crafted techno-scientific discourses, which simultaneously reflected and constructed an urban-driven set of modernization guiding principles. These were unequivocally linked to people's everyday needs, concerns and expectations. The case of Barcelona and its new radio station in the decade that ended with the International Exhibition allows us to reflect on the relation between people's construction and transformation of their everyday-life contexts, the spreading and use of media technology, and wide-ranging processes of knowledge management.

The beginnings of commercial radio broadcasting in Barcelona

The boiling urban–industrial areas of early twentieth century were scenes of intense technological development, labour disputes and social unrest. Barcelona was no exception. In the 1920s, when the first radio stations started broadcasting, Spain was immersed in complex socio-political dynamics. General Miguel

of historical reclamation'; Paddy Scannell and David Cardiff, *A Social History of British Broadcasting: 1922–1939 Serving the Nation* (Oxford: Wiley-Blackwell, 1991), p. xiii. Concerning these publications and their relations to radio stations, see the references in notes 3 and 4.

Primo de Rivera's dictatorship (1923–1930) had just started its course, one of its main goals being to control the increasing social strife.[9] In this context, strategies were developed at all levels to overcome conflict and address an escalating plethora of needs and expectations. Budding telecommunication industries and companies were eager to enter promising markets of gradually more resourceful and personalized uses of their appliances and infrastructures. Politicians struggling to reach or keep power needed to establish themselves in this new medium.[10]

Thus, concerning the development of radio broadcasting, and given the strategic conjunction of corporate and institutional interests around the need for effective commercial and political means of propaganda, it is no coincidence that several governmental offices, including the Departments of War, Navy, Interior, Education and Labour, along with the industrial radio–electric sector, promoted the June 1924 Regulations of Radio Broadcasting. Radio Barcelona was a product of the will and investment of representatives of commercial interests linked to that sector in Catalonia and Spain, alongside the expedient official support. *Radiosola*'s founders, José M. Guillén and Eduard Solà, were focused on capturing and taking advantage of the growing interest in radio technology and broadcasting. It was with this in mind that they set to gather the cream of this industrial sector at *Radiosola* headquarters on 19 February 1924.[11] The result of that meeting was the establishment of the ANR, with the unambiguous aim of 'putting within everybody's reach the great discoveries of radio communication'.[12] Later on, as expected, representatives of the military, the political and administrative establishments, and even the Church, were at attendance alongside the industrialists at the launch of Radio Barcelona at Grand Hotel Colón in November 1924.

Competition was intense. Just two years after its foundation, Radio Barcelona already experienced financial trouble while competing directly with Radio Catalana, EAJ-13, founded on 20 June 1925. On 10 November 1926, Unión Radio, EAJ-7, founded on 17 June 1925 with the aim of centralizing radio broadcasting while reducing costs and increasing commercial profit, took over Radio Barcelona, and established Unión Radio Barcelona. This led to a rupture within the ANR, as small and medium-sized retailers were gradually yet noticeably alienated from decision-making processes. Unión Radio also took over Radio

9 See: Eduardo González Calleja, *La España de Primo de Rivera: la modernización autoritaria (1923–1930)* (Madrid: Alianza, 2005); and Eduardo González Calleja and José Luis Ledesma Vera, 'Conflictividad y violencia sociopolítica en la España de la primera mitad del siglo XX', in María Encarna Nicolás Marín and Carmen González Martínez (eds), *Mundos de ayer: investigaciones históricas contemporáneas del IX Congreso de la AHC* (Murcia: Universidad de Murcia, 2009).

10 On the geo-historical and socio-political implications of radio broadcasting, in relation to issues of collective identity, propaganda, indoctrination and the exercise of 'soft power', see: Alasdair Pinkerton and Klaus Dodds, 'Radio geopolitics: broadcasting, listening and the struggle for acoustic spaces', *Progress in Human Geography*, 33 (2009): 10–27. On corporate and governmental roles in the development and regulation of early radio broadcasting in Spain, see note 4.

11 Regarding the development of the electrical sector at the time, see Chapter 11 of this book.

12 Quoted in Franquet i Calvet, 'Els primers anys', p. 19. See note 4.

Catalana on 7 June 1929, as this station was then unable to compete with Unión Radio Barcelona. Also in 1929, in the midst of closures, takeovers and fusions among different stations, the ANR promoted Ràdio Associació de Catalunya (RAC, EAJ-15), which started broadcasting three and half hours a day in Catalan on 15 April 1930. The challenge to Unión Radio Barcelona became even more compelling exactly one year later. On 15 April 1931 Francesc Macià, the president of the new Catalan government, issued a decree allowing RAC to broadcast 24 hours a day. As a result, Unión Radio Barcelona also started to broadcast in Catalan. To round off such a highly competitive context, on October 1931, the ANR launched a new periodical, *Catalunya Ràdio* magazine, which battled *RBM* for the printed radio-linked space in the city (Radio Catalana had also published its own magazine, *Radio Catalana* from 1925 until its takeover in 1929).

Radiosola and *RBM* constituted, together with these other publications, a relentless effort of popularization of radio technology and its commercial uses. Radio technology had been drawing the attention of a rapidly increasing number of people in a context where unofficial test radio broadcasts had been carried out, albeit unevenly, since 1923. Quite different concerns and expectations were somehow coming together around radio technology and, through the foundation of commercial radio stations over time, fostered the swift emergence of this new technology as a widespread medium. As we have seen, the beginning of Radio Barcelona in 1924 was in effect the outcome of a multidimensional and highly competitive endeavour around and in conjunction with the development of radio technology. Such escalating drive featured the reciprocal encouragement and joint efforts on the part of the public and private sectors at large. But it was also matched, as we shall see later in the chapter, by the interest and initiative of radio enthusiasts and a rapidly growing number of listeners.

The veins of the city and its changing skyline

The city was suddenly expanding its ways and areas of operation, its web of spatial and epistemic practices in a new turf, the radio waves. And eye-catching modifications of the urban landscape echoed these fresh dynamics. The 29-metre antennas installed on the roof of Hotel Colón in downtown Barcelona became a feature of the skyline (Figure 10.1, left). Later, on 25 February 1926, a new station equipped with a new set of 45-metre antennas with increased power and reach, was inaugurated on the grounds of the Hotel Florida of the Tibidabo mountain (see Figure 10.1, right). With these antennas, Radio Barcelona overcame a few interference problems while broadcasts started positively reaching way beyond the city limits.[13]

13 Plenty of pieces were published in *RBM* and in *Radio-Lot* between 1925 and 1927 about the installation of the Tibidabo antennas and the technical advance they meant for Radio Barcelona and its listeners. The Tibidabo mountain became a crucial location regarding the techno-scientifically driven transformation of Barcelona's urban landscape. See Chapters 5 and 9 of this book and also: Radiófilo, 'El tríptico científico del Tibidabo', *RBM*, 27 February 1926, 4.

Figure 10.1 The antennas at the Grand Hotel Colón (left),[14] and the Tibidabo mountain (right).[15]

Yet, more construction work was necessary. Live broadcasts from Radio Barcelona required a great effort in technical and economic resources.[16] For instance, a huge wire network was built in order to connect the different venues where all kind of events took place with the central studio, first at Grand Hotel Colón (see Figure 10.2, left), and later at the Tibidabo Mountain (see Figure 10.2, right). The selected venues were the vital spots of the social, political, economic and cultural beat of the city, including Barcelona's Opera House, the Gran Teatre del Liceu, the Werner and Mozart concert halls, the Palace of Fine Arts, the University of Barcelona and the Reial Acadèmia de Ciències i Arts (RACAB), the City Hall and the Urban Property Chamber, the bullfighting ring and the soccer stadiums (FC Barcelona and RCD Espanyol).

The installation of antennas and the building of an ever-increasing wire network brought the development of urban radio communication out into public view. At the same time, the entrance of radio technology in the city's thriving socio-cultural nodes opened access to public spaces, which were nonetheless

14 *RBM*, 7 February 1925, cover.

15 *RBM*, 19 September 1925, cover.

16 The calls for attracting new members and for the increase and reliability of subscriptions and contributions were relentless in the pages of *RBM*. It also became clear from the very beginning that advertising was an essential funding source (in print and on the air), readily available given the medium's commercial application: J.A., 'La publicidad por radio', *RBM*, 29 November 1924, 1.

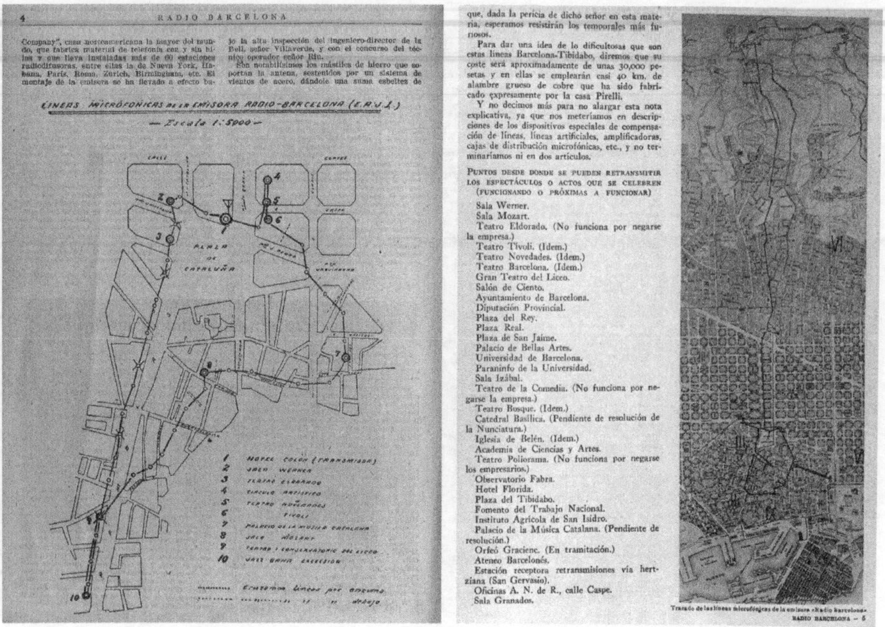

Figure 10.2 The wire networks built from the stations at the Grand Hotel Colón (left)[17] and the Tibidabo Mountain (right).[18]

prohibitive, and thus alien, for the many. As a result, the physical fabric of the city was hastily and radically changing together with people's everyday communication practices. In its unabashedly self-promotional style, *RBM* strove to put across a higher, collective meaning to the connection between the development of radio technology applications and people's practices that went way beyond the mere technical improvement:

> [...] we mulled over the momentous importance, for [the city of Barcelona] and for our people at large, of the creation of [the Tibidabo] radiotelephonic station, which daily sends its waves all over Europe and with them the affirmation of our rich and plethoric life [...] Have you pondered the importance of a powerful radio station for a nation? Have you grasped its value? A nation, sometimes, fights for air, sunk in a corner of the world. Its cultural, artistic, business life is overflowing; so rich, in blood cells, is its blood. It lives an existence disproportionate to its smallness. And in spite of all this, it must continue vegetating in oblivion, because the world is very big and that nation do not have

17 *RBM*, 16 September 1924, 4.
18 *RBM*, 19 September 1925, 5.

> the means to make its voice heard. [...] Radiotelephony waves overcome this setback and right this wrong. Radio stations waves are universal and do not need passport to cross borders. A nation full of life regardless of remaining overshadowed by circumstances, whether geographic or of any other kind, daily delivers to space its science, its music, its literature, its life as it is, and its voice is heard by all the nations in the world where its waves reach.[19]

Indeed, the construction works and the view of the finished infrastructures, as well as the pomp of the inaugurations, amounted to a sort of spectacular projection of what was happening, or what was openly endorsed to happen, at many levels of everyday practice. In this context, Radio Barcelona did not miss the opportunity to become the exclusive radio broadcaster worldwide of the grand techno-scientific finale of the decade, the 1929 International Exhibition.[20] To this end, it committed an important economic and technical effort, not only concerning the extensive wiring of the venues (see Figure 10.3), but also extending its presence to the actual events, particularly in the organization of Shortwave Conferences (Jornadas de Onda Corta) on November 1929.[21] International experts and enthusiasts gathered in the city at the time of the Exhibition to talk about radio technology. Guglielmo Marconi, the great engineer, inventor and entrepreneur, was among the renowned experts who visited the premises. However, Mussolini's senator and marquis, undoubtedly welcomed in the context of Primo de Rivera's dictatorship, clearly stated that he did not want to talk about politics or, as expected owing to his fame, radio while in Barcelona and its Exhibition.[22] Yet, the board of directors of the ANR did not miss the opportunity to talk to him and publicize the meeting.[23] So, even though Marconi's visit to the International Exhibition was rather ceremonial, he could not avoid serving somehow as sanctioning expert for radio broadcasting business and technological development. And so, Radio Barcelona actively contributed to this then ultimate and colossal landscape transformation of Barcelona, the International Exhibition. An extraordinary future seemed to be lying ahead. Just about everything looked as possible. And yet, Wall Street's Black Tuesday happened at the same time.

19 J. Ventalló, 'La importancia de la nueva emisora del Tibidabo', *RBM*, 13 March 1926, 4.

20 Franquet i Calvet, 'Els primers anys'.

21 The Shortwave Conferences were extensively advertised and subsequently covered in *RBM* from August till November 1929.

22 'Mr. Marconi doesn't want to talk about two issues: politics and radiotelephony. He is Senator in Italy and the inventor of radio. But he has already said many times all he had to say, so he does not want to talk again about these issues' (Antonio Puente de la Cruz, 'El sabio Marconi ha venido a Barcelona en su yate "Electra" para visitar la Exposición', *Diario oficial de la Exposición de 1929*, 19 October 1929). See also: *La Vanguardia*, 10, 11 and 13 October 1929, 8, 8 and 12 respectively.

23 *RBM*, 19 October 1929, 6–8; *Radio-Lot*, November 1929, 4–5.

La Exposición Internacional de Barcelona
y Unión Radio Barcelona E A J 1

Diversos aspectos del tendido de líneas subterráneas para la red microfónica de Unión Radio Barcelona y altavoces Standard Eléctrica, que se está efectuando en la Exposición.

RADIO BARCELONA — 9

Figure 10.3 The construction of wire networks in the 1929 International Exhibition.[24]

24 *RBM*, 19 January 1929, 9.

The weight of amateurs

Being the first commercial radio station in Spain did not mean that Radio Barcelona was entering a void. The world of the aficionado, the radio enthusiast, seemed to have been growing steadily for some time already, and, according to what was published in the pages of *RBM*, was crucial for the station. But who were these amateurs? They were defined and insistently acknowledged, even if self-promotionally, in the magazine, as an informally built community of independently working and self-sacrificing amateur scientists:

> None of the many branches of physics or chemistry has been as immensely researched as radio-electricity, albeit with the characteristic that this new science has not advanced owing to the work of the so-called men of science, but of ordinary aficionados, many of them lacking the basis required by any kind of research [...] Radio-electricity owes its development to the enduring and unknown aficionado who, from his modest work room, turned into a laboratory, and with the help of just a few odd parts, a hammer, a chisel and a file, builds, modifies experiments day by day, with the patience of a saint, in search of the unknown.[25]

Indeed, a section addressed to radio enthusiasts, called 'Carnet de T.S.H'. (that is, 'Wireless telegraphy ID card'), started running regularly in the newspaper *La Vanguardia* on 9 July 1924. Its aim was to solve misunderstandings generated by the newly passed regulations, as well as practical and technical issues. And, interestingly, it included the daily schedule of the radio stations (European,

25 José M. Guillén García, 'El progreso de la radio y los aficionados', *RBM*, 15 August 1925, 1. For further discussion on amateur scientists and their connections with professional scientists, see: Marc Rothenberg, 'Organization and control: Professionals and amateurs in American astronomy, 1899–1918', *Social Studies of Science*, 11 (3) (1981): 305–25; Yves Cohen and Jean-Marc Drouin (eds), Les amateurs de sciences et de techniques, *Cahiers d'histoire et de philosophie des sciencies* (special issue), 27 (1989); W. Patrick McCray, 'Amateur scientists, the International Geophysical Year, and the ambitions of Fred Whipple', *Isis*, 97 (4) (2006): 634–58. On related theoretical concepts such as boundary work and citizen science, and always in relation to science communication and media, see, for instance: Massimiano Bucchi, 'When scientists turn to the public: Alternative routes in science communication', *Public Understanding of Science*, 5 (4) (1996): 375–94; William Boddy, 'The amateur, the housewife, and the salesroom floor. Promoting postwar US television', *International Journal of Cultural Studies*, 1 (1) (1998): 129–42; Thomas F. Gieryn, *Cultural Boundaries of Science: Credibility on the Line* (Chicago: The University of Chicago Press, 1999); Gail Davies, 'Science, observation and entertainment: Competing visions of postwar British natural history television, 1946–1967', *Cultural Geographies* 7 (4) (2000): 432–60; Bernadette Bensaude-Vincent, 'A genealogy of the increasing gap between science and the public', *Public Understanding of Science*, 10 (1) (2001): 99–113; Morgan Meyer, 'On the boundaries and partial connections between amateurs and professionals', *Museum and Society*, 6 (1) (2008): 38–53; Gwen Ottinger, 'Buckets of resistance: Standards and the effectiveness of citizen science', *Science, Technology & Human Values*, 35 (2) (2010): 244–70; and Hauke Riesch and Clive Potter, 'Citizen science as seen by scientists: Methodological, epistemological and ethical dimensions', *Public Understanding of Science*, 23 (1) (2014): 107–20. On the role of amateurism in techno-scientific development, see also Chapter 9 of this volume.

mostly French at the start) whose waves could be caught in Barcelona before the launch of Radio Barcelona. Correspondingly, *RBM*, apart from the abundant practical and technical information addressed to the amateurs, persistently mentioned them as fundamental contributors, alongside corporations and institutions, to the efforts of both the ANR and the station. As Guillén wrote:

> Our prestige is not just a product of the tenacity of a few men: it is due to the collaboration we have had from the beginning of all true radiotelephony enthusiasts who, from the start, sided with us, and, also mainly, retail dealers of radio equipment.[26]

The mentioning of retail dealers suggests that the advent of radio broadcasting can be also deemed as a commercially driven transition of audio communication practices from that community of radio technology hobbyists and entrepreneurs to an actively sought-for and then steadily increasing amount of listeners.[27] In fact, CEOs, entrepreneurs, policy-makers and officials in charge of industries and administrative bodies were also, already or potentially, users and consumers of the goods they were producing or helping to produce. In our case, some of these characters, particularly those linked to telecommunication and its partner industries, could also be part of the ranks of radio amateurs. Therefore, the collaborative effort of enthusiasts at large with entrepreneurs and officials in the creation, preservation and improvement of the station was always emphasized in the contributions to *RBM*.

As mentioned, *RMB* was publishing lots of pieces on radio technology. Radio amateurs with some kind of either professional or amateur expertise could contribute articles ranging from technical information regarding stations, antennas and receivers to veritable essays on the particulars of radio communication, from theory to practice. This was an arena where amateurs and enthusiasts at large were also actively claiming their share as qualified authorities in the field. Their writings were published alongside a steadfast flow of news regarding the shock wave that represented the brisk diffusion of radio technology and the related practices. In addition, articles of all kinds from other specialized magazines, including translations, mostly from French and English publications, were also published. And finally, the increasing advertising mainly focused on

26 Guillén, 'Proyectos y realidades', *RBM*, 19 September 1925, 6.

27 'A popular theme in radio historiography portrays broadcasting as an inherently democratic use of wireless technology that developed first within the amateur community and was then captured by commercial interests. [An] examination of early radio retailing, by contrast, argues that the commercial exploitation of radio was itself instrumental in the transforming of broadcasting into a widespread cultural and technological practice'; Noah Arceneaux, 'The wireless in the window: Department stores and radio retailing in the 1920s', *Journal & Mass Communication Quarterly*, 83 (3) (2006): 581–5, p. 582. See also: Ruth Schwartz Cowan, 'The Consumption Junction: A Proposal for Research Strategies in the Sociology of Technology', in Wiebe E. Bijker, Thomas P. Hughes and Trevor J. Pinch (eds), *The Social Construction of Technological Systems: New Directions in the Sociology of History and Technology* (Cambridge, Mass.: MIT Press, 1987).

technical details and appliances, thus seasoning an end product that bore an inescapable didactic flavour.[28]

Surely, at the outset, radio technical traits responded, in a context of high industrial competition, to pressing needs of enhanced efficiency in immediate communication at several strategic levels (industrial, transportation, war). Yet, soon enough, the medium and its characteristics conceivably reflected people's penchant for self-management, whether individually or collectively, of knowledge production, circulation and consumption. Many people entered and explored the increasingly excited realm of radio waves by developing the technology, privately, in the same place where broadcasting listening, later on, was bound to happen, their homes. New and somehow startling communication highways proliferated owing to the unstoppable drive of hundreds, thousands of individual, anonymous radio enthusiasts who were soon carpeting the space with their immaterial, seemingly unnoticed exchanges. And their keenness to improve such manageable technology ran inexorably parallel to fancying a wider reach among fellow enthusiasts. Peer-to-peer exchanges gave way to communities where the sharing of self-managed content soon became a timely exploit:

> The enthusiast's contribution to the progress and advancement of radio is undeniable and recently General Ferrié, the illustrious director of the French Radiotelegraphy Services, upon giving a speech at the International Congress of Radio Amateurs, proclaimed out loud that [...] as a man of science, felt proud of counting himself among amateurs of each and every country, as he considered them as fraternal research partners.[29]

Meanwhile, different exchange initiatives that focused on the technological character of radio operation and use, such as the Radio-technical Laboratory that the ANR launched on 4 April 1927, implicitly recognized the importance of people's self-management. According to *RBM*, the Laboratory was a space where an active and fluid exchange between the station's professionals and the enthusiasts took place around the sharing of ideas and solutions concerning technical problems.[30] A little before, by the end of 1926, the ANR had promoted a project of scientific and cultural extension, which aimed to further promote radio technology, the station, and the knowledge of radio enthusiasts

28 On the technical particulars of private development of radio communication equipment and the connections with industrial and institutional developments, see the references in notes 3 and 4, and also: Hugh Aitken, *Syntony and Spark: The Origins of Radio* (Princeton: Princeton University Press, 1985); Juan Juliá Enrich, *Radio: Historia y técnica* (Barcelona: Marcombo, 1993); Gerald R.M. Garratt, *The Early History of Radio: From Faraday to Marconi* (London: IET, 1994); Gustavo Docampo Otero, *La radio antigua* (Barcelona: Marcombo, 2000); and Jesús Sánchez Miñana, *La introducción de las radio comunicaciones en España (1896–1914)* (Madrid: Fundación Rogelio Segovia, 2004).

29 Guillén, 'Progreso de la radio', *RBM*, 15 August 1925, 2.

30 *RBM* published 40 pieces about the Radio-Technical Laboratory between February 1927 and December 1929.

and the public at large of the processes of production of radio content and programmes and beyond: '[...] it is not enough to instruct and entertain [...] it must contribute to build the wireless scientific tradition we lack here today [and] raise the scientific and cultural level of our radio listener world'.[31] Next to these technological on-site educational initiatives, the ANR, also following examples from other countries, devised a much broader educational set of aims around science and technology education through broadcasting, as we shall see later in the chapter.

Yet, it is no coincidence that the acknowledgement of people's self-management of this technology was usually related to calls for financial help, mostly through subscriptions. The aficionados' unofficially created space of communication practices had to be extended to the thousands for commercial broadcasting funding purposes. And such commercial interest was inescapably linked to the broader social, political and cultural goals offered by the technical traits of the medium. The space of radio communication had to enter the private domestic setting of many potential listeners who did not require, in principle, sophisticated literacy or technical skills to join in:

> Is radiotelephony a more significant invention than the printing press? The services offered by this invention in the twentieth century exceed in part the services offered Gutenberg's invention. What book, what magazine is read by twenty million people? A political discourse [...] or a lecture may be listened to by ten and twenty million people. [And] it is not even necessary for those listening to know the alphabet. Radiotelephony is the most efficient and wonderful medium for humankind to spread ideas among men.[32]

In this context, Radio Barcelona focused on becoming a broadly communal service for the city, providing important information of practical (the weather and official time services) and socio-cultural interest.[33] The establishment of radio stations and the spread of radio as a household technology led to significant transformations in everyday communication practices.[34] In a context where radio was quickly becoming 'a vital factor in modern life and the home

31 'Radio Barcelona. Su proyecto de extensión científica y cultural', *RBM*, 30 October 1926, 3.

32 Pinochet, 'La radiotelefonía como factor de transformación social', *Radiosola*, May–June 1924, 12.

33 Guzmán, *La ciència a la ràdio*.

34 On the connection between processes of media appropriation and everyday (domestic, communication) practices, see: Hermann Bausinger, 'Media, technology and daily life', *Media, Culture & Society*, 6 (4) (1980): 343–52; David Morley and Roger Silverstone, 'Domestic communication – technologies and meanings', *Media, Culture & Society*, 12 (1) (1990): 31–56; John Fiske, *Media Matters: Everyday Culture and Political Change* (Minneapolis: University of Minnesota Press, 1994); Thompson, *Social History of Media*; Lana F. Rakow, 'The public at the table: From public access to public participation', *New Media & Society*, 1 (1) (1999): 74–82; Carolyn A. Lin, 'An interactive communication technology adoption model', *Communication Theory*, 13 (4) (2003): 345–65; Couldry, 'Media as practice'; and Bräuchler and Postill, *Media as Practice*.

which has no radio is not a home in the full sense of the word',[35] people translated radio broadcasting into a set of technological communication practices characterized by fragmented (that is, private, individualized) processes of appropriation and resignification (modes of listening). But these, in turn, were intertwined with the construction of aural public spaces (the extension of citizenship and identity by hearing the same at the same time as many others), or socio-cultural 'imagined communities' around the management and arrangement of daily social, communication, learning and working practices. The sense of immediacy, flexibility, ubiquity and simultaneity intuitively linked to the domestic adoption of radio was insistently stressed, according to commercial interests, in the pages of *RBM*, thus strongly implying that information and entertainment could be produced and consumed, circulated and (re)signified, individually and collectively, anywhere and at anytime.[36]

In an era when newspapers were already a driving social, economic, political, ideological and cultural force, and cinema was in the process of consolidating a communal (public) form of entertainment and information, radio broadcasting started breaking down the lines that physically and figuratively separated the edges of Barcelona. One remarkable fact associated with the establishment of Radio Barcelona was the entry of radio sets in hospitals, shelters, prisons and schools. This was possible through the voluntary and charitable contributions of Radio Barcelona subscribers. These initiatives constituted a direct intervention of the new-fangled communication practices in those spaces, permeating already-established power struggle dynamics in institutions that were and still are tools for the construction and consolidation of human communities. Under the designation of Radio Charity ('Radiobeneficiencia', which started in 1927), 16 radio sets had been donated and installed in such places in Barcelona by November 1929.[37]

Operating in a field of increasing industrial and business competition, Radio Barcelona strove to attract a large number of listeners. One of the most representative activities in this regard was the organization of Science–Art Festivals

35 *Wireless Weekly*, 20/20, 11 November 1932, 12.

36 For a critical overview on the construction of acoustic spaces, sonic geographies and/or aural communities, see also: Lesley Johnson, 'Radio and everyday life. The early years of broadcasting in Australia, 1922–1945', *Media, Culture and Society*, 3 (2) (1981): 167–78; Hilliard, *Radio broadcasting*; Mary S. Mander, 'Utopian dimensions in the public debate on broadcasting in the twenties', *Journal of Communication Inquiry*, 12 (2) (1988): 71–88; John Hartley, 'Radiocracy: Sound and citizenship', *International Journal of Cultural Studies*, 3 (2) (2000): 153–9; Kate Lacey, 'Towards a periodization of listening: radio and modern life', *International Journal of Cultural Studies* 3 (2) (2000): 279–88; David Hendy, *Radio in the Global Age* (London: Polity Press, 2000); Susan J. Smith, 'Performing the (sound) world', *Environment and Planning: Society and Space*, 18 (2000): 615–37; Rosalia Winocur, 'Radio and everyday life: Uses and meanings in the domestic sphere', *Television & New Media*, 6 (3) (2005): 319–32; and Pinkerton and Dodds, 'Radio geopolitics'.

37 *RBM* published more than 200 pieces (including articles, editorial and reports, and calls for fund-raising) between March 1925 and January 1930 concerning Radio Charity initiatives.

('Fiestas Científicoartísticas', also started in 1927).[38] These events were held in different venues and theatres of Barcelona and its extended metropolitan area, and one of their main objectives was to expand the station's operation. Radio artists and various musical performances brightened the Festivals, which included screenings of a film about the ANR and Radio Barcelona itself.[39] These events were reserved to members, who nevertheless could attend them with companions. Yet, the marketing for the new medium was guaranteed.

According to the pages of *RBM*, the number of listeners grew rapidly. Interestingly, while the magazine was insisting on the importance of contributing economically through a subscriber fee to its maintenance and improvement, the number of listeners was always estimated as much higher than the number of actual subscribers.[40] Piracy, or the taking of radio signals without paying a subscription, seemed to be widespread and was reported to be present in many countries (with special mentions to Germany, Italy, Russia [sic] and Great Britain).[41] This piracy accounts in part for the quickly increasing interest in the new medium as well as for the technical craftsmanship listeners showed from the very beginning, thus greatly contributing to the overall transformation of the associated communication practices. There were a number of industrially promoted initiatives to thwart piracy, including the establishment of a tax to the sales of radio sets. This measure was, as expected, vehemently contested by radio enthusiasts, and, in the end, seemed highly ineffective. As we have seen, there was plenty of information concerning amateur and aficionado approaches to the building and operation of radio sets. If control was a socio-political and industrial concern, and in particular regarding the new and highly powerful medium, access seemed a significant issue for the population at large, and many people in fact promptly became radio listeners. Many among the allegedly silent, shapeless and submissive majority of city dwellers, including scores of workers that by the thousands painstakingly repeated, day in, day out, their routine productive roles, were willing and ready to actively participate in new knowledge management processes.

These, in turn, once more as reported by *RBM*, generated mounting and unexpected conflicts, yet again, reproduced in many countries and cities. One

38 *RBM* published at least 30 pieces (mostly announcements) between January 1927 and June 1928 concerning science–art festivals.

39 *RBM* published three pieces about this film between 6 November and 11 December 1926.

40 According to *RBM*, by the end of 1927, there were only 7,500 subscribers from Barcelona's one million inhabitants, while the estimated number of listeners was around 200,000; 'Un poquito de estadística radioyente', *RBM*, 3 December 1927, 12.

41 *RBM* often reported the estimated number of radio listeners, usually comparing registered and unregistered listeners, and including comparisons with European countries. There were at least 17 pieces about or mentioning piracy between April 1926 and August 1929. 'Piracy', 'pirate' and 'hidden radio listener' ['radioemboscado'] are the terms used to address this issue. And attacks were fierce: 'The Anglo Saxons denote them without euphemisms with the sweet name of [...] *pirate*. [...] We [...] have the gallantry to soothe the term calling them simply *radioemboscado*. [...] Yet, their psychology [...] is quite complex [...] Nuances are infinitely varied albeit all come down to one common denominator: to listen and not to pay'. Crispo, 'Psicología del radioemboscado', *RBM*, 28 August 1926, 3 (italics in the original).

particularly illustrative example was the disputes developing in apartment buildings regarding the installation of antennas on the roofs. Owners and tenants quarrelled about the impact of such overwhelmingly growing metallic fields on the use, manageability, enjoyment and even the structural features and aesthetics of their dwellings:[42]

> [Radio] 'broadcasting' has reached a point in which it can be considered as one of the ordinary and normal amenities of life [...] something applicable to every tenant's normal conditions of life and the 'use' of his home [and] in the case of Barcelona [...] the immense majority of radio listeners live the sad life of buildings with many floors and even more doors.[43]

Access was essential. Choice was operative. And so, the domestic, individual and collective communication practices developed by people around the reciprocal development and diffusion of radio technology, were anonymously contributing to the steady changing of the epistemic fabric of the urban spheres they conformed, built and shaped:

> A wireless reception device must be considered as a something with high value for a tenant. It does not only provide entertainment and supplementary tuition, but market and science updates. [...] It is a veritable commercial and cultural medium. [...] Therefore, a landowner has no right to order outdoor antennas removed, if these do not bring about any kind of harm or danger either to the landowner or the rest of the tenants.[44]

Yet, to be sure, industries and governmental agencies, entrepreneurs and policy-makers who were radio enthusiasts and at the same time directly involved in the inception of commercial radio stations, were not just letting go their well-established authority. The new medium was a formidable platform to define and defend their stance on the highly and increasingly complex urban knowledge management dynamics. And so, a powerful educational programme was devised and promoted, articulated around different kinds of lectures, to compellingly enter the station's schedule.

Science on the radio and the craving for modernization

Radio Barcelona quickly developed a regular programme schedule. It started broadcasting live music from the studio. Yet, upon gradual consolidation and the knowledge of what other radio stations, at home and abroad, were doing,

42 *RBM* published at least 11 pieces about this problem between December 1924 and January 1928, yet five of these pieces were published early, between December 1924 and March 1925, showing the initial importance of this issue and the related annoyance (and pressure) on the part of early listeners.

43 Radiófilo. 'Antenas y propietarios', *RBM*, 20 December 1924, 4.

44 'Sobre el derecho de los inquilinos a montar antenas', *RBM*, 27 November 1926, 5.

programming diversified. Music kept on being the main axis, yet new subjects and sections appeared. Some worth mentioning were poetry recitals; sports and market chronicles; games and humour for children; the weather and official time services; and live broadcasts of concerts, plays and sporting events.

Listeners' tastes and preferences were crucial in this regard. *RBM* often featured pieces where listeners' participation was directly called for or shown.[45] Yet, *RBM* and Radio Barcelona served as powerful advertising and propaganda platforms for the institutions and industries that were behind the whole endeavour. And in this context, *RBM* devised its educational programming.[46] As we have seen, radio was portrayed as a key milestone in a sweeping techno-scientific utopia where the future was anything but a bright spectacle lit up with its promises of a better life:

> Radiotelephony, with its advent, has revolutionized much of our society's ways, solving a myriad of problems, some domestic, other of spiritual order; besides those concerning science, art and culture.[47]

The nurturing of that utopia was essential for those industrialists, policy-makers and a growing number of members of the academia searching for a means to control the production and management of knowledge. Modernization was and arguably still is a must.[48] The possibilities, as the potential reach of the medium, were immense; the opportunity, inescapable. Radio stations could take a very pro-active part in the education of the masses they were supposedly prone to connect

45 Two significant examples are the sections 'Nos escriben diciendo' (*Radiosola* and *RBM*, November 1923 to November 1925; 17 instalments), and 'Correspondencia' (*RBM*, March 1925 to June 1929, 184 instalments). These sections focused on technical issues and the access to the necessary technology. There were also articles featuring listeners' opinions regarding preferences concerning programming (V. 'Lo que prefiere el radioyente', *RBM*, 16 May 1925, 3–4); the quality of reception ('La emisora del Tibidabo. Tienen la palabra los señores radioyentes', *RBM*, 15 May 1926, 4–5); or, significantly, opinions about the educational applications of radio broadcasting. In this sense, the coverage of issues such as the one with the installation of antennas responded directly (as stated) to the reception of listeners' letters about the matter.

46 See: Marcel C. LaFollette, 'A survey of science content in US radio broadcasting, 1920s through 1940s: Scientists speak in their own voices', *Science Communication*, 24 (1) (2002): 4–33; Marcel C. LaFollette, *Science on the Air: Popularizers and Personalities on Radio and Early Television* (Chicago: The University of Chicago Press, 2008); Arne Schirrmacher, 'State-controlled multimedia for all? Science programs in early German radio', *Science and Education*, 21 (3) (2010): 381–401; and Marcel C. LaFollette, *Science on American Television. A History* (Chicago: The University of Chicago Press, 2012). See also note 10.

47 Ramón Pérez Vilar, '¿Cuánto paga usted, señor . . . ?', *RBM*, 27 February 1926, 8.

48 For a critical review on modernization theory in relation to the techno-scientific rationale of contemporary societies, as well as on its aesthetic, political and socio-cultural implications, see: Corinna R. Unger, 'Histories of Development and Modernization: Findings, Reflections, Future Research', *H-Soz-u-Kult*, 9 December 2010, http://hsozkult.geschichte.hu-berlin.de/forum/2010-12-001 (last accessed 15 October 2014); and Raymond L.M. Lee, 'Modernity, modernities and modernization: Tradition reappraised', *Social Science Information*, 52 (3) (2013): 409–24. Concerning radio specifically: Mander, 'Utopian dimensions'; Hartley, 'Radiocracy'; Lacey, 'Radio and modern life'; and Hendy, *Radio in the Global Age*.

with in no time. The aim was, according to *RBM*, to do away with what was seen as a highly harmful lack of interclass communication, while raising the cultural and technical level of the workers and thus increasing productivity, the nurturing basis of the allegedly so much needed and much-trumpeted modernization:

> If we could get those exemplary and selected men, now isolated in their laboratories, professorships, workshops, etc., in touch with the rest of the masses, these would automatically and infallibly ascend, with an eagerness for culture that would necessarily translate into vital improvement and perfection. Because culture is work, production of human things, doing art, science and morals, but also improving the crops, perfecting industrial methods, etc.[49]

And so, Radio Barcelona connected with the University of Barcelona on the occasion of the 1925–1926 school year opening. From then on, not only universities, but schools as well, were targeted. A remarkable campaign for radio pedagogy was launched at the same time that Radio Charity efforts helped introduce radio devices in schools among other institutions (see Figure 10.4).

Ramón Pérez Vilar, a civil schoolteacher and also the director of Radio Barcelona's educational programming, designed a comprehensive broadcasting project using the characteristics of the medium. His 'Contribución al estudio de un plan Radiopedagógico' ('Contribution to the study of a Radio-pedagogical plan') was a complete and amply reasoned outline of a radio educational curriculum published in five instalments. It combined what he defined as education, which included subjects such as religion, ethics, character formation and education of the will, obedience to superiors, school obligations and urbanity; and instruction, or the bestowing of knowledge, which included science, art and literature lectures, as well as crafts, math, languages and law classes. It also incorporated 'useful knowledge' about cooking, childcare, pets and radio technology, among other subjects, and entertaining spaces with games, clowns, jokes and the reading of literature, biographies, and texts on socio-cultural and religious matters. The programmes were supposed to last one hour and none of the sections had to run for more than 10 minutes.[50]

49 Rafael del Caño, 'La radiotelefonía y la universidad', *RBM*, 3 October 1925, 4. Similar arguments, coupled with the praise of the technical advantages or radio and statements on the formation and improvement of communities of formerly isolated listeners, were used in early advocating articles broadcasted on education, where radio was simultaneously a 'widespread element of domestic life' (S.A. Clynes, 'La influencia educadora de la radiodifusión', *RBM*, 6 December, 1924, 7) and 'the true people's Science' (Federico Jiménez Ruiz, 'La radio, Universidad popular', *RBM*, 10 January 1925, 3).

50 Ramón Pérez Vilar, 'Contribución al estudio de un plan Radiopedagógico I, II, III, IV, V', *RBM*, 3 December 1927, 6–7; 10 December 1927, 5; 17 December 1927, 3–4; 31 December 1927, 4–5; and 7 January 1928, 5. Pérez Vilar also gave one of the inaugural addresses for the radio teaching programming on 5 December 1928 (Ramón Pérez Vilar, "Radio Barcelona' y la enseñanza radiada', *RBM*, 8 December 1928, 10–12; 'La enseñanza por radio', *RBM*, 15 December, 1928, 4–7). All these pieces were part of scores of articles and news on this subject published in *RBM*.

La Escuela Nacional de Teyá y la T. S. H.

Un grupo de alumnos de la Escuela Nacional de Teyá escuchando los discursos de la inauguración de curso en la Universidad

Semanalmente es premiado un grupo de niños con la audición del jueves clásico de Toresky, con su instructivo "Colegio de niños"

Escuchando las lecciones de inglés que transmite Radio Barcelona

Figure 10.4 Teyá's public school and wireless telephony.[51]

51 *RBM*, 26 October 1926, 5.

The scope of action, with such a powerful tool in hand to reach at once an extraordinary amount of people, was seemingly limitless. Scientific, social and educational content beyond the field of radio communication, even though it amounted to a small percentage of the total programming, gradually became a key feature of Radio Barcelona. In accordance with the ANR's educational initiatives, and also following French and Anglo-Saxon models, and notwithstanding that the commercial trait soon consolidated as the main track of Radio Barcelona, different sorts of courses (of languages and typing, for instance) and actual academic lectures, however intended as accessible for a so-considered ignorant public, gradually entered the schedule.

According to *RBM*, at least 127 lectures were broadcast since December 1924, that is, from the very beginning of the station, until March 1929. These lectures were extensive regarding the range of covered topics, yet, apart from radio technology and broadcasting (24 lectures), three subject matters were perceptibly given more air time: astronomy (38), meteorology (11) and medicine–health (32).[52] Correspondingly, a significant amount of *RBM* content was related to the scientific programming of the station, including transcriptions of lectures, mostly excerpts, even though, in this case, radio technology and broadcasting led the input (17 of a total of 50 in the same period).[53]

Astronomy soon became the main topic in the lecture programming. It entered Radio Barcelona very early, on 26 January 1925, with a lecture on radiation by Josep Comas i Solà, director of the Fabra Observatory, one of the techno-scientific landmarks in Barcelona, since its inauguration in 1904, and closely related to the board of directors of the ANR and Radio Barcelona. An outline of Comas's lecture was published in *RBM* on 7 February.[54] Interestingly, astronomy had in common with radio communication a high ratio of intersection and the sharing of epistemological space between experts and amateurs, and there was arguably also a high ratio of intersection between both fields. Indeed, apart from lectures on the moon, planets, comets and other astronomical subjects, there was a section called 'Actualidades astronómicas' ('News from Astronomy') that was clearly aimed at astronomy buffs. Thus, it does not come as a surprise the importance of this topic in the station programming.

Meteorology bore a close relationship with radio communication. On the one hand, enthusiasts were concerned with the impact of meteorological phenomena in the transmission and reception of radio waves. On the other hand, meteorology was one of the original strategic applications involved in the development of radio technology, above all in relation with navigation. And

52 Guzmán, *La ciència a la ràdio*. Other topics were: academic (sciences – other than the ones mentioned – and history: 14), and institutional socio-cultural events (eight).

53 Other topics were: astronomy (three), meteorology (nine), medicine–health (six), other sciences (four), and miscellaneous content (11). For further discussion on radio and educational science content, see LaFollette, 'Survey', and LaFollette, 'Science on the air'.

54 'Extracto de la conferencia dada por D. José Comas y Solá el 26 de enero sobre "Radiaciones"', *RBM*, 7 February 1925, 8. On the astronomer Comas i Solà see Chapter 9 of this book.

when commercial radio stations started operating around 1920 it soon became clear the practical advantages this technology could offer beyond the urban realm, that is, to farmers.[55] That is how, after the inauguration of the Tibidabo station and its possibility to broadcast to the rest of Catalonia, Eduard Fontserè, founder and director of the Catalan Meteorological Service (since 1921), and also colleague (and rival) of Comas i Solà at the Fabra Observatory, actively promoted the broadcasting of meteorological information through Radio Barcelona. There was also some pressure in this direction from articles and notes published in *RBM*. Finally, on 2 August 1926, just a few months after the beginning of operation from Tibidabo, the weather report started being broadcast. A little later, on 31 January 1927, the Agricultural Meteorological Service of Radio Barcelona was created. On such occasion, Eduard Fontserè gave the first broadcast lecture on meteorology, vindicating its rank as a science. Radio was, thus, not only a channel for the practical application of a particular field of knowledge like this, but also a vehicle for disciplinary disputes in need of widespread support and which, in turn, configured an overtly pedagogical view of science and technology as a whole. In this particular case, most lectures addressed technicalities and applications of meteorological information.[56]

Finally, medicine–health was already a very popular subject in mass media when the first radio stations started operating. The press and cinema were already two extremely important means to put into operation and publicize health campaigns. These were extremely important for policy-makers and industrialists at the time, as they worked, on the one hand, as mechanisms of control of the population, above all in relation to the living conditions of workers and their families and the associated labour conflicts, and on the other, consequently, as a powerful techno-scientifically based indoctrination means to improve productivity.[57] On 9 February 1925, a lecture on care of newborns by Dr Polión Buxó Izaguirre was broadcasted and so, very early in the operation

55 'Un proyecto de transmisión de meteorología agrícola en Cataluña', *RBM*, 12 September 1925, 3–4.

56 Before his 31 January 1927 broadcasted lecture, Fontserè had published an extensive article on the matter in *RBM*: Eduardo Fontseré, 'La previsión del tiempo por T.S.H'., *RBM*, 4 September 1926, 4–6. On Fontserè, see Chapter 9 of this book.

57 See: Carlos Tabernero and Enrique Perdiguero, 'Cine y medicina: imágenes sobre la salud y la enfermedad', in Serena Brigidi (ed.), *Cultura, Salud y Cine* (Tarragona: Publicaciones URV, 2016) (in press). On the use of medicine–health content in early radio broadcasting and its implications, see: LaFollette, 'Survey'; LaFollette, 'Science on the air'; and also: Isabel Jiménez-Lucena et al., 'Un discurso sanitario para un proyecto político. La educación sanitaria en los medios de comunicación de masas durante el primer franquismo', *Asclepio*, 54 (1) (2002): 201–18. For further discussion on the connection between everyday-life media practices, regarded as inherently techno-scientific, and the circulation of scientific and technological knowledge, see: Carlos Tabernero, Isabel Jiménez-Lucena and Jorge Molero-Mesa, 'Scientific-medical knowledge management through media communication practices: A review of two opposite models in early 20th century Spain', *Journal of History of Science and Technology*, 6 (Fall) (2012): 64–84; Carlos Tabernero, Isabel Jiménez-Lucena and Jorge Molero-Mesa, 'Colonial scientific-medical documentary films in 1940s Spain as legitimating tools for Franco's regime', *Manguinhos* (2016) (in press).

of Radio Barcelona, initiated the medical–health lecture programming. Most content was about childcare, infectious diseases and dental care.[58]

Conclusion

The launch of Radio Barcelona illustrates the development and quick consolidation of commercial radio broadcasting as a fundamental urban communication infrastructure in the 1920s. As we have seen, it occurred as a function of the steady growth of experience with the medium on the part of a wide variety of social, institutional and industrial actors. The entirety of processes involved, that is, from technological developments, the establishment of stations, such as Radio Barcelona, and production of content, to application and consumption patterns, brought along transformations of a wide variety of technology mediated communication practices. Radio waves suddenly pierced and overlapped actual spatial barriers with an increase of the complexity of individual and collective communication habits.

Owing to its technical features, conducive of unthought-of communication and consumption possibilities up to that period, the transformations associated to radio broadcasting were not only of the physical landscape, but also of the epistemic fabric of communities, such as the city of Barcelona. Radio entered the most personalized of environments, the household. And people's embedment of radio in their everyday lives at all levels brought forth a new and essential space where fundamental power struggles occurred, such as the dynamics of granting and sharing authority, responsibility and values in the increasingly technological society it entered. The technical work of amateurs as well as piracy and conflicts such as the disputes about the installation of antennas on building roofs speak of people's drive for access to information, to knowledge management processes. Conversely, the devising of a thorough radio educational programme illustrates the need for social control on the part of the institutional and industrial promoters of the station.

Indeed, the beginnings of Radio Barcelona highlight how a new mass medium becomes a key turf where everyday actions and interactions between individuals, groups and institutions take place. Accordingly, people shaped fresh urban physical and epistemic landscapes, obstinately transforming Barcelona, while developing radio technology and infrastructures, through their increasing use and sophistication, and by simultaneously changing daily communication, social, learning and working practices.

And so, on a late afternoon already flowing into the evening in Barcelona in the 1920s, at around 6 pm, in a good number of homes, opulent or modest, or somewhere in between, men and women carefully manipulated a recently acquired, in some cases self-built, and possibly odd-looking device. A radio set.

58 Guzmán, *La ciència a la ràdio*.

It might not have been easy to tune it to perfection. Some persistent, delicate tweaking was surely required. Yet, the dwellers of those homes were determined. They were experienced or quickly becoming so. And on their own or gathered in groups; entire families while having dinner, with playful and possibly thrilled children; friends celebrating; or also bedridden people, sick or elderly; and those lonely or dispirited as well; all separated by walls, streets and neighbourhoods; all somehow together listen, at the same time, mostly to music, live or recorded, played from the studios of Radio Barcelona, or from one of the city's concert halls. But they also listened, either with undivided attention or while taking care of their daily domestic tasks, to some news, poetry recitals, perhaps a lecture, or some freshly devised magazine programmes. And also, inescapably, they had to go through the necessary, possibly already annoying, commercials inevitably piercing the raw information, the educational content, and the predominant, surely long-awaited entertainment. These were the first years of radio broadcasting in Barcelona. It did not last long in its daily schedule, just from 6 to 11 pm. Yet, those who could afford a radio set, or some kind of connection, most likely looked forward to it. The city was suddenly in waves.[59]

59 Details and nuances of the narrative featured in this closing paragraph are taken or inspired by the articles from *Radiosola* and *RBM* mentioned in the chapter, as well as by: Josep Broquetas, 'Nostra emissora', *Radio-Lot*, April 1926, 6–8; and Maria Cinta Balagué, 'La radio en la llar', *Radio-Lot*, June 1927, 2.

11 The city of electric light

Experts and users at the 1929 international exhibition and beyond

Jordi Ferran and Agustí Nieto-Galan[1]

In 1929, a group of scientists, engineers and entrepreneurs founded the Asociación Española de Luminotecnia (AEL). The association aimed to spread 'modern' methods for the 'scientific' use of light and its practical application to hygiene, traffic regulation, decoration and home comfort.[2] Just a year later, in 1930, the AEL published a short story that described how electricity had changed the everyday life of the average Barcelona citizen ('for the better').[3] In the advertising tale, a fairy brought a man into a technophilic world in which the evident benefits of electricity in the home – in the bathroom, dining room and entrance hall – and in the city – on the tramway, in the office and factory – were presented. The story accentuated the man's transportation into a 'dream', only comparable to that of visitors to the Exposición de la Luz (EL, Exhibition of Light) in the context of the city's 1929 International Exhibition.[4] This short story was part of the AEL's broader marketing campaign to attract new customers, as well as a sign of the enormous public impact, already exerted by the new source of power and light in Barcelona in 1929: from the Montjuïc electric magic fountain and the spectacular electric lighting on the exhibition grounds, to the promising application of electricity to street lighting and pioneering household appliances.

The first evidence of electric lighting in Barcelona stretches back to 1875, in a spotlight at the city dock. A Gramme dynamo moved by a frigate fed an arc lamp to light about 750 metres of La Rambla: from the Porta de la Pau to the Opera, the Liceu.[5] In the same year, a permanent installation was built to

1 We would like to thank Graeme Gooday and David E. Nye for their valuable comments on earlier drafts of this chapter.

2 This association, established in 1929, sought to 'disseminate the knowledge, advantages and importance of modern lighting methods and the rules of scientific use of light for private and national economy: hygiene, traffic safety, decoration, comfort, etc.'., AEL, *Estatutos de la Asociación Española de Luminotecnia* (Madrid: Gráficas Reunidas, 1929), Article 1.

3 L.S. Splendor, 'Alegorías de la Luz. Cuento de Invierno', *Novela Luminosa*, I (1), 1930. It was part of the advertisement campaign surrounding the Exhibition of Light.

4 Splendor, 'Alegorías de la Luz', 6.

5 Francesc Cabana, *Fàbriques i empresaris. Els protagonistes de la Revolució Industrial a Catalunya* (Barcelona: Enciclopèdia Catalana, 1992–1994), p. 183. See also: Joan Carles Alayo, *L'Electricitat a Catalunya: de 1879 a 1935* (Lleida: Pagès, 2007). Tomàs Josep Dalmau bought a Serrin's arc lamp in France. The test was conducted on 13 May 1875.

light the foundry in La Maquinista Terrestre y Marítima, a metallurgical factory set up in the neighbourhood next to the seafront, La Barceloneta.[6] Electricity also became an exciting novelty at the 1888 Barcelona Universal Exhibition, lighting a number of pavilions, streets and a magic fountain.[7] The ideal 'electric tour" of the 1888 Exhibition, with more than 2,000 Edison electric lamps, passed through La Rambla, Passeig de Colom (Hotel International), the exhibition, and on its grounds, the magic fountain and the night celebrations at the maritime display.

According to economic historians, 1911 was the *annus mirabilis* of Catalan electrification. At that time, rivalry between several firms steered the construction of thermal and hydroelectric power stations and lowered the cost for new users.[8] Against this backdrop, the 1915 International Exhibition of Electrical Industries and their Applications was initially conceived to celebrate the city's early electrification. In spite of its failure – World War I put an end to the project – it can be considered to be a symbolic antecedent of 1929.[9] Following the war, from 1918 to 1927, electricity production grew by 220 per cent.[10] The annual consumption per inhabitant in Catalonia in 1922 was 232 kWh, more than twice the Spanish average (92 kWh), but far from the 870 kWh per capita in Canada, 550 kWh in Norway or 485 kWh in the United States.[11] Subsidiary electricity companies such as AEG, Pirelli, General Electric, Westinghouse, Brown Boveri and Phillips grew considerably in the Barcelona area. They sold their products to Spanish markets and attracted a qualified labour force.

From the 1880s onwards, in Barcelona and elsewhere in western industrialized cities, electricity played a crucial role in the profound technological changes

6 Jordi Maluquer de Motes, 'Los pioneros de la segunda revolución industrial en España: La Sociedad Española de Electricidad (1881–1894)', *Revista de Historia Industrial*, 2 (1992): 122–42. See also: Luis Urteaga (ed.), *L'electrificació de Barcelona, 1881–1935* (Barcelona: Arxiu d'Història de la Ciutat de Barcelona, 2013).

7 Antonino Suárez Saavedra, *La Electricidad en la Exposición Universal de Barcelona* (Barcelona: Tipografia de los sucesores de N. Ramirez y Cía, 1888); Agustí Nieto-Galan, 'Scientific "marvels" in the public sphere: Barcelona and its 1888 International Exhibition', *HoST – Journal of History of Science and Technology*, 6 (2012): 33–63.

8 Two of them with foreign capital (Energía Eléctrica de Cataluña and Riegos y Fuerzas del Ebro, S.A.) and another with Catalan funds (Catalana de Gas y Electricidad). Jordi Maluquer de Motes, 'Cataluña y el País Vasco en la Industria Eléctrica Española, 1901–1935', in Manuel González Portillo, Jordi Maluquer de Motes and Borja de Riquer (eds), *Industrialización y Nacionalismos: análisis comparativo* (Bellaterra: Servei de Publicacions de la Universitat Autònoma de Barcelona, 1985), 239–52, p. 250.

9 According to the organizers, 'there have probably been very few markets which have offered the advantageous conditions now offered by Barcelona, for the electrification of its railways in all Catalonia will make the industry, trade, mining and agriculture of Catalonia of great importance [sic]'. *International Exhibition of Electrical Industries and their Applications* (Barcelona: La Neotipia, 1915), Foreword.

10 'Algunas ideas sobre el consumo de electricidad en España', *Ingeniería*, 879 (30 August 1929), 189–91.

11 Ibid., p. 190.

that progressively transformed urban landscapes and citizens' habits.[12] However, this was seldom a process of linear, uncritical progress and 'modernity'. Well into the twentieth century, electric power and lighting competed with wood, gas, kerosene lamps and candles. It took decades for electricity to fully spread into urban life, and gas even won the battle for cooking and heating.[13] What historian of technology David Nye describes as the transition from 'speculative investment' to 'established innovations' covered the early decades of the twentieth century and varied from city to city.[14] The transition called for the development of a complex socio-technical system – in Thomas Hughes' terms – [15] in which noteworthy agents such as inventors, experts and public administrations, but also private companies, research laboratories, educational institutions and regulatory agencies became crucial for the building of the electricity grid.[16]

The success of these daring urban enterprises was closely linked to private firms and their expert 'electricians' (engineers and physicists), but also to the appropriation attitudes of householders and users. As Graeme Gooday showed some years ago, beyond popularization campaigns, attempts to introduce electricity into the home were cultural projects that deserve closer examination.[17] Beyond academic experts, engineers, entrepreneurs and urban elites, any analysis of that complex technological change cannot disregard the role of electricity users,[18] consumers, amateurs and even journalists – the latter acting as mediators in the spreading of

12 David E. Nye, *Electrifying America: Social Meanings of a New Technology* (Cambridge, Mass.: The MIT Press, 1990). See also: Ken G. Beauchamp, *Exhibiting Electricity* (London: The Institution of Electrical Engineers, 1997); Mauren Dillon, *Artificial Sunshine: A Social History of Domestic Lighting* (London: The National Trust, 2002); Ernest Freeberg, *The Age of Edison: Electric Light and the Invention of Modern America* (New York: The Penguin Press, 2013).

13 David E. Nye, *When the Lights Went Out. A History of Blackouts in America* (Cambridge, Mass.: The MIT Press, 2010). See also: David E. Nye (ed.), *Technologies of Landscape* (Cambridge, Mass.: The MIT Press, 1999).

14 Nye, *When the Lights Went Out*, p. 16.

15 Thomas P. Hughes, *Networks of Power: Electrification in Western Society, 1880–1930* (Baltimore: Johns Hopkins University Press, 1983).

16 'After World War I, what is referred to as 'modern society' emerged full blown in the United States. Urban based and consumer oriented, it required huge amounts of energy to sustain a new style of life. [...] By the late twenties the use of more and more electricity, gas and oil in everyday life had become so ubiquitous as to wrap urban American in an "invisible world" of energy. Even the shock wave of the Great Depression could not halt a steady rise in household consumption of electricity, preserving the new standard of living', Harold L. Platt, *The Electric City: Energy and the Growth of the Chicago Area, 1880–1930* (Chicago: The University of Chicago Press, 1991), pp. 235–6. See also: John A. Jakle, *City Lights. Illuminating the American Night* (Baltimore: Johns Hopkins University Press, 2001), Nye, *Electrifying America* and David E. Nye, *American Technological Sublime* (Cambridge, Mass.: The MIT Press, 1994).

17 Graeme Gooday, *Domesticating Electricity. Technology, Uncertainty and Gender, 1880–1914* (London: Pickering and Chatto, 2008). See also: Graeme Gooday, *The Morals of Measurement. Accuracy, Irony and Trust in Late Victorian Electrical Practice* (Cambridge: Cambridge University Press, 2004)

18 Nelly Oudshoorn and Trevor Pinch (eds), *How Users Matter. The Co-Construction of Users and Technologies* (Cambridge, Mass.: The MIT Press, 2013); David Edgerton, *The Shock of the Old: Technology and Global History Since 1900* (Oxford: Oxford University Press, 2007).

new electrical knowledge across society. This is therefore the story of a complex dissemination of knowledge and skills that should be carefully analysed in every local context, Barcelona in this case.

International scholarship has mainly focused their attention on North American and European metropolises such as Chicago, New York, London, Paris and Berlin,[19] but smaller cities such as Barcelona in the 1920 and 1930s were not taken into account. What were the specific features of that local electricity culture, if any? A painstaking analysis of the impact of electric lighting at the 1929 International Exhibition would be useful here to address the city's electricity culture as a whole.[20] All sound data that follow in the next sections of this chapter are framed within David Nye's hypothesis on the growth of electrical landscapes in industrial cities as a sort of imitation of displays at world fairs and amusement parks.[21] In fact, as we wish to examine here in the case of Barcelona, in the early decades of the twentieth century, world fairs acted as testing grounds, as a driving force, as a model for further development, for the progressive electrification of the city lights.

We therefore begin here with a general description of the Barcelona 1929 International Exhibition, and then move on to the impressive display of the EL. In fact, the EL became a microcosm of the city's larger projects, and a laboratory for the dissemination of electrical knowledge among experts, entrepreneurs, advertisers, consumers and the general public. In addition, beyond the EL, electricity firms, electricity associations and their experts campaigned for the supposed benefits of electric lighting, and shaped the manner in which the city as a whole was lit. Shop windows, streets, public squares, showrooms, public and private buildings, neon advertisements, among many other urban places, became the new containers of electric light. Beyond the average landscape described in the AEL leaflet of 1930, this marked a fundamental change in Barcelona's culture of electricity in the early twentieth century, which this chapter seeks to examine in detail.

Popularizing 'rational' lighting: The Exhibition and the Exposición de la Luz

The 1929 Barcelona International Exhibition took place from 20 May that year to 15 January 1930 on a huge area on Montjuïc hill, which was transformed

19 Joachim Schlör, *Nights in the Big City: Paris, Berlin, London 1840–1930* (London: Reaktion Books, 1998).

20 See *Electricidad Industrial y Doméstica* regarding the need to instigate campaigns explaining the benefits of electrification. See also: Franciso Sintes Olives and Francisco Vidal Burdils, *La industria eléctrica en España* (Barcelona: Montaner y Simón, 1933).

21 In David Nye's words: 'As advertising, floodlighting, and street lighting spread through urban society, salesmen were able to ratchet up sales of electrical signs to businessmen eager to maximize the visibility of their enterprises. An unintended result of their competition was a brilliant landscape of light that by the 1920s had transformed urban space into a new form of the sublime that hid the night sky'. Nye, *When the Lights Went Out*, p. 10.

into a public park.[22] Compared to its 'cultural' achievements, the scientific content has not yet received sufficient scholarly interest.[23] Standard accounts often focus on exterior spaces, but never explore the pavilions' content, thus mirroring the photography of the event in the censored official press, which only depicted surfaces such as façades, official celebrations and openings.[24] The impressive Palacio Nacional, which held a major exhibition on Spanish art; the Pueblo Español (Spanish Village), which displayed national arts and crafts from different regions, and replicated 320 buildings from all over Spain; and the German pavilion, designed by the famous Bauhaus architect Ludwig Mies van der Rohe, have constituted the preferred topics of study thus far. This can be seen in the 1929 *L'auca del senyor Estevet*, a typically Catalan comic strip, in which the images of technology at the exhibition are limited to the magic fountain and the electrical spectacles.[25] The exhibition was conceived by the Catalan industrial bourgeoisie, during General Miguel Primo de Rivera's dictatorship (1923–1930). However, in the end, it was organized by the Spanish government with the ambiguous collaboration of the Catalan elite. Apart from being a Spanish nationalist celebration of a declining military dictatorship, it aimed to show Barcelona and Spain's industrial development, and to become an industrial forum in which the country could appropriate foreign technological achievements. In 1929, public street lighting shared electricity and gas supplies, but a large number of the electric lamps were still only provisionally installed.[26] In that fragile context, the application of electric lighting in the exhibition became a laboratory test that combined electric, practical, urban and aesthetic factors.[27]

22 See, for example, Mª Carmen Grandas, *L'exposició Internacional de Barcelona de 1929* (Sant Cugat del Vallès: Els Llibres de la Frontera, 1988); or Ignasi de Solà-Morales, *L'Exposició Internacional de Barcelona, 1914–1929: arquitectura i ciutat* (Barcelona: Fira de Barcelona, 1985).

23 Jaume Sastre-Juan and Agustí Nieto-Galan, 'Barcelona 1929: Science and Technology in a "Peripheral" International Exhibition', 7th STEP meeting Galway, Ireland 2010 (unpublished manuscript).

24 As can be seen in a folder full of newspaper clips with photographs from the press, collected by the Diputació de Barcelona [File 3384–3892, Arxiu de la Diputació de Barcelona].

25 These *auques*, and comic strips in general, are good indicators of the public perception of exhibitions (for an insightful study on humour as a source for the study of the public perception of international exhibitions – with an emphasis on the 1893 Chicago Columbian Exhibition – see Manon Niquette and William Buxton, 'Meet me at the fair: Sociability and reflexivity in nineteenth-century world expositions', *Canadian Journal of Communication*, 22 (1) (1997): 81–113). In our case, the most relevant feature is that it does not at all mention the pavilions' technological content, and when the main character visits them he only finds art or folklore. In the author's view, visiting the exhibition had nothing to do with science or technology, and in any case the only technological encounter the rural visitor had was a traffic jam on the streets of Barcelona.

26 *Barcelona: edición especial del artículo Barcelona de la Enciclopedia Espasa, dedicado a la Exposición Internacional Barcelona* (Madrid: Espasa Calpe, 1929). Until 1966 Barcelona had confined gas and electricity to its public lighting (Mercedes Arroyo, 'Los cambios en el proceso de producción y de distribución de gas en Barcelona y su hinterland (1930–1961). Entre el gas de hulla y el gas natural', *Scripta nova*, X, 218 (29) (2006), http://www.ub.edu/geocrit/sn/sn-218-29.htm (last accessed 11 December 2014). See also: Mercedes Arroyo, *La industria del gas en Barcelona (1841–1933). Innovación tecnológica, territorio urbano y conflicto de intereses* (Barcelona: El Serbal, 1996).

27 Juan Lasarte Karr, 'L'electricitat i la llum a l'Exposició de Barcelona', *Ciència: Revista catalana de ciència i tecnologia*, 36 (1930): 512–37.

There is a consensus that the installation of electricity was one of the most thrilling developments in the interwar period and that its success meant a turning point in the public image of the engineering community that had made it possible (see Figure 11.1).[28] In December 1929, after the closing of the exhibition, the journal *Electrical World* pointed out that:

> The spectacular and extraordinary effects exceed everything that have been tried in colour and mobile lighting; its splendour presents a marvellous natural view and attracts the interest of thousands of people that visit the International Exhibition of Barcelona [...] It is probably the first trial that the engineers of lighting applied to a so vast exterior lighting, a centralized control.[29]

The spectacle of light was regulated by the centralized control of local technicians and Westinghouse experts;[30] the electric magic fountain was also a good example of that collaboration.[31] The spectacle of light, colours and water tinged the illuminated magic fountain with a good dose of electrical sublime.[32] Designed by engineer Carles Buigas (1898–1979),[33] the setting displayed a spectacle in which light and water transformed the whole street from red to green in a few seconds.[34] The magic fountain provided the final touch for

28 Jordi Ferran, 'Technology for the public: Electricity in the Barcelona International Exhibition of 1929', *Annales historiques de l'électricité*, 4 (1) (2006): 31–48; Jordi Ferran, *Els públics de l'electricitat a Catalunya (1929–1936): De la Font Màgica de Montjuïc a la difusió dels electrodomèstics* (PhD thesis, Universitat Autònoma de Barcelona, 2013). In the *Historical Dictionary of World's Fairs*, John E. Findling and Kimberly D. Pelle described the 1929 Barcelona Exhibition in the following terms: 'The scene was particularly stunning at night, when coloured lights illuminated the monumental fountain at the centre of the plaza and the cascades of water descending from Montjuïc. Art deco-styled pedestals lighted both sides of the avenue, and enormous light beams burst into the sky behind the Palacio Nacional', John E. Findling and Kimberly D. Pelle, *Historical Dictionary of World's Fairs and Expositions, 1851–1988* (New York: Greenwood Press, 1990), p. 252.

29 *Electrical World* (94), 7 December 1929. Electrical 'magic' fountains dated back to the nineteenth century, being installed for the first time at the 1884 International Electrical Exhibition of Philadelphia (David E. Nye, 'Electrifying Expositions, 1880–1939', in Robert Rydell and Nancy Gwinn (eds), *Fair Representations: World's Fairs and the Modern World* (Amsterdam: Free University Press, 1994), 140-56). The 1888 Barcelona Universal Exhibition also had a magic fountain (Suárez Saavedra, *La Electricidad en la Exposición Universal de Barcelona*).

30 Charles J. Stahl, 'The floodlighting of the International Exposition at Barcelona, Spain', *Illuminating Engineering Society Transactions*, 24 (9) (1929): 876–89.

31 In this case with the German company A.E.G. (E. Posa, 'Los suministros de la AEG para la Exposición de Barcelona', *AEG al día*, 13 (January 1930), 1).

32 The notion of electrical sublime (firstly articulated in David E. Nye, *Electrifying America: Social Meanings of a New Technology, 1880–1940*) refers to a particular kind of 'technological sublime' (a notion developed in Nye, *American Technological Sublime*), which describes a semi-religious feeling of awe in the presence of impressive technological objects or performances. For a specific insightful analysis on the electrical sublime in exhibitions, see Nye, 'Electrifying Expositions, 1880–1939', 113–27.

33 For a biographical approximation of Buigas, see *L'enginy de Carles Buigas (1898–1979)* (Barcelona: Agbar, Ajuntament de Barcelona, 1998). See also: Maria Paz Silva, *22 anys amb Buigas* (Barcelona: Thor, 1981).

34 Manuel Nogareda, 'Editorial', *Diario Oficial de la Exposición Internacional*, 22, 11 August 1929. See also: David Caralt, *Agualuz* (Madrid: Siruela, 2010).

Figure 11.1 The 1929 International Exhibition. Avinguda de la Reina Maria Cristina, day and night.

the visitors' excitement,[35] and Buigas enjoyed tremendous popularity. Catalan newspapers referred to him as the 'wizard of light and waters'.[36] The entire

35 *Diario Oficial de la Exposición Internacional de Barcelona*, 6, 21 May 1929 (editorial of the day after the opening of the Exhibition). See also: David Caralt, 'Les nits de l'Exposició Internacional de 1929', in *L'electrificació de Barcelona, 1881–1935*, ed. Urteaga, 261–83.

36 Nogareda, 'Editorial'.

spectacle offered two hours of entertainment based on water, light and colour, and turned unpopulated daylight displays and pavilions into an attractive, unmissable experience at nightfall.[37] Some interviews conducted with local female staff confirm this hypothesis:

> As night falls, the lighting of the Palacio Nacional turns on. Reflectors are projected. Fountains and cascades get coloured. Lighting signals shine on the track. So many new and extraordinary phenomena take place, that's why in my opinion the night is the most beautiful part of the Exhibition.[38]

National and international newspapers described the spectacle as excellent, splendid, magnificent, amazing, impressive, highly entertaining to conquer visitors' minds,[39] and a good example for other cities that were to organize similar world fairs in the following years.[40] Not without resistance, various appropriations of all those electric lighting displays contributed to stimulating further installations on the streets, beyond the exhibition grounds.[41] At the heart of that urban 'laboratory', the EL emerged as a very particular site, which was placed at the core of local elites' cultural message to visitors and citizens at large. It strongly encouraged a rapid domestication of electric light in shops, streets and homes.

Under the auspices of the AEL and the local Comité de Difusión Luminotécnica (CDL), the EL was funded by the two major Spanish electricity suppliers, and displayed on the first floor of the Textile Art Pavilion. Productores y Distribuidores de Energía Eléctrica de España and Grandes Fábricas de Material Eléctrico spent 400,000 pesetas on the project. Additional public funds came from the Diputació de Barcelona Provincial Council (50,000 pesetas) and the City Council (50,000 pesetas). The space, light, water and scientific instruments (with a total estimated cost of 300,000 pesetas) were lent to the AEL for free.[42]

37 Carles Buigas, 'La iluminación decorativa en la futura exposición de Barcelona', *Ibérica*, 759 (1929): 2; 'La Exposición de Barcelona. La Opinión del "New York World"', *La Vanguardia*, 13 September 1929, 16.

38 *Diario oficial de la Exposición de 1929*, 29, 26 September 1929.

39 *La Ilustración Ibero-americana*, (1 (4) (1930): 63–70) published some laudatory passages in international press articles. Articles featured in *La Revue de Paris*, *Figaro*, *Daily Mail*, *Daily News*, *Glasgow Herald*, *Berliner Tageblatt*, *New York World*, *Mundo Argentino* and *El Mercurio de la Habana*, among others.

40 As has been pointed by John E. Findling: 'In 1929, Daniel H. Burnham had gone [...] to visit the Barcelona and Seville Exhibitions, and he had been particularly impressed by the night-time illumination at Barcelona, which "far surpasses all previous expositions . . . [and] was what we have dreamed of for Chicago in 1933'; John E. Findling, *Chicago's Great World's Fairs* (Manchester: Manchester University Press, 1994), p. 87.

41 'Barcelona'. *Enciclopedia Espasa-Calpe*. Apéndice I. Tomo 83 (1930), 1315. See also: Juan Lasarte Karr, 'Las obras de ingeniería en la Exposición de Barcelona', *Técnica*, 134 (1930): 17–23, and in consecutive issues.

42 The Diputació de Barcelona decided to fund this Light Exhibition because they considered it to be a national effort and that Spain could thereby be considered one more among the most technologically advanced European nations, which had previously displayed its light industry in exhibitions

The EL seminal idea was mainly due to the engineer Martín Arrúe Astiazarán (1893–1976).[43] Trained in Germany, Arrúe worked at the OSRAM company and managed to obtain scientific support from the Institut für Luminotechnik in Karlsruhe. Commissioned by the AEL, he conceived the EL as a place to 'educate' visitors on the new electricity culture.[44] In fact, the EL was devoted to popularizing everyday electric lighting applications on the streets, in shops, factories and homes. Its opening on 30 October brought together a large swathe of Barcelona's elite.

Engineer Manuel Vidal Españó (1905–1984) described details of a visit to the EL. Via a path of carefully designed rooms, the visitor experienced the virtues of 'luminotechnics', and the material culture of consumerism.[45] At the entrance, the word 'Luminotecnia' was displayed, and, at the bottom of the stairs, a lighting arrow indicated the way to the reception hall with several coloured lights. In the centre of the hall, there was a torch, blown with hidden fans. A whole section of the EL was devoted to the 'History of Light', from the biblical episodes to the use of modern electrical lamps. It described a lineal and continuous succession of human achievements, which culminated in Barcelona's electric lighting. A set of dioramas described technological progress: from the creation of light by God, primitive bonfires, torches, oil lamps, candles, gas lighting, Edison and the invention of the incandescent lamp, and then the lighting of a modern boulevard, the modern harbour, the Barcelona Exhibition and its electric magic fountain.[46] In a deterministic framework, electric lighting was presented as an agent of change from night into day,[47] as a driving force for social habits and values and was intimately linked to modernity.[48]

From the central hall, the visitor enters the theatre and its impressive stage, ready for the electrical imitation of natural wonders such as clouds, storms, sun and rainbows. After that came the highlight of the exhibit: the 'Street of Light' (see Figure 11.2). It was a reconstruction of an illuminated street with

visited and studied by the organizers of the Palace of Light (in Düsseldorf, Munich, Dresden, Frankfurt, Stuttgart, Rostock, Berlin and Paris). See 'Subvención concedida a la 'Asociación Española de Luminotecnica' para contribuir a sufragar los gastos de la Exposicion de la Luz', File 4189 Exp. 26, Arxiu de la Diputació de Barcelona.

43 Noteworthy members of the AEL included Germán de la Mora (president), Martín Arrúe (secretary), B. von Eggeling, A. Hielscher, F. Brandon, among others. Jorge Barcino, 'La Exposición de la Luz en el Certamen Internacional de Barcelona', *La Ilustración Ibero-americana*, 1 (4) (August, 1930), 36.

44 'El alma de la Exposición de la Luz. Don Martín Arrúe Astiazarán', *La Ilustración Ibero-americana*, 1 (4) (1930): 38–9.

45 Manuel Vidal Españó, 'El Palau de la Llum de la nostra Exposició', *Ciència: Revista catalana de ciència i tecnologia*, V, 36, (May–June 1930): 596–603.

46 'Los Dioramas de la Historia de la Luz', *La Ilustración Ibero-americana*, 1 (4) (1930): 42–8.

47 Jorge Barcino, 'La Exposición de la Luz en el Certamen Internacional de Barcelona', *La Ilustración Ibero-americana*, 1 (4) (1930): 36.

48 For instance, a highly favourable article published in *La Ilustración Ibero-americana* referred to the Exposición de la Luz as: 'Exhibitions like that produce real social movements since they are like historical crusades' (Barcino, 'La Exposición de la Luz en el Certamen Internacional de Barcelona', 36).

Figure 11.2 The main entrance (left) and the Street of Light (right) at the Exposición de la Luz during the 1929 International Exhibition.

its starry sky and full of well-lit shop windows. Since the effective lighting of window displays was supposed to hold a direct relation with sales figures, entire shops were reproduced in full detail. Automatic mechanisms illuminated each set, alternatively, with correct and incorrect solutions. Shopkeepers were strongly encouraged to hire one of the new expert 'electricians'. Even a contest to appraise the lighting of shop windows was organized.

Beyond the Street of Light, visitors passed through the home section, which reproduced details of kitchens, dining rooms, bathrooms and bedrooms. The same procedure was used to spread the correct lighting of factories. Finally, visitors reached the scientific section, which displayed instruments related to light, experiments from the Laboratorio de Fototécnia de Madrid, as well as optical illusions and demonstrations of optical effects.

According to the local conservative newspaper, *La Vanguardia*, the EL was 'one of the best and surprising displays at the fair'.[49] The liberal *El Progreso* described it as 'an awesome exhibition from every point of view'.[50] On the other hand, the *Diario de Barcelona*, also from a conservative point of view, pointed out how the EL caught professional and laypeople's interest, and combined technical and spectacular demonstrations.[51] The estimated visitors' figure was 600,000, and included more than 500 school visits.[52] On 13 and 14 January 1930, delegates from France, Austria, Belgium, Germany, the Netherlands, Hungary, Italy and Sweden as well as Spain attended an international conference at the EL

49 'Ayer se inauguró la Exposición de la Luz', *La Vanguardia*, 1 November 1929, 8.

50 'Se inaugura la Exposición de la Luz', *El Progreso*, 1 November 1929, 3.

51 'El Palacio de la Luz', *Diario de Barcelona*, 1 November 1929, 13. The news also announced a free technical service on lighting for the public.

52 A.T.M., 'El alma de la Exposición de la Luz. Don Martín Arrúe Astiazarán', *La Ilustración Iberoamericana*, 1 (4) (1930): 38.

to debate and deliberate over strategies to popularize electric lighting in their respective countries.[53]

The CDL had planned a broader popularization campaign on electric lighting, targeting different audiences: public lectures for experts and technicians, other lectures for industrial workers and tradesmen, but also a series for the general public that had to be broadcast on the radio. All those efforts were complemented by the publication of several technical leaflets and articles in the daily press.[54] In the local context, the CDL's executive board brought local academic scientists, entrepreneurs and journalists together. Academic scientists worked in close conjunction with engineers and popularizers. In the EL theatre, a specific series of public lectures constituted a useful meeting point for socialization and business. Broadcast by Radio Barcelona and reviewed in newspapers, they played an important role in the scientific legitimization and domestication of the new applications of electricity. The list of speakers comprised a number of engineers and professionals working on lighting in private companies, such as Arrúe himself, as the director of the EL but also OSRAM's representative.[55]

Arrúe's talk introduced several lighting problems as an object of display at the exhibition, discussed its effects in specific sites, and reproduced some of the experiments in the scientific section.[56] Arrúe was also interested in electric lighting for shops.[57] In an article published in *Electricidad Industrial y Doméstica*, he even stressed how, in some US cities and in Berlin, electric lighting campaigns had shops keep their lights switched on until midnight to encourage new habits on the street.[58] In a Taylorian framework, Vidal Españó described how lighting methods shaped the capacity to read small letters, and the

53 *La Vanguardia*, 7 January 1930, 8; 14 January 1930, 11; and 18 January 1930, 8.

54 'El Comité Ejecutivo de Difusión Luminotécnica en Barcelona. Miembros que componen dicho Comité, organizador de la brillante labor cultural que se ha irradiado de la Exposición de la Luz de Barcelona', *La Ilustración Ibero-americana*, 1 (4) (1930): 54.

55 Arrúe lectured on the concept of luminotechnics; Jaime Bachs, engineer of Compañía de Aplicaciones Eléctricas, discussed the practical aspects of electric lighting; Francisco Vidal Burdils, lawyer of Fuerzas Eléctricas de Cataluña, presented several aspects of electric lighting; engineer Manuel Vidal i Españó talked about light and the 'scientific organization' of labour; engineer Manuel Gabarró Freixas presented electric lighting in the field of sport; José Mañas Bonví, engineer and professor of the Escola d'Enginyers Industrials of Barcelona, discussed the technology of lighting; Mariano Soria, professor of ophthalmology at the Universitat de Barcelona lectured on light and the sense of vision; José Baltá Elias, professor of physics at the same university, explained the nature of light as electromagnetic radiation; Miguel Masriera, PhD in Science at ETH Zurich, discussed the mechanisms of vision and painting; Josep Comas i Solà, director of the Fabra Observatory, presented and discussed the wave-particle nature of light. 'Personalidades que han intervenido en el ciclo de conferencias organizado por el Comité Ejecutivo de Difusión Luminotécnica', *La Ilustración Ibero-americana*, 1 (4) (1930): 55.

56 'La luminotecnia y la Exposición de la Luz', 16 May 1930.

57 He was also a pioneer in this area by translating the German text: F. Putnoky, *Die Technik der Schaufensterbeleuchtung*. F. Putnoky, *La técnica del alumbrado de escaparates* (Madrid: Blass, 1927). Arrúe translated it from the German edition of 1922 and introduced a new preface.

58 Martín Arrúe, 'La luz en el comercio'. *Electricidad Industrial y Doméstica*, 2 (1930): 7.

perception of the speed of change and, in the end, enhanced efficiency at work.[59] Francisco Vidal Burdills, José Mañas Bonví and Jaime Bachs focused on the right lighting for shop windows as a new driving force for advertising and marketing.[60] They emphasized how new electric lighting technology could provide new professional opportunities and new potential markets for the new expert 'electricians'. In their views, lit shops led to further street lighting. Like many other cities, they also shaped new urban habits and daily practices in the city. Radio Barcelona broadcast the lectures, together with 15 minutes of brief clips for the general public, in prime time, from 8.45 to 9 pm.[61] In one of these clips, Arrúe stressed that good lighting was not a luxury but a 'necessity' for modern urban life.[62] The CDL was the seed for the future Comitè Català de Luminotècnia (CCL), a local association linked to AEL that organized conferences on lighting in the following years. CCL members were academics, engineers and other professionals such as designers, glaziers, psychologists, businessmen and, some years later, architects and decorators, all of them truly interested in electric lighting. Despite their lack of funding, they kept their agenda of public lectures, radio shows, and the publication of articles in the daily press and technical journals.[63] As at the EL, they reproduced several domestic spaces featuring the right and wrong lighting systems in pictures.[64] Some years later, after the exhibition, the publication of the journal *Luminotecnia y Decoración* on January 1935, coincided with the end of the CCL activities, after its persistent electric lighting popularization campaign. The journal was devoted to showing 'all things that can be done with light', and was illustrated generously with hundreds of photographs of spaces lit by the members of the committee.[65]

Since the cost of electricity became a bottleneck for its use in lighting, it had to be transformed as efficiently as possible into a basic commodity.[66] The context of the exhibition and the impressive display at the EL were designed

59 'La luz en la organización científica del Trabajo', 9 July 1930.

60 'Aspectos interesantes de la iluminación' by Francisco Vidal Burdills (15 July 1930); 'Técnica de la Iluminación' by José Mañas Bonví (13 June 1930); and 'Aspectos prácticos de la iluminación' by Jaime Bachs (2 July 1930).

61 At least three of these pieces has been found: 'Lo menos que debe saber todo el mundo respecto a la iluminación' by Martín Arrúe (1 June 1930); 'El alumbrado en la industria' by Manuel Vidal i Españó (22 June 1930); and 'El alumbrado en el hogar' by Martín Arrúe (29 June 1930).

62 Martín Arrúe, 'Lo menos que debe saber todo el mundo respecto a la iluminación', *Electricidad Industrial y Doméstica*, 3 (1930): 17. For the introduction of radio in Barcelona, see Chapter 10 of this volume.

63 For instance, they published up to 7,620 square centimetres about lighting in the daily or technical press from July 1931 to June 1932. AEL, *Memoria correspondiente al ejercicio de 1° de julio de 1931 a 1° de julio de 1932* (Madrid: Gráficas Reunidas, 1932).

64 AEL, *La luz en casa racionalmente empleada proporciona alegría, bienestar, belleza y economía* (Madrid: Gráficas Reunidas, S.A., 1933). A description of AEL publications can be found in Jordi Ferran, 'La asociación española de luminotecnia (1929–1936): La utilización racional de la electricidad para la iluminación'. *Actes d'Història de la Ciència i de la Tècnica*, 5 (2012): 51–70.

65 *Luminotecnia y Decoración* only published six issues.

66 Ferran, *Els públics de l'electricitat a Catalunya (1929–1936).*

specifically to serve these aims, and led to further campaigns for domestic appliances as described in the following section.

Domesticating electric lights and appliances: Experts, users and showrooms

From 1929 onwards, new sites and practices spread the EL model to the city as a whole. Advocates of electric lighting endeavoured to replicate the atmosphere of the exhibition in the city centre. In 1930, the mayor of Barcelona, Joan Pich i Pon (1878–1937), supported public contests revolving around the electric lighting of shop windows and buildings' facades. Pich i Pon belonged to the exhibition's government body, had worked in his own electricity firm, and was keen to pass a municipal bill for tax exemptions on lit facades.[67] In addition, the City Council organized a contest to award a price of 10,000 pesetas to the best-lit shop window. Thirty-six shops entered the contest with varied results. The press noted that the panel of judges was benevolent towards participants but unanimously praised Radio Lot's display, centred on the Frigidaire refrigerators, the most coveted electrical device of the time.[68] The aim of the contest was again to convince shopkeepers – as users and consumers – that good electric lighting contributed to better sales figures, although substantial achievements in this area did not come to fruition until the mid-1930s.[69] The project was part of a broader ambition to transform Barcelona into 'Europe's City of Light'. In 1931, la Cambra de Propietat spearheaded the proposal to organize a Light Fortnight, which aimed to reproduce the atmosphere of the exhibition's lighting. The initial programme included a shop window and facade light competition, light shows and nightly parades.[70]

However, public reluctance to the new 'determinism' of electric lighting was also frequent. The installation of a large lit column in the centre of Plaça Catalunya sparked public controversy. The satirical weekly *L'Esquella de la Torratxa*, for instance, published cartoons ridiculing the lit obelisk. It was a sign of mixed feelings, of the pros and cons of electric light. On the one hand, a character in front of the obelisk sarcastically proclaimed, 'we are already better than Paris'.[71] But some cartoons questioned its utility, aesthetics and high cost (see Figure 11.3). When it was finally dismantled, the satirical press described a critical celebration

67 *La Vanguardia*, 16 May 1930, 7.

68 'El concurs d'aparadors il·luminats', *Revista Imatges*, 1 (7), 23 July 1930, 2.

69 The goal in this case was to reproduce the North American model. It had transformed shopping into entertainment through artificial illumination (Jakle, *City Lights*, p. 253).

70 The Comitè Català de Luminotècnia, established some time after the Exhibition and inheriting the Comité de Difusión Luminotécnica, obtained resources from the Asociación Española de Luminotecnia (5.000 pesetas) but there was indifference from Electric Companies and the City Council. The event was never held. 'Una noticia muy interesante. ¿Será Barcelona la ciudad de la Luz?'. *Electricidad Industrial y Doméstica*, 5–6 (1930–1931): 26; AEL, *Memoria correspondiente al ejercicio de 1° de julio de 1930 a 1° de julio de 1931* (Madrid: Talleres Voluntad, 1931), p. 4.

71 *L'Esquella de la Torratxa*, 2602, 10 May 1929, 296.

Figure 11.3 Left: The electric obelisk in Plaça Catalunya. Right: Cartoon published in *L'Esquella de la Torratxa*, 2658, 6 June 1930, 377.

in which citizens opposed the City Council, threw stones at the electric obelisk and demanded a new urban reform of Plaça Catalunya.[72]

The growing use of electric lighting on the streets of Barcelona aroused both satisfaction and uneasiness or scepticism. At a busy intersection in the neighbourhood of Sant Gervasi, many objected to the City Council's decision to replace an electric spotlight with a new gas lamppost,[73] but new electric lighting was celebrated in the Poble Sec neighbourhood.[74] At the same time, the opening of lit facilities such as the Sociedad Sportiva Pompeya's tennis court, stimulated new social activities after nightfall.[75] Catholics considered electric light awkward for liturgical proposals,[76] and liberal urban classes perceived it as an inconvenience in theatres[77] and on the streets.[78] Nevertheless, in spite of this controversy, the core of engineers that promoted the CDL's activities assessed the progress of electric lighting. Again within a deterministic framework, they considered that electric lighting had changed citizens' lives in theatres, cinemas, restaurants, shops, in short, on the streets.[79]

72 *L'Esquella de la Torratxa*, 2658, 6 June 1930, 377; 2660, 20 June 1930, 420; 2662, 4 July 1930, 443.

73 *La Vanguardia*, 9 June 1929, 14.

74 *La Vanguardia*, 29 July 1934, 7. Regarding the electric lighting opening in the squares of Santa Madrona and Blasco de Garay.

75 *La Vanguardia*, 13 June 1931, 12. The Sociedad Sportiva Pompeya was established in the Sant Gervasi neighbourhood.

76 *La Vanguardia*, 31 March 1934, 19.

77 Manuel Cases Lamolla, 'Il·luminació', *Art: Revista de les Arts*, 9 (1934): 3. Manuel Cases Lamolla (1900–1974), architect and member of the *Fomento de las Artes Decorativas* was one of the first professionals outside the core of promoters of decorative lighting. He considered himself as a tamer of the Edison's shiny light bulb (*Luminotecnia y Decoración*, 1 (January 1935), 1).

78 *L'Esquella de la Torratxa*, 2622, 27 September 1929, 616.

79 In 1935, founders of the *Comité* and other professionals began to publish the journal *Luminotecnia y Decoración* devoted to showing lighting installations by means of spectacular pictures. A paper entitled 'El progreso de la iluminación decorativa en España' considered that not only Barcelona and Madrid had a considerable number of spectacular lighting displays in commerce (*Luminotecnia y Decoración*, 3, (March 1935)).

The pictures published in the journal *Luminotecnia y decoración* convey the atmosphere on La Rambla, where the canopies belonging to Bar Automàtic or Dancing Excelsior illuminated a significant part of the footpath and road. Electric lighting at the Pompeya music hall in Paral·lel, Hollywood Bar on the street Conde del Asalto or Cinema Mayland in Plaça d'Urquinaona attracted new customers and showed their 'modernity'. Shop windows belonging to Almacenes Jorba on Avinguda del Portal de l'Àngel were a continuous showroom of lighting in a shopping area. Marian Rubió i Bellver (1862–1938), engineer and former technical director of the 1929 International Exhibition, considered that the spectacle of Barcelona streets was simply admirable, from dusk to the moment shops close, and shop windows lit up at night. To prevent the daily sales from plummeting, the lights at night had to appear brighter.[80] Beyond the city centre, the Canódromo park facilities on Travessera de les Corts or Joyería Roca on Avinguda Diagonal spread new urban uses to other areas. At that time, the majority of the city centre's buildings had full electric wiring.[81]

After the exhibition, *showrooms*, as mediating spaces between expert and laypeople's electrical knowledge – lighting and power – rapidly spread throughout the city. Showrooms provided spaces for exchanging electrical knowledge between experts and laypeople, between salespeople and customers as visitors and potential users. However, private firms used showrooms as part of their marketing strategy for new electrical devices. Despite this technophile image of electricity in ads, showrooms helped consumers to manage it and to check out new products by themselves. They exhibited lamps, lighting procedures, vacuum cleaners, refrigerators, all aimed at piquing potential customers' interest. Around 1930 there were up to 15 private showrooms in Barcelona city centre, most of them devoted to refrigerators (10) and other domestic electrical appliances (2), and electric lighting applications (3): The *Compañía de Aplicaciones Eléctricas* in Plaça Catalunya, the *Compañía General Española de Electricidad. Lámpara 'METAL'* on Ronda de la Universitat and the *Compañía Barcelonesa de Electricidad* on Carrer de Girona.[82]

80 Mariano Rubió Bellvé, 'La psicología de la ciudad'. *La Vanguardia*, 14 May 1935, 5.

81 Horacio Capel (ed.), *Las Tres Chimeneas. Implantación industrial, cambio tecnológico y transformación de un espacio urbano barcelonés*. Vol. 3 (Barcelona: FECSA, 1994), p. 165. Unfortunately, there is no data available on electricity consumption per inhabitant at local level. Catalonia had a total consumption of 372 kWh per inhabitant in 1935 (Jordi Maluquer de Motes, 'Panorama eléctrico español hasta 1944', in Gonzalo Anes (ed.), *Un siglo de luz. Historia empresarial de Iberdrola* (Madrid: Iberdrola, 2006), 88). Lighting consumption was considered to be 15.5 per cent (Alayo, *L'electricitat a Catalunya*, p. 901). In the period between 1290 and 1935, domestic electrification rose from insignificant values to more than 70 per cent penetration in industrialized cities worldwide. Nevertheless, consumption values per inhabitant for lighting purposes presented a high level of variability among European cities in 1933: Basel 525 kWh, Zurich 471 kWh, Stockholm 236 kWh, Amsterdam 243 kWh, Paris 123 kWh, Berlin 100 kWh and Vienna 84 kWh (Luciano Segreto, 'Ciento veinte años de electricidad. Dos mundos diferentes y parecidos', in Anes (ed.), *Un siglo de luz*, pp. 29–30). Figures for Barcelona were far from those of top cities. The Spanish average was 153 kWh per inhabitant in 1933 (Maluquer de Motes, 'Panorama eléctrico español hasta 1944', p. 88). Simultaneously, Catalonia led the electrification process in Spain during this period.

82 Electric showrooms were mainly placed (14) in the Eixample, the new city, but close to the limits of the old city, surrounding Plaça de Catalunya. For refrigerators, it is worth mentioning the Sala de

Advertisements and articles in local newspapers and journals – in particular, at the *Electricidad Industrial y Doméstica* – [83] allow us to trace back details of Barcelona's local electric culture. In 1930, for instance, Jaime Bachs, the showroom manager of the Compañía de Aplicaciones Eléctricas described his audience (see Figure 11.4). In Bachs's view, among the 6,000 daily visits, there were many bystanders, but also people actually interested in installing some of the lighting solutions in their homes, factories, shops or workshops. Again, imitating the EL, kitchens, bedrooms, living rooms and bathrooms were reproduced in detail, to later show how lamps had to be properly installed;[84] the showroom of the Compañía General Española de Electricidad. Lámpara 'METAL' reproduced shop window displays with correct and incorrect lighting.[85] Lámpara 'METAL' engineer, Rosendo Pich, described his campaign for better lighting, which included visits by school children, and even portable showrooms to spread the progress of electric lighting across Catalonia.[86] The Compañía Barcelonesa de Electricidad's showroom coached professional architects and electricians on specific lighting projects.[87] It was mainly an electric power company, and did not sell appliances, its main goal being to increase electricity supply. The Compañía Barcelonesa de Electricidad had another showroom in Plaça Catalunya devoted to new domestic devices.

From a gender perspective, messages such as 'Electricity raises you to the throne of the utmost comfort'[88] sought to disseminate the idea that electrical appliances would replace the female workforce in domestic tasks. The actual benefits of the use of electrical appliances for early twentieth century housewives

Auto-Electricidad (Diputació, 234) exposing General Electric fridges; Suministros Eléctricos, S.A., (Fontanella, 14) for Westinghouse; Anglo-Española de Electricidad (Granvía, 525 and Pelai, 12) for Kelvinator, among many others. See also: *Barcelona en el año de la Exposición International de 1929* (Barcelona: Bailly-Ballière y Riera Reunidos, 1929).

83 Launched in 1930, the journal *Electricidad Industrial y Doméstica* (*Domestic and Industrial Electricity*) was devoted to popularizing new electric devices and their use in home chores. Its pages were full of ads for electricity companies, which often bought copies to circulate them among the journal's subscribers. Within a few months of being launched, the journal achieved a monthly circulation of 30,000 copies.

84 In the centre of the ceiling and on both sides of the mirror in the bathroom; a bedroom with lighting points in the ceiling, over the dresser and the bedside tables. José J. Morales, 'El alumbrado racional. Una sala de demostraciones', *Electricidad Industrial y Doméstica*, 3 (1930): 12.

85 José Jurado Morales, 'Una sala de demostraciones de alumbrado', *Electricidad Industrial y Doméstica*, 5–6 (1930–1931): 10.

86 He considered it crucial to teach children the benefits of electric lighting, keeping in mind that 'new generations will appreciate great manifestations of electricity' (Morales, 'Una sala de demostraciones de alumbrado', p. 10).

87 Advertisements of *Compañía Barcelonesa de Electricida*d presented decorative lighting as the key factor of home comfort (*Luminotecnia y Decoración*, 3 (March 1935), 1). Philips 'Design with light' campaign also stressed the link between lighting and modernity (*Nuevas formas. Revista de Arquitectura y Decoración*, 3 (1934): 110).

88 'La electricidad os eleva Señora al trono de la mayor comodidad'. Advertisement run by the *Cooperativa de Fluido Eléctrico*, published in *Electricidad Industrial y Doméstica*, 9 (1931): 4.

Figure 11.4 Top: Customers at the entrance to the Compañía de Aplicaciones Eléctricas showroom in Plaça Catalunya, 20 in 1930. Bottom: The same place lit by night.

have been repeatedly questioned by historians.[89] However, it is also the case that women became a central target of the companies' marketing campaigns in the early twentieth century.[90] In spite of the apparent lack of local users' associations, electricity firms paid special attention to women in order to increase electricity consumption. Radio, daily press and specialized magazines spread information on the introduction of electricity in houses, shops, factories and on streets. The transformation of the urban space that electricity had brought about in other European and American cities a decade earlier began in Barcelona in the early 1930s.[91]

In the showrooms, expert electricians standardized what women supposedly had to know about electricity: the vocabulary associated with electricity, the limitations of the electrical system, security notions to tackle risks and uncertainty, and, more obviously, an optimization of the cost of electrical appliances and incandescent lamps.[92] They also provided specific guidelines for consumers when buying light bulbs: the ideal voltage and power and cleaning for better efficiency.[93] Women, and users in general, often acted as mediators and active electricity agents in private homes.[94]

As experts identified the cost of electricity supply as the main critical point for stepping up the use of electrical appliances, companies' salesmen organized public shows on the consumption of electrical devices. They compared the cost of good lighting at home, in shops and in industry to well-known costs of more mundane things such as a family celebration, the publication of an advertisement in the press or the cost of wasted time by factory workers. The showroom

89 Regarding this controversy, see: Ruth Schwartz Cowan, *More Work for Women. The Ironies of Household Technologies from the Open Hearth to the Microwave* (London: Free Association Books, 1989); Cynthia Cockburn and Ruza Fürst-Dilíc (eds), *Bringing Technology Home. Gender and Technology in a Changing Europe* (Buckingham: Open University Press, 1994); Amy Sue Bix, 'Equipped for life. Gendered technical training and consumerism in home economics, 1920–1980', *Technology and Culture*, 43 (4) (2002): 728–54; Sue Bowden and Avner Offer, 'The Technological Revolution that Never Was. Gender, Class, and the Diffusion of Household Appliances in Interwar England', in *The Sex of Things: Gender and Consumption in Historical Perspective*, in Victoria de Grazia and Ellen Furlough (eds), (Berkeley: University of California Press, 1996), 244–74; Roger Horowitz and Arwen Mohun (eds), *His and Hers: Gender, Consumption and Technology* (Charlottesville: University Press of Virginia, 1998); or Ruth Oldenziel and Karin Zachmann (eds), *Cold War Kitchen: Americanization, Technology, and European Users* (Cambridge, Mass.: The MIT Press, 2009).

90 Unlike other European countries, it is not possible to find any women's association devoted to popularizing electricity. In the case of Britain, see L. Symons, 'The Electrical Association for Women, 1924–1986', *IEE Proceedings A: Science, Measurement and Technology*, 140 (3) (1993): 215–20.

91 However, it is worth noting that in Germany, for instance, full-scale electrification of households only took place in the 1950s, after World War II (Dieter Schott, 'The City and Electricity', in Mikael Hård, and Thomas J. Misa (eds), *The Urban Machine. Recent Literature on European Cities in the 20th Century*, (Tensions of Europe, electronic publication, 2003, www.iit.edu/~misa/toe20/urban-machine, last accessed 15 July 2012), 109–19.

92 Ernest Greenwook, 'Lo que la señora de la casa debe conocer', *Electricidad Industrial y Doméstica*, 16 (1931): 8. Republished in issue 37 (1933): 4.

93 AEL, *La luz en casa racionalmente empleada*, 1–2, 16.

94 J.A. Corcovan, 'Electrifique su casa. Conviértala en un hogar', *Electricidad Industrial y Doméstica*, 3 (1930): 22.

belonging to the Compañía Barcelonesa de Electricidad, for instance, displayed a set of basic experiments to assess the power expenses of different appliances.[95]

Although neon lights did not reach Barcelona's public sphere until the mid-1930s, they had already had an impact on the city centre. The company Luminosos Neón installed 75 per cent of Barcelona's 1,000 neon lights.[96] An illuminated chocolate ad in Plaça Catalunya exerted a great impact, not only for the company that funded it (Chocolates Juncosa), but also for its contribution to the shaping of the city's public image. Simultaneously, neon red lights in a brothel in the Barrio Chino transformed a narrow, dark and dirty street into an unrealistic scenario.[97]

Shop windows underwent a profound cultural change in the city, affecting public consumption and even lifestyles. Moreover, neon lights, street and traffic lights progressively built an exciting and cosmopolitan city by night.[98] Barcelona mirrored itself in cities such as Paris, Berlin or New York, which had begun transformations in electric lighting a few decades earlier, but the city's urban transformation through lighting that had begun with the 1929 Exhibition was well on its way in the 1930s.[99]

Conclusions

Our winter short story of 1930 effectively revealed how experts and firms campaigned for the use of electricity as a decisive agent in shaping everyday life in the city.[100] It was written following the impact of the 1929 Exhibition, which represented a landmark in the development of modern electricity culture. The EL, the fascinating combination of electric light and water on the exhibition grounds, the public lectures, the millions of visitors, constituted a model to be replicated by expert electricians, to be subsequently extrapolated to the city at large, through municipal campaigns, companies' showrooms, and press advertisements. In all these campaigns electrical knowledge circulated widely between expert and lay actors, and shaped the growth of the electricity grid in modern Barcelona.

The key actors in the process of popularizing electric lighting, Arrúe, Vidal i Españó and Bachs, among others, established a network of collaboration rooted in the Exhibition. Despite becoming experts serving the interests of private firms, they sought their own professional recognition. The spectacular display of the EL captured potential customers and users of new electric lighting,

95 This procedure was shown by advertisements published in *La Vanguardia* (6 June 1936, 7 and 9 June 1936, 2).

96 In the city centre there were advertisements of the next companies: Dr Andreu, Gilette, Amatller, Juncosa, Moulin Rouge, Xampany Freixenet, Jorba, Almacenes el Águila, Agua Oxigenada Forest, Odeón and Pirelli; 'Com es fa un anunci lluminós', *Revista Imatges*, 5 (9 July 1930). 21.

97 Josep M. Planes, *Nits de Barcelona* (Barcelona: Llibreria Catalonia, 1931), p. 55.

98 'Com es fa un anunci lluminós'. *Revista Imatges*.

99 Raymond Carr, *España 1808–2008* (Madrid: Arial, 2009).

100 L.S. Splendor, 'Alegorías de la Luz. Cuento de Invierno'.

whereas practical shows and demonstrations addressed the new material culture of electric appliances for new social groups. Electric lighting soon also became embedded in the work of architects and interior designers, but also in the marketing plans of numerous shops, theatres and restaurants. In addition, after the exhibition and the EL, showrooms became the new crucial sites where lay consumers were in close contact with electricity applications, and ideal places for the study of consumer behaviour. At the same time, electric lighting invaded the streets with shop windows, electric bulbs and neon lights. In spite of the lack of detailed accounts on the part of ordinary consumers and households, local consumption figures seem to indicate that the effective 'domestication' of electricity took place in Barcelona in the 1930s, probably as a knock-on effect of the 1929 Exhibition agenda.

Beyond consumers' concerns regarding electrical risks and hazards – as they have been described in studies on home electrification in previous decades[101] – the cost seemed to be the main stumbling block to the rapid increase and intensive use of electrical appliances, including electric lighting. In fact, all the public displays and shows endeavoured to convince consumers that benefits would exceed costs. Local popularizers designed strategies to convince women to become the electricity experts at home: in maintaining the lamps, being familiar with the technicalities of the home electric installation and being able to understand details in the companies' invoices. All these marketing efforts can be easily traced back through articles in the daily press, in specialist magazines or even in radio clips.[102] In practice, electricity was integrated into everyday life, and its uncertain nature was progressively excluded from debates in the Barcelona public sphere.

As described by the fairy in our short story:

> A day will come in which the laws of artificial lighting will be taught at school. Today there is no single textbook for the rational use of that little incandescent lamp that Edison invented 50 years ago, the source of the uncountable wonders we know today. The incandescent lamp, to which we owe so much, is often treated as a Cinderella, and its indisputable utility will improve, only when men learn to use it properly.[103]

In the 1930s, in spite of these frequent optimistic accounts and their public image as an agent of modernity, electric lighting and electric appliances had only made a preliminary step forward in their domestication process. The turmoil erupting on the streets of Barcelona in July 1936 dramatically interrupted that process. The city had still to wait decades to make the 1930 fairy tale come true, to become the 'ciudad maravillosa' in which electric lighting had fully spread throughout public and private settings to irreversibly shape citizens' everyday life.[104]

101 Gooday, *Domesticating Electricity.*

102 F. A. Delgado, 'Barcelona, Fanal de Luz', *Electricidad Industrial y Doméstica*, 2 (September 1930): 18.

103 L.S. Splendor, 'Alegorías de la Luz. Cuento de Invierno', p. 5

104 Ibid., p. 6.

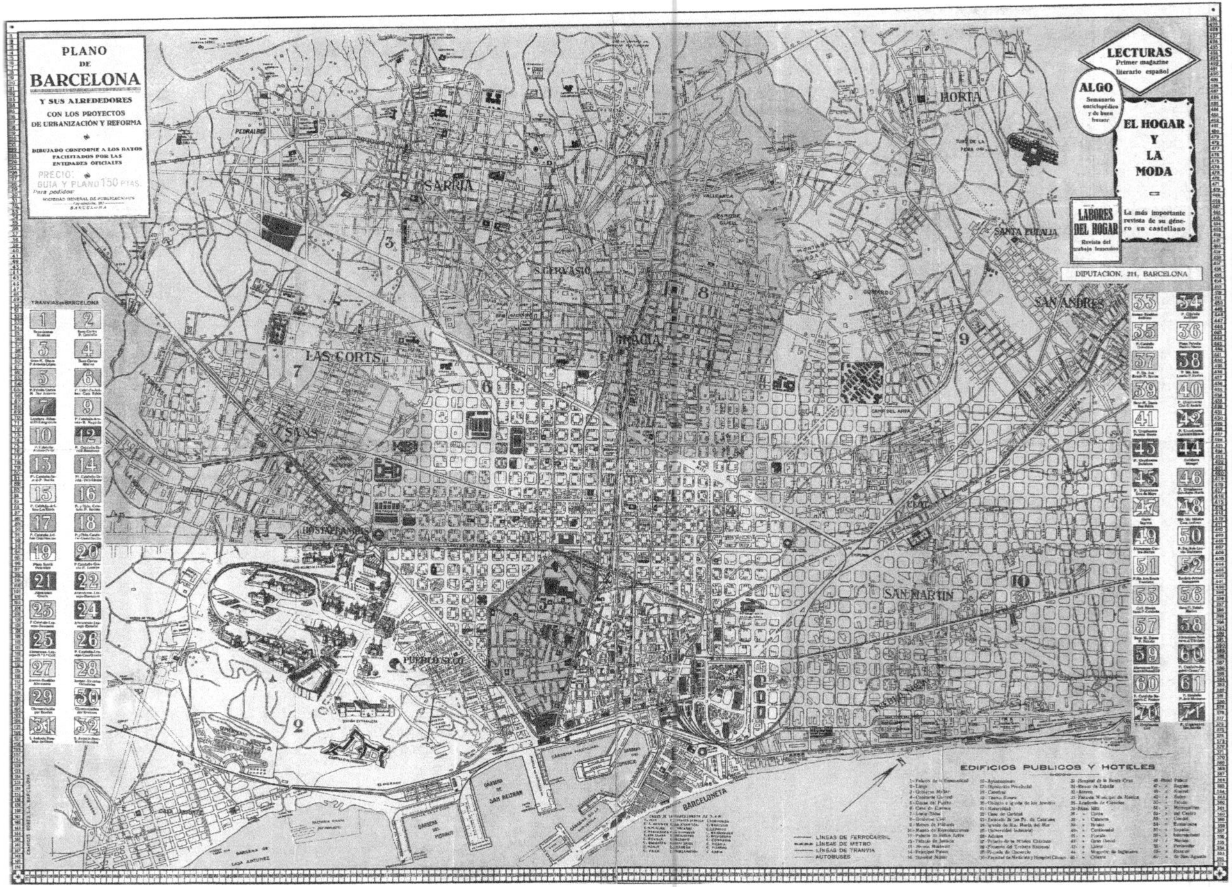

Map 2 Barcelona in 1929.

Index